CASE STUDIES

IN PROJECT, PROGRAM, AND ORGANIZATIONAL PROJECT MANAGEMENT

CASE STUDIES

IN PROJECT, PROGRAM, AND ORGANIZATIONAL PROJECT MANAGEMENT

DRAGAN Z. MILOSEVIC

PEERASIT PATANAKUL

SABIN SRIVANNABOON

WILEY

John Wiley & Sons, Inc.

Published by John Wiley & Sons, Inc., Hoboken, New Jersey

Published simultaneously in Canada

For general information about our other products and services, please contact our Customer Care Department within the United States at (800) 762-2974, outside the United States at (317) 572-3993 or fax (317) 572-4002.

Wiley also publishes its books in a variety of electronic formats. Some content that appears in print may not be available in electronic books. For more information about Wiley products, visit our web site at www.wiley.com.

"PMI", the PMI logo, "OPM3", "PMP", "PMBOK" are registered marks of Project Management Institute, Inc. (www.pmi.org). For a comprehensive list of PMI marks, contact the PMI Legal Department.

Library of Congress Cataloging-in-Publication Data:

    Case studies in project, program, and organizational project management / [edited by] Dragan Z. Milosevic, Peerasit Patanakul, Sabin Srivannaboon.
        p. cm.
    Includes index.
    ISBN 978-0-470-18388-5 (pbk.)
        1. Project management--Case studies. 2. Project management--Standards. I. Milosevic, Dragan. II. Patanakul, Peerasit. III. Srivannaboon, Sabin, 1977-
    HD69.P75C375 2010
    658.4′04--dc22

                                    2009045965

Printed in the United States of America

10 9 8 7 6 5 4 3 2 1

To Dragana, Jovana, and JR
—Dragan Z. Milosevic

*To my parents, Arun and Soisalinee; my wife, Severine;
and my children, Ananya and Yanat*
—Peerasit Patanakul

*To my father, Sabieng, my mother, Songsee,
and my lovely wife, Jany*
—Sabin Srivannaboon

Contents _____

Preface _____

Traditionally, the use of case study has been largely emphasized in many disciplines. People use cases in different manners from theory building, to theory testing, to description, or even to simple explanation. Nevertheless, learning is always one ultimate goal in which we center our attention on the gravity of the problems and issues in the case, regardless of any purpose. In particular, the learning occurs when we dissect the case, identify issues or problems in it, and then discuss or solve them.

In the field of project management, case studies as well have been one of the main sources and tools used for professional development and higher education. Over the years, the Project Management Institute (PMI) has attempted to get a large number of authors to contribute to case studies in project management. The idea is to use these cases as a means to accelerate the project management learning. This is also similar to academia where a number of cases are integrated into textbooks. A few standalone case books dedicated to project management are also available.

However, what is critically missing is a comprehensive case study book where it meets diverse needs of the readers at large. To be more specific, there is no book that has project management cases arranged in an orderly fashion that comprehensively addresses various knowledge areas, different process groups, and the global best practice standards. In particular, there are very few cases in program management and organizational project management, even though the two areas are now recognized as two standalone disciplines, and officially standardized by PMI.

We believe this book is the first of its kind to deal with the management of projects from a hierarchy perspective: project, program, and organization. The purpose of this book is to maximize the readers' learning experiences through the use of case studies, which we believe will allow our readers to carefully think and enrich their understanding of the concepts and practices in project management. In attempting to capture various aspects of project management, we have written 90 cases, each of which was triangulated by professionals with different expertise varying from engineers to industrial psychologists, to quality computer experts, to software programmers, to businesspersons' service providers, and to organization specialists. These cases are factual from real people and actual companies in different industries, settings, or cultures with diverse sizes and types of projects, although we used fictitious names to conceal their identities. Our goal is to highlight the applications and practices of project management, program management, and organizational project management in real-world settings.

The book is designed to address multiple groups of people with different needs that include but are not limited to:

- **Executives, program and project managers:** This book will help executives and program and project managers improve their management knowledge regarding projects, programs, and organizations. We present cases that discuss many best practices and lessons learned from such management in actual companies across industries.
- **Academics and consultants:** For academics, this book is a good resource of project management, and a recommended accompanying reading for their project management, program management, and organizational project management classes. The students may use this book as a reference or as a required text since the cases can well support any basic textbooks of the class, whether it is a project management, program management, or organizational project management class. For consultants, this book provides many real-world stories in which the frameworks for project and program management as well as organizational project management were implemented. They can easily incorporate a number of cases in this book, or use the entire book for their in-class trainings.
- **CAPM®, PMP®, and PgMP® candidates:** This book perfectly aligns with the standards created by PMI, and provides important details necessary for the CAPM® (Certified Associate in Project Management), PMP® (Project Management Professional), PgMP® (Program Management Professional) certification exam preparations.

For each individual, excellence in project management comes from both theoretical knowledge and practical experiences. Either one of these alone would not be sufficient in today's era of hypercompetition. After reading this book, we believe that our readers will gain such knowledge and learn from experiences shared by other project management practitioners.

All in all, this book just captures small stories. We hope, however, that these stories will serve as building blocks to drive excellence in project management, which is undoubtedly one of the fastest growing disciplines today.

Structure of the Book _____

This book offers a number of case studies that demonstrate effective use of project and program management methodologies, as well as organizational project management practices. Drawn from a variety of industries and regions, the case studies capture real-world situations, challenges, best practices, and lessons learned both from successful and not-so-successful perspectives. In order for our readers to best learn project management, we have categorized and arranged our cases into two different dimensions: case types and parts.

CASE TYPES

We classify our cases into three different types: critical incidents, issue-based cases, and comprehensive cases. The three case types differ in length and specificity, which are described as follows:

- Critical incidents are written in the form of short stories that illustrate an issue or a problem related to project, program, and organizational project management.
- Issue-based cases provide more information than critical incidents. They handle two or more issues either in project management, program management, or organizational project management.
- Comprehensive cases are the longest in length. They feature multiple issues or the entirety of the project, program, or organizational project management.

The purpose of these different levels is to offer the reader different categories of the learning skills, contingent on their experience. This way they can use this book to customize learning needs. In addition, the book has both open-ended cases, where we don't show the final outcome of the story, and close-ended cases, where the final outcomes are presented for further discussion.

While the case types are different, their structure across different parts is similar. Each case includes an introduction, main body, conclusion, and discussion items.

PARTS

In addition to the case types, we adopt the standards created by PMI, the leading global association for the project management profession, to arrange our cases. Namely, these standards are "A Guide to the Project Management Body of Knowledge" (the *PMBOK® Guide*), "The Standard for Program Management," and "The Organizational Project Management Maturity Model (OPM3®)." We follow these standards, and organize our cases and chapters into three different parts: Project Management (Part I), Program Management (Part II), and Organizational Project Management (Part III), (see Figure i).

- We organize Part I based on the PMI's *PMBOK® Guide*, which addresses the introduction, project life cycle, and organization (Chapter 1), project management processes for a project (Chapter 3), and the nine knowledge areas (Chapters 4 to 12). Added to that are the cultural aspects of project management (Chapter 2), in which we strongly feel that culture, whether it is corporate, project, or regional, plays a significant role in achieving project goals. In sum, Part I has a total of 52 cases.
- We structure Part II based on the process groups of the PMI's Standard for Program Management, including the Initiating, Planning, Executing, Monitoring and Controlling, and Closing processes (Chapters 14 to 18). We also offer cases about the themes of program management (Chapter 13), and program management in action (Chapter 18) for further discussion. There are a total of 19 cases in Part II.
- Part III focuses on issues in organizational project management, which address some of the best practices in the Organizational Project Management Maturity Model (OPM3®). This part presents cases related to strategic alignment and project portfolio management (Chapter 19), standardized methodologies (Chapter 20), and competencies of project managers and project management office (Chapter 21). We also present cases on information systems, organization, and metrics (Chapter 22) and organizational and project or program culture (Chapter 23). Cases on organizational project management in action are presented in Chapter 24. There are a total of 19 cases in Part III.

Figure i Structure of the Book

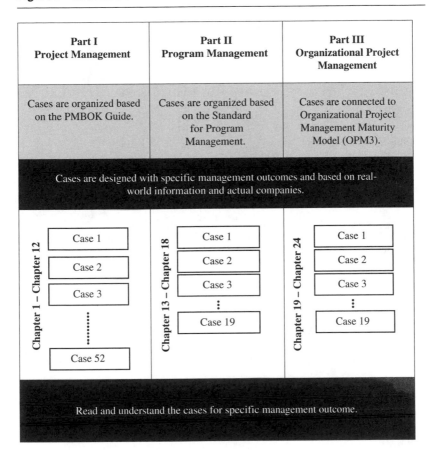

The Principles of Management _____

EQUIFINALITY

Equifinality, a term from systems science, refers to the principle through which multiple means (different inputs and processes) may lead to a same end in open systems.

CONTINGENCY

Contingency, in management terms, refers to one of several approaches one might take in dealing with a condition, situation, or set of circumstances involving uncertainty. In other words, after examining alternatives to find the most appropriate solution, another possible solution might be considered if the first one doesn't work out (a "Plan B," so to speak).

Acknowledgments _____

To complete the book, we owe gratitude to many people.

First, we'd like to thank our co-authors who helped us in writing a number of the outstanding cases or provided many valuable inputs for the case write-ups. These people are:

Abdi Mousar, Andrea Hayes-Martinelli, Art Cabanban, Bjoern Bierl, Diane Yates, Don Hallum, Ferra Weyhuni, James M. Waddell, James Schneidmuller, James Staffan, Joakim Lillieskold, Joseph Genduso, Jovana Riddle, Lars Taxen, Mani Amabalan, Marie-Anne Lamb, Mathius Sunardi, Meghana Rao, Michael Adams, Murugappan Chettiar, Nicolas Charpenel, Osman Osman, Priya Venugopal, Rabah Kamis, Russ J. Martinelli, Stevan Jovanovic, Sung Han, and Wilson Clark

Our sincere thanks to many of our colleagues, co-workers, and previous organizations or those we have been involved with in the past for the knowledge and information we gained and used for this book.

Finally, we are deeply grateful to our institutions, namely the Department of Engineering and Technology Management (Portland State University, USA), Wesley J. Howe School of Technology Management (Stevens Institute of Technology, USA), and Sasin GIBA of Chulalongkorn University (Thailand) for their support and environment, which enabled us to complete this book.

Part I

CASE STUDIES IN PROJECT MANAGEMENT

WHAT IS PROJECT MANAGEMENT?

It is well recognized that project management has been practiced since early civilization. The evidences from past history to the present are abundant: the construction of the Great Pyramids of Giza in the ancient world, the Great Wall of China construction in the 16th century, and the London Millennium Bridge in the globalization era. Without project management, these structures would not have existed.

With a competitive business environment, many organizations nowadays use projects not only to build structures, to implement changes, or to introduce new products, but also as a way to put strategies into action. Despite multiple meanings of a project, the one defined by Project Management Institute (PMI) is perhaps the most widely known definition. According to PMI, a *project* is a temporary endeavor undertaken to create a unique product, service, or result.[1] With its temporary nature, a project is often perceived as standing on the opposite spectrum of business as usual; it is often referred to as an "operation" by project management scholars. As projects differ from operations, managing projects therefore

---

[1] *A Guide to the Project Management Body of Knowledge*, 4th ed., Project Management Institute, 2008, p. 5.

requires a discipline[2] of planning, organizing, and managing resources to bring about the successful completion of specific goals and objectives. This discipline is referred to as *project management*.

The discipline of project management has evolved from different fields of application. The work of Frederick Winslow Taylor on theories of scientific management is considered to be the foundation of project management tools, such as the Work Breakdown Structure. The Gantt chart, developed by Henry Gantt, is recognized as a forefather of project management planning and control techniques. And the work of Henri Fayol on management functions is the foundation of project and program management body of knowledge.

However, it wasn't until the middle of the 20th century that project management was recognized as a formal discipline[3]; emerging from the construction of the first atomic bomb during World War II (the project known as the Manhattan Project). Since then, more and more new processes and disciplines have emerged that support the use of project management, including Time Quality Management (TQM) in 1985, concurrent engineering in 1990, and reengineering in 1993, just to name a few. As a result, more and more project management tools and techniques have emerged, including the Critical Path Method (CPM) and Program Evaluation and Review Technique (PERT) in the 1950s, and the Critical Chain Project Management in 1997.

As the discipline of project management has grown, the standards governing the field have also evolved. While each organization practicing project management may develop its own criteria, several national and international organizations have proposed project management standards. These standards are, for example, *A Guide to the Project Management Body of Knowledge* (*PMBOK® Guide*) from the Project Management Institute in the United States and PRINCE2: 2009 Refresh (PRoject IN Controlled Environment) from the Office of Government Commerce in the UK. Among these standards, the *PMBOK Guide* receives strong recognition from project management communities.

The *PMBOK Guide* suggests nine knowledge areas of project management: integration management, scope management, time management, cost management, quality management, human resource management, communication management, risk management, and procurement management. These knowledge areas are used as skeletons for organizing case studies in Part I.

---

[2]David I. Cleland and Roland Gareis, *Global Project Management Handbook*, McGraw-Hill Professional, 2006.
[3]Aaron J. Shenhar and Dov Dvir, *Reinventing Project Management: The Adaptive Diamond Approach*, Harvard Business School Press, 2007.

Chapter 1

INTRODUCTION

Chapter 1 presents examples of organizations that have recognized the importance of projects as an engine of their growth or a survival mechanism during economic turbulence. Various efforts of these organizations in response to the need for project management, therefore, were initiated.

In this chapter, there are six case studies: five critical incidents and one issue-based case. The cases generally discuss a number of concepts (e.g., organizational structures), that can be found in Chapters 1 (Introduction) and 2 (Project Life Cycle and Organization) of *A Guide to the Project Management Body of Knowledge* (the *PMBOK® Guide*).

1. AaronSide Goes to Teams
2. Cocable Inc.
3. A RobustArm Global Industries' SledgeHammer
4. Another Trojan Horse
5. Call a Truck
6. The Project Hand-off Method

These cases demonstrate different situations where companies made the transition from non-project-oriented organizations to project-oriented ones. To capture the transition efforts from multiple views and settings, we offer cases from different industries: "AaronSide Goes to Teams" is in the metal machining industry; "Cocable Inc." is in cable manufacturing business; "A RobustArm Global Industries' SledgeHammer" providesbuilding materials; "Another Trojan Horse" is in the nuclear industry; "Call a Truck" offers shipping and transportation services; and "The Project Hand-off Method" is from the field of medical equipment manufacturing.

CHAPTER SUMMARY

Name of Case	Area Supported By Case	Case Type	Author of Case
AaronSide Goes to Teams	Project Management Organization (Functional vs. Matrix Structure)	Critical Incident	Dragan Z. Milosevic, Peerasit Patanakul, and Sabin Srivannaboon
Cocable Inc.	Project Management Organization (Training by Doing)	Critical Incident	Jovana Riddle
A RobustArm Global Industries' Sledgehammer	Project Management Organization (Standardized Project Management)	Critical Incident	Dragan Z. Milosevic, Peerasit Patanakul, and Sabin Srivannaboon
Another Trojan Horse	Project Management Organization (Training)	Issue-based Case	Stevan Jovanovic
Call a Truck	Project Management Organization (Matrix Structure)	Critical Incident	Dragan Z. Milosevic, Peerasit Patanakul, and Sabin Srivannaboon
The Project Hand-off Method	Project Management Process	Critical Incident	Dragan Z. Milosevic, Russ J. Martinelli, and James M. Waddell

AaronSide Goes to Teams

Dragan Z. Milosevic, Peerasit Patanakul, and Sabin Srivannaboon

It took AaronSide, Inc. almost 80 years to grow from a small mom-and-pop business to a company that held the largest market share internationally. What made this feat special was that a single family owned the company since its inception. It is suffice to say that this success made owners, management, and all employees more than proud.

A WALL IS BETWEEN US

Operating in the metal machining industry, AaronSide's organization was perfected over time through experience and many saw this as a competitive advantage. Basically, it was an efficient, functional organization where marketing, engineering, and manufacturing with a strong quality group played a major role. The engineering department achieved the fastest 16-month lead time for a new product development project when compared with competitors. Fundamentally, product development was an operation that worked like a well-oiled machine. It started with marketing, which did market research and then threw the specification of what customers desired "over the wall" to the engineering department, which released final drawings to manufacturing, which made the quality product. The approach was called the relay race. Its secret was an efficient, functional department. Typically, if you worked in a specific department, say marketing, you would never talk to a guy from a different engineering. If you did, you might be reprimanded. Indeed, departments talk to each other, not individuals that belong to different departments. How do departments converse? Usually, only heads of departments are authorized to speak on behalf of their staff.

TO SURVIVE, CHANGE IS REQUIRED

The more intense globalization of business brought more international competition. The two largest rivals in the industry from Europe, subsidiaries of the large multinational organizations, largely expanded their operations in the U.S. market.

This is when problems for AaronSide began to mushroom. AaronSide found it difficult to compete with the Europeans, who had access to resources and new management of their rich parents. As a result, AaronSide slipped to a close third in market share, behind the European rivals. Freefall continued and by 1990, AaronSide was the distant third. Several management teams were replaced during this period, new manufacturing equipment was installed, the company was seriously reengineered, and different management was used to catch up with the leaders without significant results. So, AaronSide became ripe for a sale.

After talking with four suitors from the United States and Europe over the last several years, owners concluded that the best offer for purchase of AaronSide was one from Titan Corp, a Swedish company. So, after almost 90 years of being family-owned, AaronSide became a wholly owned subsidiary of a large multinational firm.

To facilitate the integration of AaronSide into Titan Corp's network of companies, the management team of AaronSide was retained. The first initiative of the new owner was to direct AaronSide to commission a pilot project management team (in manufacturing companies usually referred to as concurrent engineering teams), cross-functional in nature, and made up of the permanent members from marketing, engineering, and manufacturing, and auxiliary members from finance and field repair. The team was chartered to develop a new mining vehicle in eight months, twice faster than usual and as fast as the world leader. The new team was empowered to make all major decisions. The idea was to accomplish success with this team, and then use it as a paradigm along with the lessons learned from its operation to establish a company-wide project management system.

Eight months later the project was not finished, and needed eight more months to reach its conclusion. The Swedish parent asked for an immediate investigation. The investigation showed that the team did not make any major decisions. Instead vice-presidents (VPs) who were heads of the departments directed the members of their team to make no decisions, but to bring all necessary information to them and they, the VPs, would make the decision. Having discovered this, the management of Titan Corp decided to fire the CEO and all VPs.

Discussion items

1. What are the pros and cons of the relay race approach and the cross-functional team approach to product development projects? Which approach is better?
2. Who gets more power and who gets less power by shifting product development projects from the relay race to the cross-functional team approach?
3. Does the shift from the relay race to the cross-functional team approach require a significant cultural change? Explain why or why not.
4. Why do you think the VPs took the approach of not letting a pilot team make major decisions although the team was empowered to do so?
5. Was the firing of the CEO and all VPs justified? Why or why not?

Cocable Inc.

Jovana Riddle

JANE AND OBANGA

It was 6:30 on a Wednesday morning, and Jane Campbell was on her way to work. The traffic was heavy like any other day, which usually made Jane frustrated. Today, however, Jane was calm and didn't seem to mind the long drive. Why? The commute would give her time to think about what her next move at Cocable Inc. should be. Jane had some very important decisions to make in the next few weeks that would greatly impact her life.

Jane's boss, Larry Fitzgerald, recently came to her with a new project proposal. Since her current assignment was wrapping up, she had to make a choice: Should she stay with Cocable, and accept the uninteresting product development coordinator role she was offered, or should she look for a new job with a different company?

As she always did when facing a tough decision, she started to identify the pros and cons of each option. If she took a new job she would likely have to move, work longer hours, and experience a huge learning curve. Most importantly, the new job would mean less time with her precious daughter, Obanga. Obanga was now two years old and had been through a lot in her short life. Jane and her ex-husband, Obanga's dad, met in Kenya where Jane used to live. Shortly after getting married they moved to England, where Obanga was born. Their divorce came soon after Obanga's birth, just before Jane and Obanga moved to America. To provide Obanga with a better life, Jane went to Oregon Graduate Institute (OGI), to get a master's degree in Engineering Management. From there she was hired by Cocable. Thus, in her two short years of life Obanga had dealt with the loss of her father, a huge move, and the transition of having her stay-at-home mom became a full-time employee. Taking a new job, thus, could really impact the little girl negatively.

Jane's other choice was to stay at Cocable and accept the product development coordinator role, for a project that would likely last for at least eight to nine months. This new role would not be very challenging and, more importantly, would not increase her moderate salary. As she finally arrived at the office, Jane knew she had to make a decision soon, one that she couldn't regret.

BACKGROUND

Cocable is a company based in Chicago, which makes interconnecting cables for industry gear. Their annual sales are around $78 million and the company currently employs 720 people. Jane's role thus far at Cocable has been in operations and product development. The new project she has been offered would require Jane to coordinate product development for gear, similar to her current assignment. Therefore, the learning curve would be nonexistent, and would likely not be challenging or exciting for Jane.

The product that needs to be developed is brand new and would require the formation of a 25-person team. Jane's role in this 25-person team would be to coordinate product development. The members assigned to the team thus far are from all product development functions and various groups. Most of them have spent their entire careers working on similar products and have not been challenged or motivated in their jobs in a very long time.

In an attempt to encourage Cocable employees, the company decided to hire an external consulting team to train employees on project management, the area that had never been formally known in the organization. The aim was to help facilitate the product development process at least on a temporary basis.

Jane and this 25-person team were requested to attend this training before they would start on the new project.

TRAINING

The newly formed team would have to undergo a two-day training session, which would be divided into two 5-hour modules. According to historical attendance of training sessions, it could be expected that 22 of the 25 people on the team would be present for each session. The main purpose of this training would be to get all the team members on the same page, and to encourage learning best practices and project management knowledge. Most importantly, this would be a first step in building a cohesive team while providing all members with vivacious and interesting in-classroom training. The team building would also allow the leaders of the group to stand out. The five most prominent leaders would be asked to make up the "Cascade team" that would help design and deploy the project management manual that would be used later in the company.

TRAINING BY DOING

It would take Cocable two weeks to organize and coordinate the required training session. Upon the conclusion of the training, the Cascade team would focus on designing a manual, using the project management training they just received.

To test whether the manual was feasible, Jane's new project would be the first to use its draft. If it was effective, then the manual would be deployed to all projects within the company. During the first three months of the project, the team would have to consult the manual and apply its guidelines to each task. By using the manual for each task the team would learn the ins and outs of the project and the application of project management. This typical practice, which is very common in project management, is referred to as "training by doing."

After the three-month anniversary the team would be up for their "three-month review," during which the application of the designed manual would be analyzed. The review would be in a workshop format and attended by the 25 employees/trainees and the supervising consultant. The purpose of the three-month review is to identify any mistakes the employees make while applying the manual to their tasks. For example, each employee would present a real-world example of how they used the manual in everyday tasks and potential mistakes in application would be identified by the team.

PRE-IMPLEMENTATION

The pre-implementation phase is the period of time between the three- and six-month anniversaries of the project's start. During this time, the trainees continue to apply all the rules in the manual. The "six-month review" is a workshop-style format attended by the supervising consultant and the 25 trainees. Again, each employee is asked to present a real-world issue they have encountered and any potential mistakes/deviations from the manual implementation are identified by the larger group. This workshop would also conclude the pre-implementation phase of the project.

CUT-OFF

The company would then transition from following the old handbook for all its tasks to using different project management rules presented in the new manual. The application of the old handbook would be abandoned and each step would be transitioned to operating in a new way. The transition date is also known as a cut-off date, the point at which a company chooses to transition to a new way of doing things.

This would be the first time in company history that a standard manual was introduced and applied to projects regardless of their size. Things seemed to be changing at Cocable, and Jane felt that she might want to stick around for a while.

Discussion item

1. What are the advantages and disadvantages of introducing project management to new product development the way they did in the Cocable case?

A RobustArm Global Industries' Sledgehammer

Dragan Z. Milosevic, Peerasit Patanakul, and Sabin Srivannaboon

READY, STEADY, NOW . . .

"We have a corporate mandate to start the management of our projects by means of Standardized Project Management (SPM) processes by the end of the year. Considering that most frequently projects are done in the Engineering department, its head, Blaine Peters, will be in charge," said Tim Robison, the general manager of the RobustArm Global Industries (RGI) plant in Duckville, Oregon, to close the meeting of the management leadership team. And the SPM race began.

BACKGROUND

RGI company was a global multi-million-dollar business that served the planet with the most cost-effective building materials. The company had several branches, and the one in Duckville was one of the biggest plants which employed 220 people, and had annual sales of around $2 million. The company strategy led the plant to cater their products to Western United States and China.

What made RGI a distinct company was the three-pronged corporate culture: *improve continually, standardize processes,* and *purse change*. In the early 1990s RGI won the Baldridge award for quality, and embarked on a never-ending race for continuous improvement. All employees relentlessly searched for the next production method to enhance or to remove a bottleneck. Most importantly, the Baldridge award changed their worker mindsets forever, making them feel like owners of processes, rather than just regular employees.

One of the major aims of this corporate culture facet was standardization of processes. The new corporate mandate of pursuing SPM has the purpose of bettering ways of running projects, and raising efficiency and reducing costs in order to eventually provide greater value to the customer. At RGI, they even standardized one facet of meetings: To convene any meeting, you had to send a standardized agenda ahead of time.

10

"Nothing is permanent except change" was the motto at RGI. This was not a "one-major-change-initiative-at-a-time" type. Actually, this was a whitewater type of change where a multiple of change initiatives were unfolding at the same time. So, while they were preparing to launch the SPM initiative, there was a serious effort related to the six sigma initiative, targeting the improvement of production facilities, and several teams used the best of their knowledge to upgrade processes in production, HR, IT, R&D, engineering, etc.

GETTING TO WORK

As soon as Blaine left the conference room, he began to run the SPM race. He did so by following the corporate guidelines for the major change. According to the guidelines, he established the project and named the change initiative: *SPM Implementation Project Sledgehammer*, and assumed the job of Project Manager. Blaine then chose his project team members, from the Engineering department, who needed to be properly trained in project management first before proceeding to the next step. It took him a week to find the right project management trainer, and another two weeks to arrange the three-day project management training. So far, project affairs went smoothly and were expected to continue that way. In three weeks, the team would need to produce an SPM manual, which had 10 standard items with a template for each one, including:

- Scope statement
- Work Breakdown Structure (WBS)
- Responsibility matrix
- Schedule
- Cost estimate
- Quality plan
- Risk plan I (P-I matrix)
- Risk plan II (Risk response plan)
- Progress report
- Closure

To finish the Sledgehammer with flying colors, the project team would need to organize separate training, designed specifically for the use of the SPM manual, for all members of RGI's Duckville plant. When this was done, the project team would announce the cut-off date by which all new projects had to be managed with the SPM manual. It sounded that simple.

Discussion items

1. How important is the SPM process to the company? Explain your answer.
2. What are the pros and cons of the approach that RGI used to develop the SPM process?

Another Trojan Horse

Stevan Jovanovic

MEET THE TROJANS

John Lackey can't hear the noise of the hundreds of people cheering in congratulations for him. His thoughts have drowned out the noise. He just received a reward for Best Project Management of a Utility Plant in the United States. The plant is Trojan Nuclear Plant, which was officially shut down in 1996. The year is now 2000.

John is about to give a speech in front of hundreds of people. He is expected to give the typical acceptance speech and thank everyone for their hard work and thank the award committee for their recognition. This is not, however, the usual utility plant management story. John is consumed with thoughts of how a common speech could possibly convince the world that a nuclear plant, having produced zero power in four years, deserves an award praising its project management. The story of a project that involves decommissioning of a nuclear plant cannot be short. This project is also very unique with equally unique circumstances.

The whole project plays out in his mind. His thoughts start from the very beginning . . .

THE SIMPLE TASK

The task sounded simple enough, to shut down a nuclear plant, decommission it, and make the plant, or at least its location, safe for the future. The plant was near a major metropolitan area, which made the first line of reasoning simple; get the dangerous stuff away from the dense population and reduce risk for human harm. Trojan was a large plant—it was the largest and most powerful nuclear plant of its day. It was built on a huge site that supported tons in equipment. Although most people seldom consider decommissioning, as part of managing a power plant, it is actually a necessary part of the plant lifecycle.

As one would expect, most of the equipment was used to handle extremely hazardous nuclear material. Nuclear material stays dangerous for very long periods of time, for thousands of years. The varying degrees of danger involved for

12

the different parts of the huge plant ranged from slightly dangerous to extremely poisonous. Every inch of the 600-acre complex had to be thoroughly examined and a plan had to be made to deal with each and every inch. All of the equipment was hazardous and had to be handled with extreme caution. The project would take years.

TRAINING

Decommissioning of a nuclear reactor is unique in that every single detail of the project must be planned out and reviewed long before any work can begin. Error was unacceptable, especially for a project such as this that would be scrutinized not only by the authorities, but by the general public as well. Nuclear power has always been a hot topic, generating extreme public opinion. The Trojan Nuclear Plant, even during its operating life, was always a source of controversy for people on both sides of the nuclear power debate. Every aspect of its operation, especially bad news, became fodder for the media. Any negative news immediately became front page material. John and his company definitely did not want their name associated with negative "nuclear news." The only way to ensure that this did not happen was to be as prepared as possible.

Naturally, the first step was getting up to date on the latest in project management techniques. For this they chose the four-day training course by Scope Management (statement, WBS, process, changes), Cost Management (estimates, earned value, cost baselines), Time Management (jogging line, TAD, milestone charts), and Risk Management (risk events, PI Matrix, Monte Carlo analysis). The course was structured per the *PMBOK Guide*. Now armed with tools and concepts needed for the work ahead, they felt prepared for any size project.

LEARNING

One of the biggest challenges for this company was the change in mindset from an operating company to a project management company. An operating company does use basic project management techniques but spends much of its time and energy executing operations. Naturally, however, a project management company spends all of its energy on project management. Although the transition from an operating company to a project management company may not sound challenging, there can be difficulty when a person has spent their entire career focusing on project execution. The tendency for that type of person might be to revert to old ways. On a project this complex, however, skimping on project management could be disastrous.

The first step in decommissioning the plant was to break down the project into smaller projects. These "smaller" projects were by no means easy, but were more adaptable for applying and refining their recently acquired project management skills. They were essentially used as learning blocks and every "smaller" project management project was carefully reviewed and each lesson learned was clearly outlined.

One of the projects was the removal of the central reactor. This involved transferring the entire reactor hundreds of miles from the plant location. Such a project had never been attempted before, which means no historical information upon which to rely. The entire decommissioning project had many "firsts" and of them all the reactor removal was by far the biggest and most complicated.

The reactor, a concrete and steel maze, was the size of a basketball court and weighed many, many tons. The safest way to remove the reactor was in one piece. The reactor was not originally designed and built, however, with the intent of being picked up and carried around in one piece. It was a system of structures, pipes, and mechanical equipment. Damaging and breaking open any part of the reactor system would have allowed poisonous material to leak out. This was the worst possible scenario and could not be allowed to happen. The project was the pinnacle of the team's project management use. They managed to safely lift the reactor, transfer it to a ship, move it up river (including going through four dam locks), transfer it to a land vehicle, and "truck it" to its final destination. They managed every detail of the move. To the great relief of everyone, the move was a success. The team and the company had made their mark in the pages of project management history.

Discussion item

1. What do you learn from this case?

Call a Truck

Dragan Z. Milosevic, Peerasit Patanakul, and Sabin Srivannaboon

CAT, Inc. was a powerful company. They provided their customers with the goods to transport, the food to eat, the gas to drive, and a place to sleep overnight. In fact, they provided everything that a driver would need on the road. What kind of company was that? CAT, Inc. was a broker between companies that transport the freight and the owners of the freight. And they were the premier broker of the market.

In the 1980s, the technology wasn't very complicated or advanced. "Computer" was a fairly new term that was just recently becoming known to this business. Most requests and information were handled by fax or telephone. CAT, Inc. was one of the many companies that used this technology.

THE NEW CEO

CAT, Inc. just got a new CEO, James Carter. James was voted and selected by the majority of managers who were former drivers. To the surprise of the majority of the employees, the new CEO was not a former driver. He was a businessman who viewed CAT, Inc. as a dinosaur, an about-to-be-obsolete business that used fax and telephone.

The new CEO had a new vision. The vision was to be in touch with customers by a new means that radically changed the way people communicated. The means was known as a computer. This required massive investments. Luckily, CAT, Inc. was in a position that could afford the changes. So, to make his vision possible, he launched a number of operative and strategic changes.

Among the operative changes, James ordered a stop to many fax-based communications between his people and customers and changed them to be fully computer-based. In that manner, CAT, Inc. became the first company in its industrial history that was able to provide everything a driver would need on the road, ordered by request through a computer system. However, because not many people had computers at the time, the company also maintained a fair use of the traditional means of fax and telephone, but limited them to minimal usages.

The information through the traditional methods was kept in the digital format, nevertheless.

In addition to the operative changes, strategic changes were many and in-depth. One example was the new infrastructure. CAT, Inc. built the infrastructure that supported the new system and better facilitated customer requests. Despite a number of positive changes, consequences were inevitable. And one conse-quence went to the technology development group, where the old faces were all laid off. In place of them, the company hired a new software group from a tech-nology company's technology group. Immediately after that, a training program with a new content that supported the computer system was established.

In addition, a call center was set. Project groups arranged around their cus-tomers were also formed for the first time. The projects related to new products and services that were performed in the call center were led in the matrix way by project management. In other words, cross-function teams were initiated to respond to the customer needs. But, there was no standard procedure yet for managing projects.

There were so many things that the company would need to do in the next several years. But now, CAT, Inc. was a new company inside-out.

Discussion items

1. What were other operative and strategic actions, not mentioned in the case, which might have been executed to accomplish James' vision?
2. How do the changes affect the strategy of CAT, Inc.?
3. What role did project management play?

The Project Hand-off Method

Dragan Z. Milosevic, Russ J. Martinelli, and James M. Waddell

Hospi-Tek is a medical equipment manufacturing company who has historically used a project hand-off approach to develop its products. They are currently under intense time-to-market pressure from their primary competitor, forcing senior management to reevaluate this approach.

A HAND-OFF METHOD

Under the project hand-off method, the Hospi-Tek product development effort began with the architectural team who developed an architectural concept and derived the high-level requirements of the medical device from the work of the product marketing team. The architectural concept and specifications were then handed-off to the hardware engineering team, who assumed ownership of the project. The engineering team developed the hardware requirements, engineering specifications, and the product design, which were then handed-off to the manufacturing team, who assumed ownership of the project. The manufacturing team developed the manufacturing processes, retooled the factor, and produced the physical product. The product and project ownership were then handed-off to downstream engineering teams, such as the software development team. The software team developed the software stack, then handed-off the combined hardware/software product, as well as project ownership, to the validations and test team. Finally, the validations and test team performed product- and component-level testing to ensure the product achieved the functional, quality, usability, and reliability requirements.

Management of the project was accomplished through a project management-only model, with multiple project managers in control of the project as it progressed through the development life cycle. Thus, a project manager with the functional expertise specific to the phase of development the product was currently in assumed ownership of the project.

JUDGEMENT

The hand-off method of development is common in smaller, less mature, and technically focused companies in which true project management value is usually not well understood and the engineering functions reign king. Unfortunately, this method is not scalable, and as a company begins to succeed and grow, product and process complexity requires the management team to look at alternative methods to structure and manage its product development efforts. This was the case with Hospi-Tek.

Discussion items

1. Why, by implementing the hand-off method, are there multiple project managers in control of the project as it progresses through the development life cycle?
2. Do you agree with the judgment that the hand-off method is popular in companies where true project management value is usually not well understood? Why or why not?
3. What are the pros and cons of the hand-off method?

Chapter 2

CULTURAL ASPECTS OF PROJECT MANAGEMENT

This chapter discusses the cultural issues of project management, which basically impact how projects are performed. "Culture" involves how people do things, the methods they use, that influence a project's ability to meet its objectives. It is a collective programming of the minds, generally used to understand basic values of a group, and is used (or sometimes abused) by management to direct the behavior of employees to achieve better performance. Two cases are featured: one is issue-based and the second is a critical incident.

1. Engineering Culture at Beck
     Engineering Culture at Beck is an issued-based case. It discusses the functional culture of an engineering department.
2. The Jamming
     The Jamming is a critical incident. It specifically discusses a culturally compatible strategy, called the jamming.

CHAPTER SUMMARY

Name of Case	Area Supported by Case	Case Type	Author of Case
Engineering Culture at Beck	Functional Culture	Issue-based Case	Dragan Z. Milosevic, Peerasit Patanakul, and Sabin Srivannaboon
The Jamming	Culturally Compatible Strategy	Critical Incident	Dragan Z. Milosevic, Peerasit Patanakul, and Sabin Srivannaboon

Engineering Culture at Beck

Dragan Z. Milosevic, Peerasit Patanakul, and Sabin Srivannaboon

IF JIM SAYS SO

Jim Traddell, director of a PM Office, and the project initiator, led the meeting with the consultant. He made every effort to look decisive, as he dictated the conclusions of the meeting to the scribe.

Raja James was Jim's partner. He was helping Jim on this project, particularly on the aspects of the culture. The team wanted to learn more about the engineering culture at Beck. To obtain the information, the meeting with an engineer at Beck was arranged at a local restaurant around the corner. Raja was about to leave the office to meet his informant.

BACKGROUND

Beck operates in an environment that abounds with rapid and discontinuous change in demand, competition, and technology. Even worse, that information is often not available, obsolete, or inaccurate. The company is founded in 1946, and its primary income comes from projects. Annual income is approximately $800 million, with 4,000 employees. Disequilibrium, along with perpetual and discontinuous change makes all competitive advantages temporary. Organizational units are loosely coupled with a lot of entrepreneurial behavior. Any advantage is temporary, contributing surprise and flexibility to Beck strategy. In this industry, risk is viewed as a factor upon which to capitalize, and the trick is that succession of fleeting advantages lead to higher performance.

ENGINEERING PRIDE

Tippie Anne, the 34-year veteran and engineer, showed up at 11 AM in front of Bombay restaurant for the lunch meeting. After pleasantries were exchanged, Raja invited Tippie inside where the aroma of curry dominated the air. As they helped themselves to an abundance of Indian food, Raja took the first step: "We'll have to cut to the chase. Let's start the interview now, as we both are very busy, and I would appreciate finishing by 12 noon." Raja announced his intent to focus on the professional part of the lunch. Tippie agreed with his idea, commencing a conversation. "Jim Traddell, one of my good friends, asked me to tell you my story of Beck's culture. Personally, I don't like to begin interviews with the word 'culture' because it tempts interviewees to become theoretical. But this interview is an exception." Raja replied, "Thank you. Then, let me ask you, how do you do things around Beck's now and how did you used to do them?" Raja hoped that using this simpler language would inspire Tippie to speak about culture in easier terms. The conversation went as follows:

Tippie: Well, we have a habit of calling ourselves "engineers," which we are. Perhaps this is in protest to having no such name 20+ years ago when Beck was small and had no real financial power. Today, for comparison, the new Beck is 4,000 employees strong and has $900 million cash at hand to purchase other companies. Nowadays, everybody respects us as engineers. If you talk to senior managers, engineering managers, and workers, you'll see for yourselves, how much these folks esteem us. Not that other specialists are not held in high regard as well, but when in the past our company would negotiate products, engineers would not be always seen as an important part of the negotiating team. Surprisingly, at old Beck our product sold without engineers around. But not for long. The old Beck valued MBAs. If one desired to get ahead, he had to get an MBA degree. Engineers were engaged almost like mercenaries. Yes, I am telling you the truth. Don't laugh there, Raja. The newly minted MBAs, mostly second-rate engineers, would prepare business plans for new products, the NPV (Net Present Value) and PI (Profitability Index), mostly what we call today derivative products—routine stuff. Of course, they embellished the products, hired engineers like "mercenaries" and business thrived. As I said, though, not for long. Once the substance was sold out, the derivatives didn't sell and there were no new breakthroughs (products new to the world) and platforms (products new to the companies), the business sank into the red, resulting in years of suffering.

The new Beck was in the making for years and the senior managers spent a lot of time creating project culture. Those platforms and derivatives products helped employees come about. Managers as well as workers understood that Beck would not survive if it didn't have high-value products. Those high-value products must be designed and developed by engineers and, they were. That's why they are respected and held in high regard, which is why they are feeling engineering pride.

DESIGNERS TO COST

Raja: I heard the terms "design to cost" and his cousin, "cost to design," mentioned many times at Beck. Would you please explain them to me?

Tippie: These proud engineers from the new Beck are very different in one respect from the engineering guard of the old Beck. Namely, current engineers are much more cost conscious, so their skill is much higher and the cost attitudes are essentially more developed. For example, the new ones always design products with the cost in mind. So, the finished product price must be one-fifth of a set of corresponding prices of spare parts. These proud guys have a different beginning policy. Theirs is "design to cost" as opposed to the old Beck's "cost to design." These two differ as two philosophies of life but their names are similar.

The former means that first the target product prices must be established; then the product is designed backwards. The desire is to first obtain the real product, component, and feature prices, then determine spare part prices. So, we have developed a tool called Kano's maps, which tells the price of each feature, thus we know which combinations of features really make money, and which don't.

The latter cost-to-design approach implies that engineers first design the product, then, they figure out the price, which the customer may consider overly high. Once product price is a known component, feature and spare prices can be calculated. While the former approach is proactive and customers love it, the latter is reactive and customers consider it obsolete.

Sometimes, customers view the cost-to-design approach as a way to rip them off. Skills requiring the design-to-cost approach are markedly higher than those demanded by the cost-to-design approach. Plus, customers like the former approach because it immediately tells the price, and gets their business from others. Sometimes, customers consider the cost-to-design approach not suitable for knowing the target product price sufficiently early. This skill, that engineers at new Beck have and those at old Beck did not, offers our organization strategic market opportunities. In particular, we see new markets and customers become available, and we think this is another good reason to feel like proud engineers.

Not so long ago, we got a call from a wholesaler who could turn around 1,000 pieces of one of our products with certain features. We consulted our Kano's map, and we learned that the asking price was right.

CUSTOMER-CENTRIC

Raja: I would like to learn how customers impact your engineering culture.

Tippie: Our engineers try very hard to be customer-centric. I know it sounds like a buzzword, but it is very precise in what it wants to denote. As in the old Beck, and in many other companies, we didn't ask the customer what they wanted in new products. We assumed by doing that, we would become customer-oriented. Rather, we developed whole arrays of processes and tools that pointed toward seeking out what the customer exactly wanted. The bottom line of being customer-centric is being able to help translate what customers precisely want of the product design. For that, we first used tools of survey design, customer segment, and custom visit documents to understand what customers wanted before the work began. Note that sometimes customers do not know what they want. So, our engineers designed all of the tools and processes to help them figure out what customers asked for. Then, in the design stage, we may use Quality Function Deployment (QFD) or rapid prototyping, to give design features customers long for.

They have a long experience of working with customers' people. For instance, they sit on joint development teams with customers, or use lead users' concepts to reveal what future products would look like. Our engineers are trained to stand in front of customers. You know, psychologists get trained how to talk to patients and elicit meritorious information; they are only college grads who get this education. No engineers get this kind of training. We at Beck spend thousands of dollars to prepare engineers to be comfortable talking to customers and becoming customer-centric.

RUN TO THE END

Raja: What else distinguishes Beck culturally from others?

Tippie: One thing I should mention as unique to us is our motto of "run to the end." Namely, in old Beck, our first priority was to secure new business, and a concern for producing a product at times fell through the cracks. So, as a consequence, acquisition of the business really mattered, and implementation of business suffered. Obviously, more attention was paid to the front end of the order process, rather than to the back end. At times, we did not know where this product was in the production cycle. As a further consequence, all kinds of promises were made to customers to bring the business in, usually forgotten easily in the production stage, and that was considered "normal" and ethical. Customers, anyhow, felt cheated.

The new Beck does not tolerate this. We worked very hard to show our engineers that all products are made equal. So, the customer has every right to know how long down the manufacturing line their product is. And, hence "run to the end." We put a lot of effort into eradicating this ugly habit—not only to know the product production status but where it will be the next day, in the next two days, until its completion. We, in other words, care about any product in production and show it.

Raja: Give me an example.

Tippie: We had a big wholesale customer in China. Thanks to this approach, we followed the product and its life cycle. So, this customer called us, and told us that market shifted away and two planned features didn't sell. We were at the third quarter of the project's life cycle but we managed to drop the two features.

Raja: What other ways of doing things around Beck are unique?

Tippie: We are a typical high-velocity company around the area. So, we have many ways that we share—the rewards, matrix organization, decision making, etc.—but those I described make us who we are.

Raja still had several questions he wanted to ask, but an hour had passed. He had no choice but to thank Tippie, and leave the restaurant. He had another appointment at 12:30, the second lunch meeting he needed to have for today.

Discussion items

1. List major definitions of the culture mentioned in the case. Analyze them and tell what is unique in each.
2. What do you like and dislike about Beck's culture?

The Jamming

Dragan Z. Milosevic, Peerasit Patanakul, and Sabin Srivannaboon

SCENARIO 1: JAM WITH THE COUNTERPART

An executive five-member team was formed to manage a small but global company. Because they were allowed to choose where they wanted to live, the team spread across Finland, Denmark, Sweden, and England. Although each member was multilingual, they spoke in English during their weekly teleconference. Every month the team met at one of the company's divisional headquarters and spent the next day with the managers from that division. Members were encouraged to be part of every discussion, although their individual roles were very clear, so that interaction on a day-to-day basis was unnecessary. Even though the team never went through a formal team-building process, its emphasis on an agreed team mission, shared business values, and high performance goals for all members made it a true model of a well-jammed multicultural team.

SCENARIO 2: THE NPD GAME

When the team members first went to work on a product development project in a small high-tech company in the United States, it appeared that they would forever be at odds over every aspect of managing a project. A few projects and many fights later, however, a German, an American, a Mexican, and a Macedonian looked as cohesive as any other team. As they marched through their projects, they acquired an in-depth knowledge of each other's cultures and project management scripts. Not only did they know each other's religious holidays and eating habits, but they also reached a point of accepting American concern for cost tracking, German obsession with precise schedule management, Macedonian dedication to team spirit, and Mexican zeal for interpersonal relationships. The road to their masterly jamming was not paved by deliberate actions. Rather, it evolved from patient learning, many dead ends in their interactions, and the need to be successful in their work.

JAMMING

The situations described here can be called "jamming,"—a strategy that suggests the project manager and the counterpart improvise, without an explicit mutual agreement, and transform their ideas into an agreeable scenario for their work. In this sense, they are like members of a jazz band following the loose rules of a jam session. "Jazzers" jam when they begin with a conventional theme, improvise on it, and pass it around until a new sound is created.

This strategy implies what is apparent in the executive team (scenario 1)—all team members are highly competent. Such competency enabled them to fathom the counterparts' assumptions and habits, predict their responses, and take courses of actions that appealed to them. Another condition was met for jamming to work with the executive team, in particular, understanding the individuality of each counterpart. A counterpart's fluency in several scripts clearly meant that he or she might propose any of the scripts' practices. Knowing the individuality then meant anticipating the practices. That the counterpart was analyzed as a person with distinct traits, and not only as a representative of a culture, was the key to successful jamming.

However, there are intrinsic risks in the use of the jamming strategy. As it occurred in the initial phase of the high-tech team (scenario 2), some counterparts did not read the jamming as recognition of cultural points, but rather as an attempt to seek favor by flattery and fawning. Although the team never faced it, it is also possible that jamming may lead to an "overpersonalization" of the relationship between the project manager and the counterpart, characterized by high emotional involvement, loss of touch with and ignorance of other team members, and reluctance to delegate.

Jamming's basic design may not be in tune with all cultures and may not even be appropriate for the execution by teams composed of members with varying levels of competency in other people's project management scripts. While in its early stage of development the high-tech team members' varying levels of competency were a significant roadblock, their further learning and growth got them over the obstacle. Still, the number and intensity of cultural run-ins that the team experienced before maturing supported the view that this strategy tends to be shorter on specific instructions for implementation and higher in uncertainty than any other unilateral strategy. However, its plasticity may be such a great asset to multicultural project managers that many of them view it as ideal in the development of a culturally responsive project management strategy.

Discussion items

1. In what situation would the jamming approach work well?
2. What are the pros and cons of the jamming approach?

Chapter 3

PROJECT MANAGEMENT PROCESSES

This chapter presents case studies of issues related to concepts in Chapter 3 of the *PMBOK® Guide*, namely the project management processes in which a process is defined as a set of interrelated actions and activities performed to achieve a predetermined product, result, or service.

There are four issue-based cases in this chapter.

1. Special Session
    Special Session is an issue-based case that generally describes different project management process groups, defined in the *PMBOK Guide*, including Initiating, Planning, Executing, Monitoring and Controlling, and Closing.
2. Waterfall Software Development
    Waterfall Software Development is an issue-based case which centers on the waterfall model for software development projects. In this case, six common phases of the model, including its limitations, are discussed.
3. Extreme Programming
    Extreme Programming, an issue-based case, talks about a methodology that is intended to improve software quality and increase the responsiveness of the software development organizations to changing market demands. The process has the iterative nature of the development, which facilitates continuous feedback to maintain management visibility of the project status and budget consumptions.

4. Do You ZBB?

This issue-based case presents one technique of project estimation and the selection process, known as zero-based budgeting (ZBB). It provides an example of applying the ZBB concept to everyday life; in this particular case, getting a degree while still balancing the work needed with life. In comparison, a strategy is expressed in terms of a bucket of money that a business unit wants to spend making the strategy work. In the process of selecting the projects, their estimates therefore should be equal or not exceed the bucket sum.

CHAPTER SUMMARY

Name of Case	Area Supported by Case	Case Type	Author of Case
Special Session	PMI's Project Management Process Groups	Issue-based Case	Sabin Srivannaboon
Waterfall Software Development	Software Development Model	Issue-based Case	Osman Osman
Extreme Programming	Software Development Model	Issue-based Case	Mani Ambalan
Do you ZBB?	Select Project	Issue-based Case	Rabah Kamis

Special Session

Sabin Srivannaboon

One fine Monday in spring 2008, Thomas Glacia was waiting for his students in a classroom at the 4th Avenue building to come back from a 15-minute break. Thomas was a senior lecturer of a local institute in Eugene, Oregon. This term he was teaching two classes, which were already completed. But his assignment wasn't yet over. He still had to teach this special session, which covered project management concepts. Why special? It was because there were only five undergraduates in the class; Jill, Kelly, Alice, Mackey, and Octavio, who failed the final exam, and needed to retake the exam in three days. As Thomas was looking at his lecture notes, his students returned to the room for the second half.

SPECIAL SESSION

Thomas: Okay, guys. Let's continue. We have discussed project management concepts and the differences and relationships between project management and their related principles. We also talked about the triple constraints which every project must meet: the time given, the budget, and the acceptable quality standard. Now let's move on and talk about the project management process.

Now someone tell me how many processes do we have, according to the PMBOK? And what are they? You guys already learned about it. Come on.

Alice: There are nine processes. They are time management, cost management, quality management, risk management, etc. Right?

Thomas: Well, nice try, but not quite. Those are the knowledge areas, Alice. We have only five process groups, not nine! What are they? Anyone?

Kelly: Initiating, Planning, Executing, Monitoring and Controlling, and Closing process groups.

Thomas: Exactly. These processes overlap, and are used as a guideline for applying appropriate project management knowledge and skills during the project.

They are iterative, and many processes are repeated during the project. The first process is initiating. Like its name implies, this is when we initiate a project. This is when we get some ideas of the project scope, and identify the purposes of the project.

Kelly: What are we really looking for in the scope and purpose of the project?

Thomas: We need to clearly and explicitly define what the project is intended to achieve and what its scope of interest will be. One output of the process is a project charter. And the charter is . . .?

Jill: I just read about it. The charter is a summary of project team members.

Thomas: Well, that could be one element in the charter. The charter is actually a document that formally authorizes a project or a phase and documents initial requirements that satisfy the stakeholders' needs and expectations. Now for the second process, the planning process—what do we have to do there?

Mackey: We can break a project into a number of smaller pieces, and arrange them in a top-down structure, which is called a work breakdown structure. Also, we should define what resources and time commitments are required to carry out the project through the work breakdown structure.

Octavio: Also, we need to create a project plan that involves resource plan, financial plan, quality plan, acceptance plan, and communications plan.

Mackey: But how will we deliver a project and present it to our customer for their acceptance?

Thomas: Not too fast, Mackey. That's the third process, the execution. The execution process is undertaken to perform the work defined in the project plan to achieve the project's objectives. The activities here may include creating project deliverables; managing tools, equipment, and people; managing risks; and creating project data such as schedule, cost, technical, and quality progress.

Alice: I think the most challenging task for me is not executing, but rather monitoring and controlling the tasks executed.

Thomas: That's the fourth process. We monitor and control projects to identify the potential problems in a timely manner so corrective actions can be taken promptly. Monitoring and controlling projects will benefit the project performance since it is observed and measured regularly to identify variances from the project plan.

Octavio: Could you give some examples of monitoring and controlling a project, please? I mean what exactly do we need to monitor in a project?

Thomas: What about risk? When risk becomes a problem, it can greatly affect a project's cost, schedule, scope, or quality. So, it needs to be closely monitored and controlled. One way of monitoring and controlling risks is to create and deploy a Risk Management Plan which anticipates any project challenges. Usually, the progress of monitoring and controlling is provided through the project status report to all stakeholders of the project.

Jill: Monitoring and controlling are the same, aren't they? I am not quite clear on these two terms.

Thomas: They are related. Monitoring is collecting, recording, and reporting information concerning all aspects of project performance that the project manager or others in the organization wish to know. In our discussion, it is important to remember that monitoring should be kept distinct from controlling and from evaluation.

On the contrary, controlling uses the data supplied by monitoring to bring actual performance into approximate equivalence with planned performance. Evaluation is performed through judgments that are made about the quality and effectiveness of project performance.

Mackey: How do we monitor our projects?

Thomas: Usually, the first task in monitoring projects is to identify the key factors in the project action plan. The factors should focus on results rather than activities. Then, we need to collect several important data. Data can be frequency counts, numbers, subjective numeric ratings, indicators, and verbal measures. Then, we can generate project progress reports after data collection has been completed.

Alice: That must be difficult. I bet there would be problems with the project reporting.

Thomas: Yes, there are actually three common project reporting problems, which are too much detail, poor correspondence to the parent firm's reporting system, and a poor correspondence between the planning and monitoring systems.

Jill: What kind of analysis do we use to monitor the entire project?

Thomas: There are several techniques. The most well-known one is the Earned Value Analysis. Earned value analysis is abbreviated as EVA, measuring overall performance by using an aggregate performance measure. The earned value chart depicts scheduled progress, actual cost, and actual progress to allow the determination of spending, schedule, and time variances. In practice, there are a number of commercial software packages that can help in monitoring project status, but the common desirable attributes that most project managers prefer are user friendliness, schedules, calendars, budgets, reports, graphics, networks, charts, migration, and consolidation.

Mackey: What about controlling?

Thomas: Control is an act of reducing the difference between plan and reality. Project control focuses on three elements of a project: performance, cost, and time. Is the project delivering what it promised to deliver or more? Is it making delivery at or below the promised cost? Is it making delivery at or before the promised time? There are two purposes of control. First, it is to regulate results through altering activity. Second, it is to conserve the organization's physical, human, and financial assets.

Kelly: What about the control report? I learned about this, but I forgot. What should be included?

Thomas: The control report should include project objectives, milestones and budgets, final project results, and recommendations for improvement.

Jill: I see. A good controlling system must help. So what kind of controlling system should we establish?

Thomas: It should be flexible, cost effective, and truly useful. More importantly, it should operate in an ethical manner, in a timely manner, and be sufficiently accurate. Anything else you want to add?

Kelly: Well, it should be easy to maintain and simple to operate.

Alice: It should be fully documented and extendable, as well, I think.

Thomas: That's good. One of the problems in controlling projects is the control of change. It is difficult because it causes uncertainty, increases sophistication in a project, and has to modify rules applying to the project processes.

Let me recap. We already discussed four project management processes; project initiation, project planning, project execution, and project monitoring and controlling processes. Now let's move on to the final process, the closing process. How many ways can a project be terminated?

Jill: I think it can be terminated by the extinction. Sorry, I don't remember the rest.

Thomas: Come on. The exam is coming in three days. In addition to the extinction, a project can also be terminated by addition, integration, and starvation.

Octavio: I didn't know a project can be stopped by addition! What exactly is that?

Thomas: It's in the book, Octavio. You should read it again before the exam. It means that project personnel, property, and equipment are simply transferred from the dying project to the newly born division. It transforms a project into a division of the firm and then, if real economic stability seems assured, the new project in that division can be created.

Kelly: How do we know when to terminate a project?

Thomas: Making a decision to terminate a project is difficult, but a number of factors can be used to reach a conclusion. For example, the factors that are considered in terminating projects are technical, economic, market, the customer's satisfaction, the impact of the project on the organization, etc.

Alice: What should be included in the project final report?

Thomas: You tell me.

Mackey: I think the report should include the process knowledge gained from the project.

Octavio: It should include what we have learned. For instance, it should incorporate project performance comments, administrative performance comments, personnel suggestions, etc.

Thomas: That's good. Time is almost up. Make sure you guys review the materials today again before the exam. This is your last chance. And I hope that I won't see you in class again next year. Good luck guys!

Discussion items

1. What project management tools were mentioned in this case?
2. Which of the process groups do you think is the most challenging one? Explain your reasons.
3. How are the knowledge areas and project management processes related?
4. In your opinion, what is termination by starvation?

Waterfall Software Development

Osman Osman

Robert Adam, the project manager of Renegade, just received a notification that his project was in an idle stage for four hours due to a specification conflict. The conflict came from the disagreement of how to accomplish one of the project tasks. Dangerously, the disagreement was built around specification deviations, thus delay of the project was likely inevitable.

Robert was well aware that his project would make no progress until this issue was resolved. Consequently, if the project did not meet the specifications during the review, it would not progress at all until a correction had been made or modification to the specification document had been approved and recorded. This irritated Robert. So, he called for an emergency meeting with the involved parties.

RENEGADE

Renegade is a health insurance provider in the southeast region. The company has employed the "waterfall" methodology for their software development projects. The waterfall methodology is based on the assumption that projects can be managed better when segmented into a hierarchy of phases, stages, activities, tasks, or steps. The waterfall methodology therefore directs a software development project to progress in an orderly sequence of development steps. This also means concurrent development steps of a project are prohibited.

Renegade has experienced a rapid growth over the years. In just a short five-year span, the numbers of subscribers has increased from two hundred thousand to more than one million subscribers. Management predicts even more growth. This number is only the subscribers, and their dependents are not included in this number. Therefore, the rapid growth of subscribers means extensive data to manage for Renegade. The company has been using Microsoft Access as their data management system. Renegade realizes the limitations of MS Access in managing large amounts of data, now that they have to deal with their aggressive demands. Therefore, an upgrade to more sophisticated data management systems is a must.

After some research, top business managers decided to convert the database management system to Oracle, which should have been be able to provide flexibility, ability to manage huge amounts of data, expandability, and referential integrity.

THE PROJECT

To respond to this need, Renegade initiated a project whose purpose was to implement:

Oracle database with proper connectivity to the existing system, and

Safe data transfer from MS Access to Oracle

Project Budget: $200, 000
Project Duration: 10 weeks

The pharmacy department (PD) was responsible for the project and its cost. The PD selected Robert Adam as the project manager and Leila Rakoba as the business analyst, both of whom were very experienced individuals.

Robert was the main person in charge of any problem/issue/concern regarding the project. He was responsible for all aspects of the project, including project schedule, development matrix, budgeting, and issue escalations.

Leila was the major point of contact for the user. She was responsible for collecting all the requirements from the users and documenting them in the business requirement document.

WATERFALL SOFTWARE DEVELOPMENT

As mentioned, similar to other projects this project followed the waterfall software development model. The project must complete each phase and go through a final walkthrough (FWT) before it could move forward. The formal walkthrough committee usually involved the phase completion check-off and presided over all sign-offs. The committee consisted of managers from different functional areas. Robert was also involved in the FWT, but was not part of the committee. So his signature wasn't required for phase completion sign-off.

The purpose for the FWT meeting was to ensure that specification requirements of the particular phase were clearly met and documented. Requirements must be validated and exit criteria must be satisfied before the project could progress. In other words, the waterfall model created disciplined project management and ensured the adequacy of documentation and design reviews. It set all requirements, schedules, and expectations before the project kick-off.

Figure 3.1 shows a preliminary outline that represents the overall process of software development life cycle at Renegade.

Figure 3.1   Software Development Life Cycle

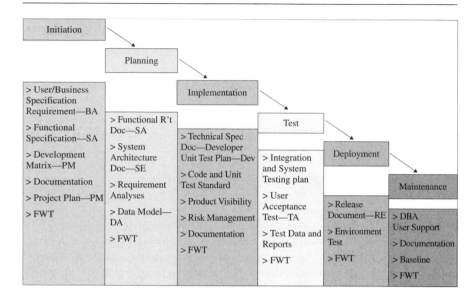

INITIATION PHASE

As the project manager, Robert was responsible for the development of the resource requirement document and project plan. To prepare these documents, he needed to get help from different functions. In particular, Robert needed assistance from the Solution Engineer (SE), Systems Analyst (SA), Developer, Test Analyst (TA), Release Engineer (RE), Data Architect (DA), and Database Analyst (DBA).

Robert also needed to write a development matrix document that consisted of the deliverables and the responsible members for the deliverables. Additionally, he needed to produce a resource-planning document that included the roles and requirements of the resources. For example, Robert requested the database analyst who had a minimum of 3.5 years of work experience in related fields, and possessed good Oracle and MS Access technical knowledge.

FIRST FORMAL WALKTHROUGH MEETING

It was November 13, 2007, 10 days prior to the kick-off meeting. The PD already allocated all required resources and assigned them to Robert for the project. It was his responsibility to use the resources optimally according to the schedule and budget. With Leila's help in facilitating and conveying all the project requirements, the meeting was sure to go well. The team was more than ready.

PLANNING PHASE

As expected, the team passed the first FWT meeting, and moved onto the planning phase. Robert and his team went on, and started laying out the project plan. But when the team worked on the project plan for a week, a conflict arose. That's when Robert called for an emergency meeting.

EMERGENCY MEETING

In the meeting, Robert wasn't very excited to hear issues or complaints from Luke, who was the systems analyst. Robert knew that Luke was very experienced but stubborn and liked to make last-minute changes. The role of the systems analyst was to design how to implement the Oracle system, and the developer (Jason) and the solutions engineer (Jasmine), should agree to the plan before the next FWT.

Robert: So, what is going on? Could someone explain to me what is happening?

Jasmine: We could not agree on how to handle *flat file bit* conversion. I am not sure if you're familiar with the *flat* file.

Robert: No, please explain.

Jasmine: Okay, it is easier if I draw. (See Figure 3.2.)

Jasmine: The new system must be able to receive the 16-bit from the existing system to ensure that input feeds are not misconfigured. So the new system must be programmed to handle it. The small box within the Implementation box is the conversion process that is going to be integrated into the introduction of the new database. This conversion block is Luke's plan and the point of conflict. Jason and I are opposed to it, as it is beyond the scope and not part of the project plan.

Figure 3.2   The Flat File

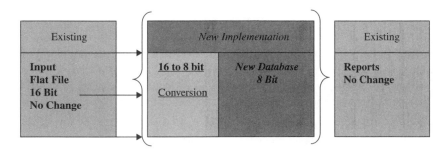

Luke: Robert, what I am suggesting is that this conversion process will satisfy the need to implement the project and it will not take long to develop. I do not see any reason to do the more complex work of programming the new DB to 16-bit capability. *Why break a wall when you can walk around it?*

Robert: Luke, I realize the complexity of programming the 16-bit DB capability but your solution is suggesting a scope change. Again, we cannot spend more time going back and forth; once the plan is set we should abide to it. Therefore, scratch the conversion plan. Stick to the original script. Any questions?

Luke: It does not make sense that we have to reinvent the wheel. I have learned a lot about this system during the development of the *referential integrity* (thousands of tables in the database connecting through a key or specific reference number). The conversion will take 18 hours less than developing a 16-bit database interface.

Robert: I do realize and appreciate your enthusiasm but specification requirement documentation indicates that the new database will be 16-bit capable of interfacing the input of the 16-bit flat file.

Even with discussions after discussions, Luke seemed not willing to concur. Robert looked for someone who could assist in solving this issue. Luke was a senior systems analyst and had been around for a while. He reported to a more-senior manager who was not easily accessible. Therefore, Robert decided to talk to his own manager, Steven, who was the program manager. Steven sent Luke an email informing him of the direction the project should take and carbon-copied Luke's manager.

Robert learned the next day that Luke conceded and the project was progressing ahead. Shortly after, with some leftover tension still in the room, the next FWT meeting went smoothly. The project successfully moved ahead to the implementation phase.

NEW PROBLEM

Since then, the project had been going along as planned except until it reached the test phase. Sam, the test analyst, found incorrect implementation done during the development phase. He came to Robert to consult.

Sam: Robert, I cannot continue with the test. It is failing for populating the table and running appropriate data in the database tables. I already explained this to Jason. He roughly looked at it, and estimated about a week to fix the problem. But you know we don't have a week. Could you please talk to him?

Robert: When did you talk to him about this?

Sam: This morning. I have sent an email to him and talked to him two times already. He told me he has to complete another high-priority assignment and then he can get into this. I cannot wait any longer. I have to complete the user acceptance test and develop the test reports in two days.

Robert: Well let me handle this. I will talk directly to his manager to speed things up. Can I use your phone?

Robert grabbed the phone, and made a call to Jason's boss, John, who was a development manager.

Robert: John, this is Rob. I would like you to talk Jason. This is about completing or correcting the Oracle implementation code he and Sam discussed. We really do not have much time left and we cannot delay this anymore. I understand he is busy but the project is on hold. And this project is critical to our company, you know that.

John: Oh, I wasn't aware of this. Let me check with him, and I will get back to you. I am sure he could postpone what he's working on now. He must have misunderstood the project priority. I'll handle it.

Soon after Robert hung up, Jason called and informed him that he was ready to work. Surprisingly, it only took five hours for Jason to fix the error. Now the project was able to move on, however, it was behind schedule. At that point, Robert knew he would need to make a lot of phone calls and send several more emails to get this project done.

Discussion items

1. What are the pros and cons of the waterfall methodology?
2. How would you change the waterfall model to better respond to the time-to-market pressure?
3. Do you agree or disagree with Robert's ways of handling the problems? Why?

Extreme Programming

Mani Ambalan

eFusion Inc. was a Portland-based eCommerce company. In eCommerce business, one of the limitations that online consumers used to face was discontinuity in communication with the online store or company. This enforced discontinuity had been a significant cause for "shopping cart abandonment." PushToTalk™, a "Voice over IP" (VoIP)-based product innovation from eFusion, was a new breakthrough idea, which attempted to address this discontinuity. This innovation was based on an existing mature technology at eFusion called the EAG (eFusion Application Gateway). PushToTalk™ was a new communication paradigm, where online consumers could talk with a customer service representative of an eCommerce portal without having to shift their frame of reference from the website to a telephone.

Luis Machuca, the COO of eFusion, looked at this overwhelming innovation as the "defining signature" for the company. Changes in the underlying market conditions caused the product to go through a series of ownership and management changes—from eFusion, Inc. to ITXC Corp to eStara, Inc.

During the early stages of its evolution, early adopters and potential customers were companies that were "born online." The prevailing project structure was unique; it was functional within the organization and matrix structure within the company. The "Intel development culture" was prevalent; this, in combination with the fact that there was significant product learning that was still needed, resulted in a space shuttle methodology and solution—a highly overengineered product design approach was adopted.

This, in conjunction with a rigid development methodology, introduced latencies (*latency* is a delay, which is a period between the initiation of something and its occurrence) and resource costs in the system that made it unsuitable to respond to the rapidly changing market needs and conditions. The project methodology was slow and cumbersome and the product in its early form was too complex for the customers. This resulted in the entire initial target market being wiped out. This led management to consider and introduce a different development philosophy. They introduced a new model of software development and project management called "extreme programming," and ensured that there was a paradigm shift across the functional teams. There were many challenges that both management and the teams faced in adapting to the new paradigm. The company was under pressure to regain the market momentum that it had lost, and the teams were anxious to make any significant progress.

INTRODUCTION OF EXTREME PROGRAMMING

The extreme programming methodology is a software development/project management methodology that promotes agility and simplicity, typically involving pair programming and a cycle of frequent testing and feedback. It supports frequent "releases" in short development cycles, leading to productivity improvement by introducing checkpoints in which new customer requirements can be adopted. The methodology was introduced as an experiment to increase the responsiveness of the development organizations to changing market demands. Alignment of goals and a realistic comprehension of expectations were key to implementing the methodology across the organization. A critical element in the change was a reassessment of project timeline expectations and the management of schedules, tasks, and the scope to meet those timelines.

THE PROJECT

Midway during the project, there was a layoff due to cost-cutting measures, which reduced the number of people available to do QA, and increased dependence on the operations team. This affected the overall product delivery date. The team mitigated such potential slips by having detailed daily meetings to ensure that overall QA at the end was minimized. The team also cross-QAed each other to create a higher degree of reliability. However, a critical problem emerged: very low user adoption of the product itself. The engineering manager and the development team were involved in the analysis of this key problem and were in search of a solution. Project plans changed as a result of the problem.

There were product changes to enhance user adoptions. These changes were introduced based on feedback from the engineering manager and collaboratively ratified by management and product management. The extreme programming framework allowed for changes such as this to be absorbed without significantly impacting the development plan. Also, the product was designed in a sufficiently flexible manner to absorb the impact. In retrospect, the changes did improve adoption marginally; however, adoption seemed to be influenced by more fundamental business reasons. So, as a pure cost benefit analysis, the changes were not worth the impact.

This project was not organized in the typical way. There were budget constraints that required cost control measures. A new engineering manager was given the leadership of the engineering team. This manager also took on the responsibility of the project manager and the integration and release management team and minimized dependence on operations management. Furthermore, the engineering manager also reduced the dependence on explicit quality assurance teams and incorporated much of the quality assurance as a part of the software development. All of this was done by adopting the extreme programming methodology, a new paradigm of software development. This methodology facilitated a highly iterative, rapid development, which only delivers what the customer needs while creating better software.

Top management's involvement was largely along setting expectations around time-to-market issues as well as team motivation. There were organizational problems like relocation, adapting to new the paradigm, change of roles, change of management members, and so on. All of these happened a couple of times. Mr. Machuca, eFusion's COO, mentions that "top management was aware of the organizational problems, and relied on the engineering manager to mitigate the issues. The organizational difficulties stemmed from the cost control measures." The new engineering manager introduced a radically different development methodology. This seemed too risky for his peers, who played a supporting role in the development, such as operations. Furthermore, there was an implicit threat of job risk associated with such a methodology. These organizational problems were clearly communicated during the weekly status meetings in a very open manner. Both engineering managers and some of the team members felt that the project did not suffer much from turnover, rather, it suffered from a fear of loss of jobs.

In the traditional method, the project manager was responsible for facilitating the *PMBOK Guide* process. In the new model, the iterative nature of the development facilitated continuous feedback so that management always had visibility into the status and budget consumptions.

The WBS followed the traditional extreme programming model of breaking down the project into several small stories. This was done by the product manager. Each of the stories was then broken down into tasks. This was done by the engineering manager. Each task was taken up by an engineering team, and iterations were planned around these. Tasks were usually no more than two days in length. Such a simple deconstruction facilitated a continuous scheduling, budgeting, and planning. The very nature of extreme programming was built around change in the scope of the project, so there was change at the story scope and priority. However, once a story was accepted, it usually proceeded without much change.

RULES OF EXTREME PROGRAMMING

The continuous visibility that extreme programming provided created a constant communication between the product manager and the engineering manager. The product manager had the final say on the priority and the scope of a story. The engineering manager had the final say on the delivery time of the story. This strict separation of responsibility usually resulted in very little conflict. The small team size and continuous "build and iteration" model usually resulted in the critical path issues to be synchronized with the release plan; i.e., once a story was taken up, it was on the critical path; once the story was completed, it was out of the critical path. This was facilitated by the fact that any subsequent modification by another team of components that affected another team were guaranteed to always work due to the test-driven nature of extreme programming; i.e., all automated tests had to pass for components to be accepted into the code tree. And once the tests passed, the critical path components could be consumed by other teams without concern on the impact of the change.

The engineering manager did not use any project buffer time, as there was continuous visibility across the organization into the project progress. The operations manager introduced buffer time to reduce the risk of deployment. This buffer was about 30 percent of the expected time, and usually was helpful. Also, according to Eli, an engineering lead, "The risk-taking attitude increased dramatically over time, and rightfully so. At first we were trying to have 99.999 percent up-time, which requires a lot of strategizing before making changes and inhibits innovation and change. When we came to realize that our new market did not require this level of reliability, risk-taking increased. At the end of the project when the company was fishing for some way to make money, risk-taking was extremely high."

The design methodology prior to extreme programming was detailed software design based on traditional models (UMLs, etc.). In the new model the following was the methodology adopted:

Simplicity: The design was kept as simple as possible, and avoided introducing any capability that did not have an immediate need.

Class Responsibility Collaboration (CRC) cards: Traditional CRC cards, a brainstorming tool used in the design of object-oriented software, were used for the detailed design sessions.

Spike solutions: A simple actual code test of whether the proposed design was sound was conducted to mitigate design risk. No functionality was added early.

Improvement: Code was refactored (improving source code without changing its overall results) whenever and wherever possible. This model employed design to cost and manufacturability.

In the earlier model, the requirements were frozen well before development efforts were undertaken. In the new model, the requirements of a story were frozen before development was undertaken, however, the requirements of the other stories were allowed to change.

In both models, before the freeze, information that was mandatory involved:

a. Operational requirements
b. Performance requirements

c. Administration requirements
d. Management Requirements
e. Security Requirements

In the new model, requirements were always in a fluid state until development was ready to act upon them. Product management was the only department that was allowed to change the requirements. This was usually informed by customers or sales.

a. Usually there were no more than three iterations.
b. Design was frozen prior to development. But the design itself could be refactored if the development required it.
c. Yes, these were part of the plan before executing a story.
d. The freezing decision was made by the team lead, if the design met the needs of the story. There was a document that formally designated a frozen design.
e. Based on the extreme programming model, there were several automated tests for each task. These were part of an overall test framework. Tests were concomitant with development.

Design reviews were a collaborative team effort and were part of the everyday activity in extreme programming. These were part of everyday "stand up" meetings. Any issue that required further attention would then be focused upon by the appropriate resource. The entire development team participated. On routine issues, the design was resolved by the involved team members. The complex ones were signed off by the engineering manager. CRCs were the normal method of documenting designs.

The extreme programming model required the use of continuous integration. This meant that all development was always integrated. Furthermore, the model also required that the development methodology required constant refactoring. So the system evolved from simple to complex, and this permitted the multiple subsystems to evolve in an iterative manner.

In the new model, risk mitigation was achieved as a deliberate process of development. Specifically, the following were the risk-mitigation plans that were in place:

a. The product manager was always available to the team for clarification on the requirements.
b. Code was written to common coding standards.
c. The automated tests were coded first.
d. All code was written by teams of two. This, in conjunction with point b, mitigated the risk of a team member leaving.
e. Only one pair integrated code at a time.
f. Code was integrated often.
g. There was collective ownership of code.

h. Optimization was left to the end. This resulted in the team focusing on the requirement first.
i. All code had automated unit tests.
j. All tests needed to pass for the code to integrate.
k. When a bug was found, new tests were created to track this in the future.

Ongoing risks were identified as part of the daily stand-up meeting. There was no explicit risk analysis phase in the beginning. The individual risks that emerged were either due to engineering identifying gaps in requirements, or product management deciding to alter the priority and scope of a story under development. The above risk mitigation tactics were usually sufficient to cover the risks. Yes, these were part of the ongoing daily activity and implicitly contributed to the project plan. The primary troubleshooting mechanisms were the automated tests.

The project was monitored by tracking the velocity with which the stories were completed. This was communicated daily to product management as well as the engineering manager. Product management used the project velocity to assess whether the project was still on budget. Specification control was ensured by product management and the engineering manager ensuring that the tests accounted for the requirement being satisfactorily met. The primary method of identifying a problem was the nonconformance of the software to a test that was agreed upon. Control decisions were made by resolving the conflict of requirements and the tests. These were usually part of the daily development activity. The primary means of communication was email and phone.

The product management was treated as the customer by the development team. They were embedded with the development team, and were continually available for clarifications. The stories and the project velocity tracking metrics were the primary content of the communications. Communication was open. Most communication was informal.

There was a continuous customer involvement in conjunction with the product manager. After the first launch, the team members as well as the management realized that the customers were not interested in much sophistication in terms of usage and they preferred something simpler. Once past the beginning stages of the project, the customers' feedback was considered seriously and was used in decision making. The customers' primary influence was on the usability of the product. To encourage a customer to use this product, there was a deliberate campaign to educate the customer on the new version of it. This campaign also set expectations in terms of timeframes and deliverables. Product management created product documentation, as well as canned demos to highlight the new aspects of the product. These were initially rolled out to a few customers and their feedback was incorporated to enhance the communication of the message.

Vendor relationships were largely handled by the business development group and they were involved in the requirements phase of the project. The primary impact that the project had on vendors was to recreate new interfaces to facilitate the enhanced product. Vendors were treated as extraneous to the project and they were required to attend any team project meetings.

The project did use common project management tools for scheduling, budgeting, and the like. Microsoft Project was used as a tool for scheduling, resource planning critical path management, etc. It was also used in a continuous manner for tracking the activity around the project.

The original project management style and the associated engineering methodology were overly time-intensive for the nature of the solution. Specifically, the solution demanded a more iterative model of development; whereas the style was more a waterfall model (a sequential software development process), perhaps more suited for large systems with several subsystems. The newer extreme programming–based approach appeared to deliver greater short-term success. This model was different from all the other project management styles in the organization because all the others were the classical waterfall model.

Discussion items

1. Analyze the applicability of the extreme programming methodology to non-software-development processes.
2. Analyze the feasibility of mixing traditional development methodology with the extreme programming methodology.
3. What do you like/dislike about the extreme programming methodology?

Do You ZBB?

Rabah Kamis

A DREAM

Bob was really excited about his previous year's job performance and could not wait to go through the process of getting recognized by his boss in his review. He could not tell if he was getting a promotion or a nice hefty raise or a combination of the two. He thought he set a great role model for working with peers and customers alike to get the job done. He remembers getting three spontaneous recognition awards for helping customers and other groups across the company to launch a critical new product. Bob worked days, nights, and weekends to make sure the product was up to the highest quality and functionality standards.

He can imagine it now: His name will be called to come on the stage during the department update meeting to get a department or a company-wide recognition award, the audience will clap, and his name will be well recognized by many groups. He will have to decide where he is going to hang up the plaque to be visible to people passing by his cubicle. There is no doubt about it; he will be sought after.

Bob moved from one daydream to another. How much will his new salary be? Is he getting a promotion this year? He had seen the first draft of his review and it looked pretty positive. He did not have to worry about his performance review. He got up from his chair and walked to the conference room to get his review from his former manager, Sally, and his new manager, Shea. When he arrived, his ex-manager was meeting with a peer of his in the room and they overran their time. Shea was not there yet so he had to wait a little longer. When the time came, after he entered the room, Sally greeted him and they chatted for a little while waiting for Shea to show up. Not too long after that, Shea arrived and took a seat.

Shea: Here is your review. I would like you to read it and then we can discuss it.

Bob was surprised that Shea handed him the paper because he expected Sally to do it. He took the paper and started reading. The first page listed his accomplishments for the previous year and it seemed unchanged from the first draft he had reviewed with Sally. He noticed that the second page, which listed his areas of improvements, had changed radically to a more negative look. Bob did not get a promotion or a raise. On the contrary, Bob was given a negative rating.

One of the comments he found on his area of developments was: "Bob always makes commitments he does not meet and he does not update his stakeholders with the status of his deliverables. He should show discipline and communicate his ZBB list to stakeholders." Bob was shocked!

WHAT IS ZBB?

ZBB stands for Zero Base Budgeting which is a principle of prioritization used by many companies and government agencies and individuals. It is a tool that helps employers, departments, and employees to set out their priorities, project costs, and identify which projects will get funding and support. It is a critical and necessary process to manage the limited resources, especially time.

THE ZBB PROCESS

Different articles and literature agree on the concept of ZBB but may differ on the number of the process steps. Here is one common step-by-step process for creating a ZBB for an individual contributor:

- The individual comes up with a list of the projects and deliverables assigned to him by his superior. The projects or activities must be aligned with the team and organization goals or MBOs. (MBO refers to Management by Objective which is a method for driving results by setting the objectives of the team periodically and delivering on these objectives).
- The next step is to quantify the amount of effort or work needed to carry out these projects or activities, in hours, days, weeks, or months.
- The list of tasks is then ordered or prioritized from the most important or critical task to the least critical. The manager, supervisor, or project manager can help the individual to prioritize. As a matter of fact, an individual must have his manager agree on this list and together priorities are set. The task or project priority must be aligned with the group's priority.
- Now, it is time to draw the employee's personal ZBB line. This is the line above which the employee has enough time and resources to complete his/her tasks. The tasks below the line will not get done or will be done if one of the above-line tasks gets cancelled or done earlier than expected. The ZBB line is positioned at the Zero hour, i.e., the last hour consumed by the above-the-line tasks.
- The list with the tasks prioritized and the ZBB line drawn must be reviewed with the boss. The boss can concur or ask for modifications from the employee.
- The ZBB list needs to be communicated with peers, customers, and business partners. This is important so the expectations are set and understood

by the stakeholders. They can feed back their concerns to the employee to do modifications.

- As new projects arise, the ZBB list can be modified to include these projects or new activities and the ZBB line can move up or down depending on the importance of the new project(s).
- The process is repeated periodically. Some individuals have weekly ZBB lists while others have quarterly ZBB lists.

EXAMPLE

The following is a demonstration of creating a weekly ZBB list for a full-time student with 12 credit hours.

The student sets up the goal, objectives, and deliverables for the quarter which must be aligned with his/her job as a student. The long-term goal is to get a degree and graduate from college. Every quarter, he/she needs to complete at least 12 credits successfully. For this quarter, the student needs to complete 3 classes/12 credits by the end of the quarter. So his/her deliverables are completing the classes and balancing the work needed with life.

1. Task list: Study/do homework; exercise, sleep, shop, eat, clean house, and do laundry and dishes; party/hang-out, watch TV, and listen to music; camp, visit family, and stay connected to the world (read news, check the weather). The time available is 7 days x 24 hours = 168 hours.
2. Quantify time to be spent on these tasks: Study/do homework 4 hours per credit; exercise 2 hours, 4 times in a week; sleep 8 hours a day; shop, eat, clean house, and do laundry around 10 hours weekly; party/hang-out for 4 hours with buddies; watch TV and listen to music 1 hour a day; camp or go on a trip when possible for 8 hours; visit family or call for 2 hours; and stay connected to the world (read news, check the weather) 1 hour daily.
3. Prioritize the list:

a. Sleep	8 hours a day	56 hours
b. Study/Homework	12 credits x 4 hours	48 hours
c. Eat in/out	2 hours a day	14 hours
d. Exercise	2 hours 4 times a week	8 hours
e. Shop groceries/other	3 hours a week	3 hours
f. Clean house/Laundry	10 hours a week	10 hours
g. Stay connected/News	5 hours a week	5 hours
h. Visit family/Stay in touch	2 hours a week	2 hours
i. Watch TV/music/movie	7 hours a week	7 hours
j. Party/hang-out	4 hours a week	4 hours
k. Play games	4 hours a week	4 hours
l. Camp/trip	8 hours a week	8 hours

4. Drawing the ZBB line: The total available time in the week is 7 days x 24 hours a day = 168 hours

Task	Hours	Weekly Hours	Time Available
a. Sleep	8 hours a day	56 hours	112 hours
b. Study/Homework	12 credits x 4 hours	48 hours	64 hours
c. Eat in/out	2 hours a day	14 hours	50 hours
d. Exercise	2 hours 4 times a week	8 hours	42 hours
e. Shop groceries/other	3 hours a week	3 hours	39 hours
f. Clean house/Laundry	10 hours a week	10 hours	29 hours
g. Stay connected/News	5 hours a week	5 hours	24 hours
h. Visit family/Stay in touch	2 hours a week	2 hours	22 hours
i. Watch TV/music/movie	7 hours a week	7 hours	15 hours
j. Party/hang-out	4 hours a week	4 hours	11 hours
k. Play games	4 hours a week	4 hours	7 hours
------------------------------------ ZBB LINE ------------------------------------			
l. Camp/trip	8 hours a week	8 hours	−1 hours

5. Since no boss exists here, this step is skipped.
6. The student needs to communicate this list to his stakeholders who can be his friends, class team, instructors, and family.
7. The task list or task priorities can change on a weekly basis. For example, if a really close friend or family is getting married and the student is the Maid of Honor, then that task is added and prioritized for that week.
8. The student should review the list at least once every academic quarter.

WHAT ABOUT BOB?

After Bob recovered from his shock, he resumed the review discussion with his manager.

Bob: Sally, you were my manager last year. If you thought that my performance was impacted by no "ZBBing" why didn't I hear any feedback from you? All I heard were positive comments and smiley faces from you.

Sally: Bob, for the grade level you have, it was my expectation that you own your ZBB list and you should have come to me for help.

Bob: But Sally, you have been copied on all positive comments I received from our customers and peers I helped get going. They all appreciated what

I did and awarded me many spontaneous recognition awards. I impacted the company in a noticeably positive way.

Shea: Bob, while you did impact the company in a good way, you let down some of your teammates.

Bob, getting angrier to hear Shea who did not manage him last year, resumed his conversation to Sally.

Bob: Sally, I have not heard any feedback from my peers or you all year last year. You are giving me this feedback now, in my review time? Don't you think this isn't the right time or place for it?

Sally: Bob, you own your career in this company and you should come to me for help. But here are your options: You can contest the review to my boss or write a one-page rebuttal to the review and include it in your file.

Bob ended up contesting the review outcome with HR which carried out an independent investigation to see if Bob was treated unfairly. Bob had to submit documents demonstrating his performance. HR gave Bob's first manager, Sally, a fair chance to explain why she thought Bob was underperforming. After evaluating both arguments, HR discovered that Bob and his manager were both correct but because the manager did not communicate to Bob his ongoing performance during the year, they felt obliged to reverse the rating. Bob got his raise but not a promotion.

Discussion items

1. Explain in your own words: What is ZBB?
2. How does ZBB benefit the company?
3. What are the disadvantages of ZBB?

Chapter 4

PROJECT INTEGRATION MANAGEMENT

This chapter presents case studies related to Project Integration Management, which is Chapter 4 of the *PMBOK® Guide*. The project integration management includes the processes and activities needed to identify, define, combine, unify, and coordinate the various processes and project management activity. There are two comprehensive cases and one issue-based case in this chapter:

1. The Abacus Project

    The Abacus Project is a comprehensive case portraying a story of a successful project. The case details major events according to the project's life cycle. Various project management issues are presented, namely project organization, project life cycle, project strategy, project leadership, project spirit, learning, and excellence in project management.

2. The Ticketing System

    The Ticketing System also presents a story of a successful project. Different from Abacus Project, the Ticketing System is a small, internal IT project. This issued-based case portrays the issues in the early phases of the project such as understanding problems and requirements, searching for options, and making decisions. The case also discusses a project management approach that the project team used, which is quite different from typical project management methodology.

3. WRQ Software Development

   WRQ Software Development details project interaction—unforeseen problems of one project that will impact the other projects. This comprehensive case also discusses project team structure, feature-driven software development processes, dimensions of success, and team learning.

CHAPTER SUMMARY

Name of Case	Areas Supported by Case	Case Type	Authors of Case
The Abacus Project	Organization, Life Cycle, Project Strategy, Leadership, Spirit, Learning	Comprehensive Case	Peerasit Patanakul and Jospeph Genduso
The Ticketing System	Requirement Gathering, Project Management Approach	Issue-based Case	Mathias Sunardi
WRQ Software Development	Team Structure Software Development Process, Success Dimensions, Team Learning	Comprehensive Case	Peerasit Patanakul and Michael Adams

The Abacus Project

Peerasit Patanakul and Jospeph Genduso

It has been a week since the team energized the transmission line to the residence of the Seattle/King County region of Washington State, the day the project team members dubbed, "Day Zero." Trian Moore, the Director of Planning, recalls the day zero, which occurred on December 31, 2003. Five hours before turning to the New Year, the team switched on the new transmission line. They barely made it but they did and it was very exciting to get that done. Beslie Lelleher, a Lead Construction Environmental Engineer, said to Trian proudly that "even with the rough start, we brought this project down to a one-year delay instead of having a two-year delay. We have to tell ourselves that we have done a great job. And, we have to praise our project manager, Tom Lennon." That prompted a whole avalanche of Trian's memories of the project details. And, at the right time because several of them—"them" meaning some of the key members of the project team—were to give debriefings on Abacus project management, followed by a project management (PM) workshop to DBA employees with various levels of experience in PM. Joining on for the assignment are Tom Lennon, a project manager; Misha Coffman, a lead construction manager; Tene Mynard, a project environmental lead; and Beslie Lelleher, a lead construction environmental engineer.

PROJECT BACKGROUND

DBA is a federal agency headquartered in Portland, Oregon, that markets wholesale electricity and transmission to the Pacific Northwest's public and private utilities as well as to some large industries. DBA Transmission Line Project was initiated in 1999 according to the load problems in the northwest electrical grid based on the report from the planning department. In addition, the contract signed by President Kennedy in 1960s as part of the Canadian Treaty, or the Canadian Entitlement, was expiring. Per the contract, Canada would build three storage dams and would immediately sell all the power to the United States for a period of 40 years. Since the contract was expiring and the demand for electricity in Canada has been higher, the United States must return the power to Canada.

PROJECT ORGANIZATION

Debriefings began with project organization. The transmission business line of DBA is a typical matrix organization. The Abacus Project has the team, consisting of a core group and support staff, from cross-functional and cross-governmental agencies. In particular, after the project initiation in early 1999, Tom Lennon was assigned as a project manager and the team was formed. Originally, the team started with four to five key members from the environmental, engineering, and planning departments. The total personnel count at the end of the project was almost 200, including contractors and subcontractors. Also, the City of Seattle, the customer, got involved as part of the official project organization. See Figure 4.1 for an outline of the players involved in the project.

HOW THE IMPLEMENTATION UNFOLDED

After project organization, attention was focused on project life cycle (PLC). Abacus is a construction project with well-defined phases. They are project initiation, environmental evaluation, design and preconstruction, construction, and cleanup and transfer to operation phases.

PROJECT INITIATION PHASE

The first phase was initiated by the planning department and did not involve the project manager. In early 1999, the planning department ran their electrical load forecasting models to get an idea of the state of the load on the grid. The results from the models showed the potential load problems, which were then prioritized based on their forecasted negative impact. Out of those problems, the brownouts or blackouts in the Seattle/King County area had the highest priority. The system planners immediately drafted alternate solutions to solve this problem. These solutions were then input back to the forecasting models and the preferred alternative was determined. The Abacus Project was initiated. Trian Moore, the Director of

Figure 4.1 The Abacus Project

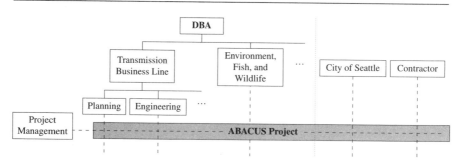

Planning, remembered clearly that at this point, the project manager was selected. Tom Lennon was a good candidate because he had done a similar project before. Tom started forming a team, developing a schedule, alternatives, and estimates of what the project could cost.

ENVIRONMENTAL EVALUATION PHASE

In this phase, the lead environmental engineer was tasked with producing an Environmental Impact Statement (EIS), clearly a turning point. Tene Mynard, the environmental lead, hired five different firms to compile the information into an EIS. These five firms evaluated the planning group's solutions as to their impact on the environment. Armed with that information, the environmental subteam put together a draft EIS. Once the draft EIS was completed, it was released to the public for review. Since the creating of the EIS is an iterative process, with all parties involved including the public, there were many addendums and changes to the original document.

Tom: At first blush, it looked like it wasn't too difficult a thing to do because there weren't that many homes in the area. But there was a watershed for the City of Seattle. So, your group, Trian, wanted this fixed right away.

Trian: Yes, we wanted it fixed ASAP.

Tom: So I generated a very fast schedule, and got a team of people together initially. We then started our location and environmental process on March 28, 2000, after we got a Notice of Intent to prepare the EIS and to conduct a project-scoping meeting published in the federal register.

Tene: I remember that we were looking strictly at three alternatives across the watershed. Since there is also an existing line across the watershed, one of the alternatives was to be immediately adjacent to that existing line. That was the one we focused on the most and went through the environmental process. But during the process, we had pushback from the City of Seattle. I think it was April 11, 2000, when we had our first scope meeting.

Tom: They did not want this transmission line through their watershed at all. We also received pushback from the environmental community. During the time we started this process, they had completed their own environmental analysis of the whole watershed. They wanted to keep the entire watershed for environmental purposes like a wildlife refuge and minimize erosion; they wanted to keep their water source as pure as possible. They did not want any additional clearing, but we needed to do initial clearing through there.

Question of Audience: Did you consider only one or more options?

Tene: We looked for more options. In fact, we had to look at other alternatives outside of the watershed. The environmental community agreed. We made a conscious decision to explore more options at that point knowing that was going to delay our process by at least a year, more likely two years. The other options impacted a whole lot of houses and residences. So no matter where we went, there were a whole lot of people involved. We looked at three main corridors outside the watershed. We had a very large pushback from those corridors as well.

Tom: People were literally screaming at public meetings. So we had to have meetings for every one of these alternatives, and that did take us another year to get through that process. At the same time we were negotiating with Seattle on going through the watershed. It took about two years to get through the negotiations, starting from October 2000. Finally, we were able to get to a point where we struck a deal with Seattle. It would allow us to go through the watershed. So we kept focusing on that one alternative that's adjacent to the existing line and kept going back to that. At the same time, we were able to do the surveying that we needed to do to determine where the towers needed to go. We even went to the point of ordering the materials and purchasing additional lands. So, all the focus was starting to be on that one alternative. By the time of the record of decision where DBA came out and said that this was the alternative we were going to go with, we had an agreement with the City of Seattle. We were able to pull everything together so that we had a record of decision in July 2003.

Trian: We all agree that we have done an impressive job here. We have done so many things in parallel. It paid off toward the end. I wonder how long it would have taken us if we had done everything sequentially.

DESIGN AND PRECONSTRUCTION PHASE

As the final part of the environmental phase finishes, the design phase begins.

Voice from Audience: Please summarize to us the scope of the design phase.

Trian: Tom, could you summarize to us the scope of the design phase?

Tom: Sure. What we did in this phase are multiple things. By the end of the phase, we wanted to get what we call a specification package. Normally, the phase starts with surveying the line and determining what structures to use. We would then be able to see where to locate those structures, which type within a tower series that we wanted to have per tower sites, and determine what the sags would be—that's all part of the design. Toward the end of the phase, our team had to go out into the field and start locating road access to each site. We also had to determine how many trees would be taken out.

Trian: And all of those things had to be recorded in a specification package?

Tom: Right.

Question of Audience: When did you start contacting the landowners because we had to buy the rights to be on their property, right?

Tom: Good question. We started contacting landowners after the surveyors got their information together. Normally, we have to negotiate with landowners, which can be a lengthy process, and it was no different with the Abacus Project. We had to condemn some landowners to put the line in. We had to go through a court process to get the rights. After all that was done, and the tower information was determined, we ordered the material. It took months to get the material onboard. Based on my experience, a standard transmission line project can take up to seven years. The Environmental and Design phases alone can take up to three years.

Trian: When did we send a request for bid?

Tom: Right after we got the specification package. The construction started after we had the notice to proceed from senior management.

CONSTRUCTION PHASE

Trian: Beslie told me that we would be able to catch up on some of the delay in the construction phase.

Tom: Right. In this phase, Misha (lead construction manager), along with Beslie (lead environmental construction engineer) worked with the contractor in order to build the transmission line. The reason that we could catch up with some of the delay was that the team was able to position themselves to start immediately after the record of decision was signed by senior management. In fact, the contractor on the job was already assembling the towers in anticipation of that event. But we also had some problems with our contractor.

Question of Audience: I heard that they were so slow afterward.

Misha: In addition to the contractor being slow, the special type of tower footing that they installed took too long to build and caused as much damage as another quicker method. We awarded the contract to them because they claimed that the footing they installed would cause less environmental damage than the standard footing.

Tom: Right. After negotiation, we released them and awarded the contract to another contractor, who specializes in the more standard footing. After that change, work got done much faster.

CLEANUP AND TRANSFER TO OPERATION PHASE

Trian: As of today, we are not totally done with our job, are we?

Tom: No. We still have to do some cleanups and transfer to operations. In this phase the site is continually evaluated for erosion and corrected for regrowth of foliage.

Beslie: I have to make sure that restoration is complete and we have stabilization at the site. I'm responsible for the storm water control permit and we have to prepare a storm water pollution prevention plan. And DBA regulates it. Where we are now: We have to do a lot of replanting and seeding, so I probably won't be done until July.

Question of Audience: How long will DBA be responsible for monitoring this site after it turns it over to Seattle?

Beslie: DBA will be responsible for monitoring this site for three years at which time it will turnover the site to Seattle. Once the line is energized, it is handed off to operations and maintenance, but we still have those responsibilities.

PROJECT STRATEGY

Discussing strategy was not easy because most project managers did not discuss it often. The Abacus Project had one main goal when it began: It had to keep the lights on for the residents of the Puget Sound area in Washington State. With the greater demand for electricity, the stability in the grid had to increase. By increasing system reliability within DBA's grid, the Abacus Project would boost the value *to the whole agency*. With a stable grid, the power could be distributed with less cost, due to the better stability. In addition, the reputation of the Agency would also improve.

The project and its product were defined based on the load forecasting models of the planning group. Trian Moore recalls, "The project was triggered in 1999. We received the new load forecasts and then identified the problems. We looked at the worst problems and focused on them. We agreed that the potential of losing the Puget Sound area to a blackout during extreme cold weather was the worst problem. So that was where we focused our attention." This forecasting is aligned with the business strategy of high line reliability. DBA's vision statement stresses reliability. "The Pacific Northwest electric power and transmission system has led the nation in four key values: low rates consistent with sound business principles, high reliability, a light environmental footprint, and accountability to the region's citizens."

The Abacus Project developed in conjunction with the DBA grid that enhances reliability. Since DBA brokers this power to many sources, including utilities, it is a priority and *advantage* to have a reliable grid.

The Abacus Project *strategic focus* was on time, but did not compromise on the environmental impacts of the project. It was clear since the initiation phase of the project that DBA's position was to increase the reliability in the grid and be environmentally friendly as much as possible during the project implementation. The project strategic focus was therefore to abide by several policies imposed by both the federal government and the City of Seattle. As a result, the team involved members from the Environment, Fish, and Wildlife group (internal within DBA), and adhered to the strict policies with regard to the environment. By abiding by these policies, the team had to follow several additional steps, including preparing the public reports and reviews. The construction had to be inspected by internal DBA departments, other government agencies, as well as the public.

Question of Audience: Explain slowly your strategy, using our methodology, and having in mind that we are not familiar with the industry terms.

Tom: The focus changed as we went along. We added to it. It was initially to get it done as quickly as possible. After we had our town hall meetings with the public, our focus began to change. We not only had to worry about the time but we had to start worrying about dollars and the water quality for Seattle. Also, the whole area inside the watershed is under a habitat conservation plan; we had to worry about that also.

Trian: I agree, but the dollars were not that strict, right? Senior management seems to understand that the environmental aspects of the project were the priority. I did understand that we attempted to get the project done as soon as possible but did not compromise on the environmental issues.

Tom: You are right. What started out as a $10.5 million project, became a $35 to $40 million project and we are still getting invoices in. Some of the activities that should have been done are not yet completed, like the roads haven't been fixed back up again. Several of the erosion control measures are not complete, so we will have to go back in the spring and complete those activities. So, I guess, you are right. Environmental issues did come first, while keeping in mind the need for getting the project done as soon as possible, especially before the winter when the demand on the load is usually at its peak. It was quite amazing, wasn't it, that we could energize the line by the year end even though we got the go-ahead in July.

PROJECT LEADERSHIP

Trian: Let's move on to project leadership issues. Any volunteer to start?

Tom: Let me start with evaluating myself. I think my style was leading by example, focusing on communication, and being hands-on when I needed to get involved. Like that time when I got a call saying that the micro-pile footing wasn't working out. I had to take a look at it right away. My first thought was that I needed more information. So I asked the team to start gathering some facts. As they were doing that I was heading out to the site to see what they were talking about. I immediately saw what was going on. I told the team to scrap what I had said earlier and go straight to Seattle and tell them that we would like to change the footing design. Fortunately, it went well and we went right over to the Department of Health and a couple of other places and told them we were changing our design and wanted to get their approval. I think this is a typical case where I would get involved and start negotiating to make sure that we were headed in the right direction.

Tene: There wasn't much experience building a 500-Kv line. We hadn't built one since the 1970s. As far as Tom's style, he has the most experience. He worked on an amazing number of projects at once and had been here a long time. He's good to work with because you can learn a lot from him. So his style is direct and he is good to work with.

Misha: I agree. Tom is a great leader on the job. I say this not because I want to please you, Tom. I just want to state a fact: Tom is very good. He's there when you need him, and stays out of your way when you don't need him. Then if something goes wrong, he will pull you off to the side and tell you how, in his opinion, you did it wrong. He doesn't embarrass you in front of people and supports what you did. He's top-shelf; one of the best out there, in my opinion. Beslie Lelleher was also a good leader with her team.

Beslie: I think for you to be successful, your team has to have the same goal. And you have to empower them. And you have to know when to step in and make a decision. Tom does. You are free to run and go, go, go, but if a decision needs to be made you know who the decisionmaker is. And that's important. That's how things get done efficiently. But Tom's definitely not a microman-ager. He expects you to do whatever and if you need help, come to him. So you are free to manage in whatever style is yours. The key is knowing when to step in and make a decision.

Trian: This is good. In my opinion, all of you have done a superb job. So, we agree that the key leadership issues here are setting a clear goal and empower-ing the team to accomplish that goal; focusing on communication; and getting involved when needed.

PROJECT SPIRIT

This project created a great sense of spirit on the team. By overcoming such adversity the team grew closer and worked extremely well together. An exam-ple of the spirit of the project was the tenacity to find a way to get the project completed through the watershed without much affecting the public. The odds and forces against the project reinforced the spirit. With all sentiment going against it, the project benefited from having a great support structure. Top management supported the project all the way through.

> **Tom:** One nice thing about this project was that management let me be very creative with the team. Our team members put their heads together to figure out ways to get things done with minimal impasse and I think we were really successful in that.

> **Beslie:** I think having the same goal brings a team together. Everyone worked well and that brought them together. Everybody wanted to be successful. It was the most challenging project anyone had worked on. We had known the players for many years, so that helped too. The environmental monitor, Steve, he commented to me over and over again how he so much enjoyed working with DBA because it was like a family. He said we are top-notch and high-quality people, and the fact that everyone kept their sense of humor helped a lot. He said he would have never expected this of a federal agency. It was quite a compliment, don't you think?

> **Trian:** So, the project spirit was high. This sentiment roots from a good team, good leadership, support from management, a family-oriented working environment, a sense of responsibility to the public, and a drive for success.

LEARNING

Preproject Learning

This project benefited from past projects. It had a plan and process to start the project. The project manager had worked on similar projects previously so he knew the general processes and procedures that would need to be followed from the beginning. This information was gathered both by the documentation of the previous projects led by the project manager as well as public information from other projects.

Ongoing Learning

> **Tom:** Abacus was a great project for maintaining ongoing learning. With all the changes and issues that came up, the team had to create and learn on the fly. The team constantly learned from working with the public. They learned about the watershed—how to go through a watershed and what to do with respect to not contaminating it. They learned some construction techniques—working with micro-piles and then having to switch over to plate footings; and they learned about tower structure—since the towers were of a new design, each one had to be "tweaked" to work in a specific location. These lessons learned were put in public documents, emailed to team members, and documented in meeting notes.

Beslie: I had to learn to create and implement plans because of the watershed. We had communications plans; traffic, safety, sanitations, and water quality emergency response plans. DBA hadn't worried about communications in the past. It was a totally new process to us. It wasn't the "straight flow down construction" communication model. We had communications plans between the leads, but nothing between DBA and the contractors or the contractor and the subs. We realized that that was a huge, valid plan that needed to be looked at on a project such as this.

Tom: In terms of the construction, to minimize erosion we started out with a micro-pile footing. Normally we have a plate footing, or grill footing, which require large holes. With the micro-pile, the hole is only six inches in diameter. So the thought there was the impact would be a lot less. As it turns out, to get the micro-pile in the ground you had to disturb a lot more ground moving equipment around. So we wound up stopping the process and going back to our conventional way and making sure that it was done very well.

Tene: I learned that the environmental process needs to begin early and the more time the better. We need to address environmental issues early and quickly. By not having enough time, poor decisions can result in a job not done cost-effectively or well.

Tom: The team met throughout the project to go over the lessons learned in previous phases. The whole team was involved so that the lessons could be passed throughout the whole organization. The project manager had to present the findings to several committees that were made up of managers and senior managers from all over the organization. This communication added to the spreading of important information and lessons learned from the project.

EXCELLENCE IN PROJECT MANAGEMENT

Behavioral Excellence

Tom: For the Abacus Project, behavioral excellence was strived for within the team as a whole. The team members were looking to be excellent in all aspects of their jobs. Each team member felt free to share opinions and ideas on the project. This promotes excellence in many areas including communications and innovation. Senior managers set an excellent example by supporting the project manager in his decisions.

Excellence in Project Monitoring and Control

Tom: Even though the project monitoring and control were very informal and casual, there were times where formal control was needed within the team.

When the formal control was implemented, it was at a level that didn't restrict the creativity or innovation of any particular team member. The monitoring within the team was mostly done at status meetings with the team and through emails. The subteams had weekly meetings and if the project manager could not attend, details were emailed to him.

Excellence in Innovative Thinking

Having the ability to think innovatively was very important to the Abacus Project. By creating ways of doing tasks that hadn't been thought of before or completing tasks in parallel, the team created ways to save time and money.

Tom: We positioned ourselves with the team so that we were able to start construction immediately after the record of decision. This is very unusual. We usually don't start surveying until after the record of decision. Prior to that we had a contractor on board and they were already assembling towers even before we obtained the record of decision with the anticipation that we were going to get it then immediately began construction on the ground. That was exactly what happened. Instead of having a two-year delay, we brought it down to one year. That time savings a result of one of our innovative ideas.

Misha: Right away we ran into a lot of constraints with atypical construction methods. We weren't allowed to drive over the ground off-road, we had to stay on roads, which makes digging holes very difficult. When we moved material we would have to pile it on plywood sheets, and then put the materials back so nothing was touching the ground. We had to steam all the equipment vehicles before they went into the watershed. The vehicles were inspected to make sure there wasn't any grease or leaking oil, and once we parked inside on the roads we had to put a diaper under the vehicle underbody. The hydraulic systems of the larger equipment had to be changed over to vegetable oil instead of hydraulic oil, in case they had a spill. Both spills are bad but vegetable oil is a little more palatable for a cleanup than hydraulic fluid. We had to implement emergency response plans inside the watershed. These were pretty controlled and constrained construction methods. So, we came up with the innovative idea of using helicopters. To minimize impact to the watershed, we used a helicopter to take all the trees out. That was very expensive. We wound up using helicopters to put the towers in place—again very expensive. Even though expensive, the use of helicopters was innovative and saved time and extra work. By eliminating the logging and tower material trucks coming though the watershed the team saved time by not having to repair the watershed.

Trian: Great job guys. We have shared a lot of valuable information today. Let me ask the audience: Do you have any comments or suggestions? Would you highlight what have you have learned from today's session? Otherwise, let's break out into groups and please develop a plan for a project similar to this. Let's assume that our next project is Grand Coulee-Bell Transmission Line.

Discussion items

1. What have you learned about the issues in the case?
2. Project organization in the case is described as the matrix. Please explain major features of matrix organization. Based on which information in the case, can you tell that the matrix is strong, balanced, or weak? Explain.
3. Describe your opinion about the pros and cons of both hands-on and hands-off styles of leadership. Which of the two styles do you prefer? Why?
4. Describe in your own words what project spirit is.
5. This case offers some examples of project management excellence. What is project management excellence to you? Define it in your own words and give examples.

The Ticketing System

Mathias Sunardi

The Office of Information Technology (OIT) of the Silicon Forest State University (SFSU) uses a ticketing system (software) to report bugs across different departments in OIT that maintain the information system at SFSU level; from the front end (user interface) to the back end (database, server, network). At some point, Remedy ticketing software was chosen. However, that decision was somewhat "flawed" in the sense that not all departments that needed the ticketing system supported Remedy. Remedy was Windows-based client software and it worked fine with most of the front-end side of OIT, but OIT's back end was mainly maintained under UNIX. This led the UNIX team, and several other groups that maintained the back-end resort, to use a different tool (Request Tracker) which was Open-Source software, and worked under the UNIX system. This case discusses project communication management, especially software bugs reporting among multiple departments involving in software development.

BACKGROUND

The OIT is a functional organization. It is under the supervision of the Vice President of Technology, Pike Gresham. OIT has four major departments: the Computing and Network Services Department, which manages the network system at SFSU; the Information System Department, which manages and implements all administrative systems for the university including databases on UNIX servers; the Instruction and Research Services Department, which provides equipment, hardware, software, resources, training, and support to students, faculty, and university staff; and the User Support Services Department, which is essentially the interface for users: students, faculty, and staff to the rest of the OIT when help is needed with technology-related issues. The ticketing system is used throughout these departments to communicate all the technology-related reports, issues, and requests.

Given that some of the departments use different ticketing systems—Request Tracker, or RT, (UNIX-based) and Remedy (Windows-based)—there exists a "gap" in the way information is being processed. The front-end side, which uses Remedy, must manually translate/convert the message from the Remedy format to the RT format in order for it to reach the back-end teams. This causes several problems. First of all, the process is time and resource consuming. It means that someone had to be on standby to do the "translation," and it took a lot of time to do, which made the response time relatively slow. Second, Remedy is more of an enterprise-level tool that provides many features which are not needed by OIT, such as finely detailed forms that request much information which does not fit with an organization like OIT. Those features make the reporting process even more confusing to the users. And finally, some information might be "lost in translation"—when the message is translated, there may be some information missing or misinterpreted. This issue has been on everyone's nerves for three years.

IT STARTED OFF, FINALLY

Ron Bashley has been working with OIT for three years and just recently has been promoted to Desktop Support and Project Coordinator. While he enjoyed his new position, he missed his previous office, where he had a nice big window to look out from the second floor of the building. His new office is located in the basement of a different, older building. Regardless, he is enthusiastic about his new position as it is the type of work he has been looking for.

It was July of last year, when Ron was going through his daily email-checking ritual that he noticed an email from his manager, Baken Dryhed, the Director of User Support Services (USS) Department. The subject read: "We need to fix the ticketing system." "Finally!" Ron screamed in his mind, "We're going to do something about the ticketing system." The body of the email was an invitation from Baken to all the users of the ticketing system for a meeting to discuss a solution for their problems. The meeting was to take place in their usual weekly meeting.

The meeting was attended by everyone who had been invited in the email, which includes: Baken, as the head of the USS Department; Ron, who manages user interface with the ticketing system; Harry Bonnett, the Director of the Information Systems Department; and the directors of Instruction and Research Services Department, Computing and Network Services Department, and some of their managers who work under them; and Bob Biyon, the Technology Manager from the School of Liberal Arts and Sciences. Bob is not working under OIT, but he uses the ticketing system to maintain the computer lab at the school. Together, they formed a committee for the project, as is usually required in most projects at the university.

TIMELINE

In the meeting, all attendees agree that they have a problem with the current ticketing system and they need to find a solution for it. The university has been on a tight budget, and OIT is one of the departments that experienced the most severe budget cuts. The cost of licensing Remedy at $20,000 per year, along with earlier disappointments by several other third-party software companies, leads the team to decide to find an open-source software solution. Baken proposed that the project be done in a year. Based on resources availability, Ron and Harry think the project could be done sooner than that. They proposed that the project be done in six months.

For the following few weeks, the team collected requirements from everyone in the meeting: what they wanted the software to be able to do; what they hated about the current system; etc. Ron and Harry and their teams were responsible for investigation into the search for options. Through their email mailing list called Listserv, the users passed along their new requirements, in addition to the ones already mentioned in the first meeting. These requirements were collected, and, together with the options they found, were brought to their weekly meetings to be discussed.

DECISION

The whole information-gathering process took about three months before they reached a decision. After considering all the requirements and the available options, they decided to replace Remedy with Request Tracker—the tool already being used by some of the departments in OIT. This was a good thing for them—since the software was already in use, they wouldn't need to spend a lot of time learning how to use it, and the change would only be implemented in the few departments that previously used Remedy.

MAKING IT HAPPEN

For the migration process, Ron became the team leader, and his team consisted of himself, Harry from the UNIX department, and one programmer from Harry's group. In addition to being the team leader, Ron was responsible for creating the user interface, and Harry and his team were responsible for replacing the Remedy system—patching, updating, and so on. From this point forward, the committee from the previous meetings did not meet any longer; only Ron and his team met weekly and discussed progress. Ron described the whole process as being "very informal." The action items sometimes came up on-the-fly, and were assigned to whomever would volunteer to take action according to their capabilities and availability.

One major concern for Ron was communications. This was a cross-departmental project—Ron reported to Baken, his supervisor, and Harry and his team reported to someone else, not Baken. With the way the organization was structured, Ron and Harry were basically at the same level. According to Ron's experience, it was usually difficult to convince other departments on the same level to make time to work on this kind of a project, since they would be mostly focused on the department's main responsibilities, and assign low priorities to those projects that come from other departments. Although everyone on the team knew each other quite well, Ron proceeded with caution.

The project was maintained through emails and the ticketing system (RT). The UNIX team, and the other back-end teams used RT to manage their projects; they mainly sent out a "ticket" if they needed an action item to be done. The person who would do the item was either assigned or they volunteered to pick up the action. After the person completed the item, a "reply ticket" was sent back through RT, so everyone—especially the requester—knew who worked on the item, and when it was done. However, Ron's department and the rest of the front-end team did not use their ticketing system (Remedy) to manage projects in the same way. Rather, they mostly used emails to communicate and manage projects with the team. So in this project, when Ron had some requests and/or bug reports, he would send emails through the team's mailing list to Harry's team, and Harry and/or his team would send a ticket regarding Ron's requests which was communicated with the rest of the team.

Fortunately for Ron, the project ran smoothly with no major issues. The whole OIT team was used to working through emails, tickets, and such electronic media means. No formal forms or records were used other than the tickets and email records. The departments had a high degree of autonomy, and there were little or no interferences from upper management. The interaction between Ron and Harry's team was almost seamless, and Ron's concern about the project being cross-departmental and convincing the other department to spare some time to work on the project was unnecessary; everyone hated what had been going on in the ticketing system, and had been anxious to have something done about it, and finally they got the chance. The transition was done by November of that year, one month sooner than the expected schedule of six months.

Currently, the OIT uses RT throughout the organization and they continue to manage their projects using the software, as has been done by the back-end teams since the beginning. Although Ron knows that everyone admits that RT is not perfect, the fact that it is open-source, meaning it will be relatively easy for them to customize it, they were pleased with it, and they would happily say, "Hey, at least it's not Remedy!"

Discussion items

1. Based on the information provided in this case, develop a Work Breakdown Schedule for Ron's project.
2. Identify the risks associated with this project and propose response plans.
3. What are the key success factors of this project?
4. Propose project management methodology that is appropriate for a project of this nature.

WRQ Software Development

Peerasit Patanakul and Michael Adams

Six months into the development of the R-Web software project (Release 7), Peter Adams, the project manager, rushes into product manager Tim Johnson's office to inform him of bad news.

Peter: Hi Tim. Sorry for rushing in today. But I have bad news about our R-Web project that needs our attention right way.

Tim: That's alright, Peter. So, what's up?

Peter: About a month ago, we got a report from our customers that they have a problem with Release 6 which we released back in June.

Tim: I know. We decided that it was not an issue.

Peter: Right. A week ago, another customer reported the same problem. This time, I asked John to run the tests. And man, it is a critical problem related to the security vulnerability.

Tim: What? This is serious then.

Peter: Right. I have to admit that the first report from the customer sort of fell through some holes.

Tim: What do you mean?

Peter: Well you know that we didn't really have a process defined for dramatically raising the visibility of a serious problem, or security vulnerability in this case, which is pretty critical in this type of software.

Tim: Right. Right. Right. We definitely have to look into that issue. But now what are we going to do next? We have to fix the security vulnerability problem right away. Man, this is a serious problem. Oh, boy. Do we have to put the development of Release 7 on hold? Management already promised the release date to the customer.

This is happening in WRQ, a company specializing in software development. Let's see it closely.

WRQ Inc.

WRQ is a developer of PC-host emulation software that integrates legacy computer systems into ubiquitous LAN and Internet environments. Operating in the industry that is best described as Enterprise Application Integration Software, or software for managing, consolidating, and coordinating disparate computer applications and systems, WRQ has been in existence and in the market since 1981. Their product lines, including the original Reflection software, are sold in over 50 countries worldwide and are in use by four out of five Fortune 500 companies. The closely held company's sales were estimated to be $100 million for each year, with one year sales growth estimated at 11 percent. Thirty-five percent of its revenue is generated internationally. Each year, 20 percent of the revenue goes into R&D while 10 percent of the revenue goes into marketing.

WRQ's business model is predicated on achieving annual product maintenance agreements with their customers, where a customer realizes product support services—annual major software releases and patches—included in the annual maintenance fee and in competitive displacement of their competitors in corporate and government IT environments. It is inherent in the organization to build and maintain customer loyalty to effectively achieve strategic objectives, and WRQ's culture of highly trained and experienced technical support, product management, and software development teams are continually focused on customer needs and challenges.

WRQ products have kept pace with continual technological change, and the IT needs of their customers for businesses and government agencies that have evolving technology environments from work automation to information management, and into the business transformations of the Knowledge Economy. Of course, each product, and the project leading to the product, this one included, must be clearly understood.

PROJECTS AND PRODUCTS

WRQ's Reflection for the Web, better known as R-Web, is a software product developed in Java, residing on a web server, so that users can access legacy platforms from the Internet. Along with the traditional emulation capabilities, the software incorporates secure Internet access, Windows authentication integration and other third-party components, among other capabilities. WRQ's culture instills a product-centric view of the annual software releases. Each year represents another single, major release of the product on the overall products roadmap, and the team effort is focused on defining requirements, software design and development, and meeting the number one requirement of the project, the release date to the customers.

The scope of the Release 7 project was to define requirements, design and develop, and release software capabilities to be integrated in a customer's IT environment, on the customer's workstations and web servers. With Release 7, several improvements and new features will be added to the existing product. They are web-to-host terminal emulation file transfer, security proxy server, usage metering server, and administrative web applications. The team also pursues FIPS 140-2 cryptography certification from the National Institute of Standards and Technology (NIST) for the R-Web Release 7 project. FIPS 140-2 cryptography certification is the first-ever certification effort for the project team.

PRODUCT STRATEGY AND COMPETITIVE ADVANTAGE

The product strategy is comprised of a number of elements. Those elements address the issues of (1) meeting an annual time-to-market product release goal, (2) accurately defining customer requirements and meeting those requirements through development of software components that add value to the existing product line and other WRQ product lines, and (3) meeting annual revenue goals.

Product direction and competitive advantage is defined in a Product Requirements Document which is developed and maintained by the Product Manager, Tim Johnson. The document formalizes the scope of the release effort, and represents a list of prioritized features derived from customer requests and competitive analysis. The feature priorities are communicated to the Software Engineering Manager, Peter Adams, who assigns development teams to determine the feasibility of developing the components, and to perform the development work.

ORGANIZATIONAL STRUCTURE AND CULTURE

WRQ is traditionally an engineering-driven organization. The typical project team structure can best be described as a functional/matrix hybrid. For Release 7, Peter also acts as a Chief Programmer/Project Manager of a given feature set in the product release. Peter has Software Engineer direct reports and Technical Product Manager direct reports, as well as indirect reports from Development Services of a Program Manager, Localization Manager, Test Engineers, Technical Writers, and a Configuration Developer (see Figure 4.2).

WRQ has a strong and well-defined organizational culture, where the customer and product-centric focus have been traditionally expressed by managers in their actions and communications with employees. The employees, who have many years of experience with the company, relate strongly with this engineering culture and have a strong belief that they represent an elite, dedicated group.

BACK TO RELEASE 7

Tim: Peter, do we have to put the development of Release 7 on hold?

Peter: Not exactly, but we have to allocate some resources to fix this problem.

Figure 4.2   Partial Organizational Structure of WRQ

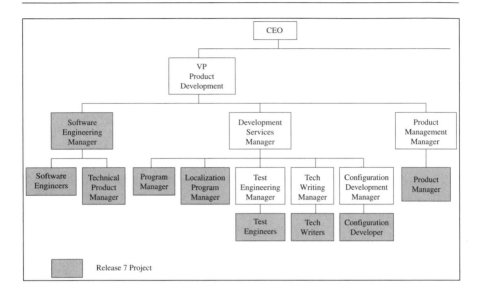

Tim: Do you have any idea how much disruption it will cause on the development of Release 7?

Peter: My estimate would be 1.5 months of one full-time equivalent of software engineers, test engineers, tech writers, etc. I have to discuss this resource allocation with John, our Development Services Manager also.

Tim: Wow. It is pretty costly. But we have to do what we have to do.

Peter: Right. I would say this is a huge disruption that will suck up weeks of time in calendar terms. I still have to figure out whether we should issue a security advisory or whether we needed to notify customers. If so, I need to determine which customers and how we would do it. We have to discuss this with our VP.

Tim: Let me know when. I am free this afternoon.

The Release 7 of the R-Web software represents the contractual maintenance release of the product. This software development project generally takes about one year. The product release date was set by management and was indicated clearly in the product's roadmap. However, even before an official project start date, the team accepts the ebb and flow of requirements analysis, software design, and further requirements analysis based on design feedback in the development process, so that development team members can explore the feasibility of features.

In addition, even though a software release has met quality criteria, after an official end of the project, the team still has the responsibility of feature development if customer problems are encountered.

RELEASE 7 DEVELOPMENT PROCESS

The Release 7 development process is Feature-Driven Development (FDD). This process was implemented to improve the management of the R-Web software development process. In the past, the development process of the first few releases were very ad-hoc. Then, in the next few releases, the development process changed to a methodology pretty well-aligned with IEEE standards with staged delivery driven by two or three overlapping waterfalls.

FDD is a lightweight software development process that has been used on small to large software and computing system projects. FDD can be described as an approach to managing software development projects that need to accommodate short business cycles, and where results are immediately visible to the project team, because of two-week development cycles. FDD is composed of five processes, with defined entry/exit, verification, and task elements for each phase (see Figure 4.3). Advancing to the next phase is not contingent on completion of the previous phase; phases can occur simultaneously given the project scope, management style of the project manager, and experience of development team members.

- Process 1—Develop an Overall Model: Domain experts; chief architect and chief programmers gather to define the domain area, and incorporate into the overall model.
- Process 2—Build Feature Sets: Chief programmers define major feature areas decomposed from the domain area. Each area is further decomposed into feature sets (activities).
- Process 3—Plan by Feature: The project manager, development manager, and chief programmer produce the development plan, and determine the order of feature implementation based on dependencies, development

Figure 4.3   Feature-Driven Development Processes

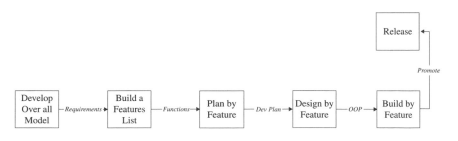

load, and feature complexity. In the case of WRQ, in addition to the software engineering manager who serves as a project manager/chief programmer of a feature area, there are other chief programmers at WRQ who also serve as project managers of other feature areas.

- Process 4—Design by Feature: Being responsible for core object-oriented programming activities, chief programmers select work packages and assign developers for programming activities. Design inspections are held.

- Process 5—Build by Feature: As to orchestrating core object-oriented programming activities, working from artifacts of the design phase, developers implement features through code, unit test, code inspection, and, after successful inspection, then promote to the software configuration management system.

The iterative, short timeframe of developing feature sets (two weeks) enables the project team to refine requirements based on changes garnered from development exploration, process results, and customer input. In theory, each of the post-processes can contribute to requirement changes that lead back to making changes to the domain model in the first process, and the overall development effort (see Figure 4.4). At WRQ, requirement changes do occur in different processes that require the changes of the domain model and development effort. However, there are cutoff dates for initial change order (ICO) and final change order (FCO) that would limit requirement changes later in the project.

Several project management artifacts are derived from the project planning in the third process, Plan by Feature. They provide visibility into the project schedule and status for the software engineering manager, chief programmer(s), and management. Those artifacts are work breakdown structure, project schedule,

Figure 4.4   Feature-Driven Development—Changing Requirements Process Flow

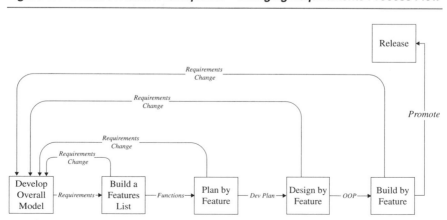

progress totals summary report, features completed versus weeks elapsed, features status report, and feature set status dashboard, etc. For Release 7, the software engineering manager created a distributed information system where chief programmers and other project team members can update the status on the Internet, and where project reports are automatically updated from the chief programmer information, and to collect plans and development documentation from the project. The system is based on a Wiki, where users are allowed to freely update web pages.

The software reaches the release stage after the required integration testing has been completed. Although this integration testing is not explicitly called for through FDD, WRQ includes this testing as an intermediary activity between Process 5 and Release. Managers discuss the processes they use.

Tim: Peter, I know that you introduced FDD as our development process when we worked on Release 6. And we had lots of problems then because our people were not used to it. How is it with Release 7?

Peter: Well, we're still in the phase of experimenting or the "fumbling around" stage to some extent with FDD, but I think the general pattern is well understood. We try to adhere to the pattern as much as possible.

Tim: Okay. But you have to remember that the FDD process is still not a standard process in WRQ. The R-Web team is the only one using FDD. We have a couple of teams using Extreme Programming (XP) and we have a couple of teams that still use our WRQ traditional seat-of-the-pants method that they have used for many, many years, and they have just learned how to make that work.

Peter: I know that. As I said, we are experimenting with FDD. If it serves us well, it will become a standard in WRQ someday. One thing that I'd like to see improve before it becomes a standard is a code review process. I realize that there were a lot of codes that weren't reviewed as early as I would have liked. Part of that was because people just don't like to do code reviews. What I want to work hard on is defining a much more effective inspection and code review methodology for the team that actually makes it less onerous—one that has less overhead and is more asynchronous. I would like to come up with processes that enable code reviews to be better, easier, and more effective.

PROJECT SUCCESS DIMENSIONS

Since the release date was set by management, the ability to meet the release date is considered an important success dimension. With the set release date and a fixed level of resources and budget, the project team has to prioritize the customer requirements and respond to those requirements as much as possible. Even managers are very much involved.

Peter: Now Tim, with a big disruption from the Release 6, we may not be able to respond to all of the customer requirements with Release 7.

Tim: I realize that. We have to really sit down and look at some of the features that we can cut out. As you know, Peter, I have a long list of what I would like to see in the product. And as usual, I have to live with what I can get in this timeframe.

Peter: I understand. And you know that I always want to launch valuable software to our customers. I will try to do it again this time despite the disruption from Release 6.

Tim: As of today, we commit to four new features and a cryptography certification.

Peter: Right. They are web-to-host terminal emulation file transfer, security proxy server, usage metering server, administrative web applications, and FIPS 140-2 cryptography certification.

Tim: Are we able to include all of them with Release 7?

Peter: I still hope so. But we have some problems with a few of the features.

Tim: I heard about the problem with the usage metering server.

Peter: Right. By adding this feature, Release 7 is also a project where we have more than the usual amount of interaction with other development teams. This feature not only enables us to meter our team's product, this release also became capable of metering other products made by other teams. In this case, we were defining new protocols and doing stuff that had to integrate into multiple product lines. That's usually not a problem for us technically. The problem is that we have to deal with license compliance and other legal team concerns. So there are inefficiencies and challenges there that normally we don't experience when we are working on a purely technical level. That actually makes the requirements gathering for the metering stuff much, much more complicated than what we usually have.

Tim: And we didn't plan for such an issue.

Peter: Not exactly. But so far, we are doing okay on this feature. We should be able to deliver it even though some of our team members would love to have more time to make it better.

Tim: Yes. But we have the deadline. How about the FIPS 140-2 cryptography?

Peter: Well, the problem with the FIPS 140-2 certification is unexpected as well. To get certified, we have to work with an independent government-approved laboratory. The lab will work on the documents with us, inspect and validate the code, and prepare the report to NIST (National Institute of Standard and

Technology). NIST then will issue the certificate for cryptographic software once it has been validated.

Tim: I know that.

Peter: When we started this project we were very much hoping that we would have a validated crypto module by the time the product is shipped. At this point, I don't see that it will happen.

Tim: Really? Why?

Peter: So far, we have a tremendous delay dealing with the lab. They have a lot of customers and since we do not have a track record with them, we are not their priority. What we can do is get our job done as soon as possible to build some slack time for the lab. By the time we ship our product, let's hope that we will get on a pre-approval list.

Tim: That's not bad. Being on a pre-approval list is an enabler really of a lot of sales.

Peter: Right. But still, we won't be officially certified then. I heard that once the lab is done with their job, it will take about a month for the final report to go up to NIST. Then, it may take 12 months for NIST to be responsive. So, we are looking at years ahead to get certified.

Tim: Well, we were so naïve to set such an unrealistic hope when we started this project. But, if we are now shooting for a pre-approval list, it still gives us some benefits. As I said, we will be able to sell more.

Peter: Right. We will try to get done pretty much what we have to get done. The external issue is beyond our control. I think this will be the only part of this project with which I will be disappointed. However, I think we will learn about how difficult it can be to schedule something based on not just one external organization but kind of a chain of external organizations.

TEAM LEARNING

At this point, the challenges presented provided experiences that the project team can learn from in a number of ways, both in incorporating methods into the software development process that were not defined explicitly, and managing internal and external relations with other teams, and third-party organizations. There is also a consensus with the project team members that requirements engineering could be better formalized, when the R-Web team interacts with other WRQ development teams, and external vendors.

Peter: When I asked myself, "What did we learn so far?" I answered myself that next time, when we have to interact with other groups that we usually don't interact with, we probably need to build a little more variance into the plan, and anticipate a little more wisely.

Tim: That's a pretty typical answer.

Peter: In the case of usage metering stuff, as painful as it is, we really need to spend time developing really solid requirements. Even though we try to be as agile as we can, I think we've done a pretty good job figuring out where we should do that and where we should be a little bit more exploratory and reactive.

Tim: In my point of view, it is not only necessary that we have solid requirements, but also we have to prioritize them correctly.

Peter: You are right.

Tim: Peter, one thing that I like very much is that you incorporate a retrospective communication method for the project team members. So far, this method serves us well as a continual, anytime review of the project rather than a project postmortem learning approach.

Peter: We've stopped the postmortem approach and instead use multiday project retrospectives. A couple of the things we covered were the tendency of work to fill all available time until a final cutoff, and the challenges that arise when features are promoted but not sufficiently testable. In the next project (Release 8) we have been addressing both of those to an extent by defining intermediate milestones that represent significant bundles of testable functionality from a test engineering perspective.

Tim: With all that being said, what would you do differently if you had to lead a project like this again?

Discussion items

1. List all the problems involved in the R-Web software development project discussed in the case. Discuss solutions for the problems, and propose some preventive actions.
2. As a project manager of a multiple-release software development project, what can you do to alleviate the impact of unforeseen problems from the previous release on the success of the new release project?
3. Discuss the advantages and disadvantages of the FDD process.
4. Propose a way to continuously capture team learning?

Chapter 5

PROJECT SCOPE MANAGEMENT

This chapter deals with the processes required to ensure that the project includes all the work required, and *only* the work required, to complete the project successfully. This is also known as the Project Scope Management, which is covered in Chapter 5 of the *PMBOK® Guide*. There are four cases in this chapter—three critical incidents and one issue-based case.

1. Workshop: Project Definition

    This critical incident discusses an example of a scope statement used in practice. Detailed explanations of the components made up of the project definition in general are discussed. Please note that *Workshop* is a series of critical incident cases, where further discussion is presented in Chapters 6, 7, and 8 on various subjects.

2. Work Breakdown Structure as a Skeleton for Integration

    This is an issue-based case that discusses the WBS construction and potential concerns that might arise if the construction is not validated with major parties of the project.

3. Project Anatomy

    Project Anatomy, an issue-based case, centers on the project decomposition issue. The team desires to decompose every major project's effort and make sure that the project is on strategy. Logically, the project anatomy might be equivalent to the WBS with some differences.

4. Rapid Prototyping

   Rapid Prototyping is a critical incident that takes on a situation where the scope of the project isn't clearly defined. As a result, the project ends up being late with cost overrun.

CHAPTER SUMMARY

Name of Case	Area Supported by Case	Case Type	Author of Case
Workshop: Project Definition	Scope Definition (Scope Statement)	Critical Incident	Dragan Z. Milosevic, Peerasit Patanakul, and Sabin Srivannaboon
Work Breakdown Structure as a Skeleton for Integration	Development of WBS	Critical Incident	Wilson Clark and Dragan Z. Milosevic
Project Anatomy	Project Decomposition	Issue-based Case	Joakim Lillieskold and Lars Taxen
Rapid Prototyping	Scope Verification	Critical Incident	Stevan Jovanovic

Workshop: Project Definition

Dragan Z. Milosevic, Peerasit Patanakul, and Sabin Srivannaboon

With expertise in project management, Konrad Cerni was a senior consultant at Ball, Inc., a very well-known company in the region. He graduated a Ph.D. in Engineering Management from one of the leading universities on the East Coast, and turned himself to a practitioner role since. Konrad, who preferred not to be addressed as "Dr.", had worked in the field of project management at a wide range of companies in different industries from a traditional manufacturing firm to a very complex aerospace operation.

His recent client requested Konrad conduct a workshop specifically designed for project management tools, for about 30+ project and program managers in the company. Because of the participants' busy schedules, the workshop was requested to be one eight-hour session. With a variety of project management tools (generally speaking 50+ tools are available in the practice and literature of project management), Konrad had to pick only those important ones, and cover them in as many areas as possible. One of the tools he included was called a Scope Statement.

WHAT IS A SCOPE STATEMENT?

Typically, a scope statement is a document succinctly describing the project objectives, scope, summarized costs, and resource requirements. The details of the document may vary from one company to another. Fundamentally, however, the document answers the crucial question of "What do we produce in this project?" The answer thus creates a big-picture view of what the project is all about, setting the scope baseline to follow in whatever is done during the project.

A PICTURE IS WORTH A THOUSAND WORDS

Konrad presented an example of the scope statement in practice. The particular company discussed has the project definition that captures six major items: strategic goals, tactical goals, milestones, constraints, assumptions, and specifically excluded scope.

1. Strategic Goals: This element acquaints you with the business end to be attained; the origin of the task; the owner or customer; and the type of product fixture, structure, assembly, or study, etc. In this section, state the overall purpose of the job and include any meaningful background that aids in describing the purpose. For example, the overall purpose of this project is to develop a project management software that will capture 40 percent of the market share in two years. The project is originated in the corporate strategic plan.

   The goals should also include major commitments established with the customer, a subsection called the project goals. This section is written to clearly delineate goals for schedule, cost, and technical performance. An example would be: "This effort will be completed within one calendar year, will cost no more than $500,000, and will result in a report per World Bank guidelines." Make sure that you prioritize the objectives. These priorities will serve as decision-making criteria in the trade-off situations.

2. Tactical Goals: This element should describe the major tasks of the project, for example, conceptual development, detailed design, delivery of a complete and tested steam system; preparation of operating manuals; training of the owner's crew, etc. An example would be: "We will design, procure, install, and commission the manufacturing plant." This section may have four to six major deliverables such as detailed design, prototype, and training. These deliverables become level one in the WBS and will be further broken down into more detailed WBS elements/deliverables such as documentation, installed facilities, services, contractual end products, etc.

   They should be described in terms of how much, how complete, and in what condition they will be delivered. The work scope and the WBS can be worked in an iterative manner to assure that the WBS displays the scope in its entirety, and to provide direct reference from any WBS element back to the work scope narrative.

3. Milestones: Identify and define key milestones, including required completion dates and completion criteria. Key events such as fab completion (a manufacturing plant which fabricates items), assembly completion, test completion, document package sign-off, or customer acceptance may be included. List all contractually fixed events and any other major schedule milestones that are critical to completion of the work. Often these milestones are dates related to deliverables/end products from the previous section.

4. Constraints: List special technical requirements, codes, and standards such as ASME or ISO. Describe facility requirements for fabrication, assembly, testing, or other facilities to accommodate the work. Define functional/operational requirements, data requirements, and special instructions. Identify design criteria. Describe technical constraints, if they exist. Schedule constraints may include interface with progress or completion of other work. In some cases, scheduled delivery may be contractually very rigid. Financial constraints are

Figure 5.1   *Project Definition Form (Scope Statement)*

Project Definition Form (Scope Statement)

Strategic Goals: Improve customer service for our products and services by 5 percent via deployment of a new customer relationship management software.

Time: Finish by September 15, 2010   Cost: $150,000   Quality: Per service level agreement

Tactical Goals: Analyze workflow, configure software, develop prototype, and release software.
Major Deliverables: Workflow analysis, configure settings, prototype, training, release

Key Milestones

- Workflow analyzed by March 15, 2010
- Configure complete by April 15, 2010
- Prototype complete by August 15, 2010
- Training complete by August 15, 2010
- Release by September 15, 2010

Major Constraints
Our key developers will not be available in June because of their visit to our European ally.

Major Assumptions
Configure software to meet our workflow; we will not change our workflow to meet software.

Specifically Excluded Scope
This project does not include training on customer service skills.

often related to funding and should be identified. Facility requirements may be better planned when the financial constraints are known.

5. Key Assumptions: With every task comes a set of assumptions and frequently unresolved uncertainties. Identify and list those assumptions. If some needed information is not yet available when the work scope is being prepared, use your experience and best judgment, or ask others who have been involved in related work. For instance, assumptions may be that a software testing will be done by the external resources, or that design activities take place in

accordance with a design manual. If assumptions are made with reasonable judgment, the work scope can be sufficiently complete to develop the other elements of the project plan.

6. Specifically Excluded Scope: Describe what is not to be included in the task, what is contractually excluded, or not included for other reasons. If the customer waives the need for operational testing, for instance, that should be stated. There are many examples. State specifically what is sometimes related to a similar task, but at this time is not included. This will help planners, engineers, management, and customers better understand the scope of work.

Konrad provided another example as presented in Figure 5.1.

Discussion items

1. What are the pros and cons of the project definition?
2. When should the project definition be used?
3. Should small projects bother using it? Why/why not?

Work Breakdown Structure as a Skeleton for Integration

Wilson Clark and Dragan Z. Milosevic

Matt: I keep looking at this piece of paper and can't believe my eyes, or I must not understand it.

Percy: What paper are you looking at? The one I brought?

Matt: Yes, let me check if I understand it well. The sheet of paper I am holding shows the work breakdown structure (WBS) for the timely termination of the Opto-Mechanical group's portion of work for Project Lada (see Figure 5.2). Is that right?

Percy: Yes, right.

These beginning moments of the conversation between Matt Boon, the engineering manager of the Opto-Mechanical Group, and Percy Bedge, project manager of Lada, do not promise much cooperation. Rather, the atmosphere in which this conversation in Matt's office takes place smells of the open conflict and apparent tension between the meeting participants.

MICROMANAGEMENT?

Matt: Well, on the second page, it shows an estimate of the person-hours and needed calendar time for each of my engineers involved in Project Lada, from now until the end of it.

Percy: Yes. That's the termination plan for Lada. What's wrong with it?

Matt: I don't want to sound negative, but all of it is wrong—from details to philosophy.

Percy: Give me details.

Matt: Your Lada planning indicates at what days on the calendar and how much you need each of my guys from Opto-Mechanical Group. That's based

Figure 5.2   Project Lada WBS for the Opto-Mechanical Group

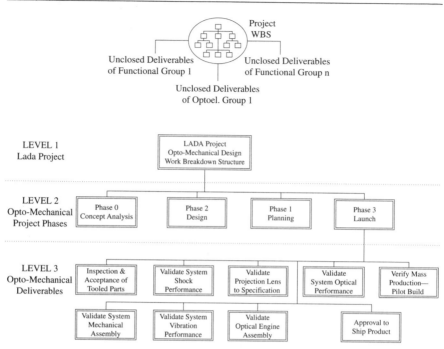

on the assumption that they are engaged in work on only one project, and that's *your* project, Lada. Those are wrong details. As a matter of fact, they are shared among six projects. Since you scheduled them for certain time periods for which they are already scheduled by other projects with higher priorities, those higher-priority projects automatically obtain my guys.

Even if you planned properly, there is another issue for which I would not agree with your WBS and related termination plan. I thought we already resolved the ownership issue. You guys, the project management team, own deliverables from the top three levels of the project WBS. And we, the functional groups, own all deliverables in the lower levels—the fourth and below. "We own them" means we plan them, we execute them, and we control them, which means we monitor and report them to you. You have no right to meddle or control activities on our deliverables, unless we screw up. To remind you, I found the company's PM charter. It shows that this is the current and ruling division of ownership signed-off on by all engineering managers and project management, and senior management.

But you seem to forget or ignore the PM charter, just as you tried to ignore it during Lada's Map Day. With the termination WBS, it appears again that you would like to micromanage us.

Percy: I see where you come from. Frankly, my intention was not to micromanage you. Give me a few days to go over it, and I'll get back to you.

Discussion items

1. What are the pros and cons of having many levels of WBS?
2. Do you prefer a few or many levels of WBS? Why?

Project Anatomy

Joakim Lillieskold and Lars Taxen

The project was heavily delayed. The new project management team was trying to grasp hold of the situation. The scope of the project was clear: Develop a new central processor for the AXE system. The processor was to be state-of-the-art. It would mean a four-time increase in capacity of the system in terms of speed, bandwidth, and quantity of transmission. Further, the processor included new capabilities, new mechanics, new firmware, a new real-time master and slave ASIC, lower electrical consumption, and a reduction in size and weight. At the moment, the risk analysis showed the project would be a nightmare. There was no simulator that could manage the ASIC. As a matter of fact, several of the large ASIC-producers could not support the development project at all. It was considered the most advanced and complex ASIC ever developed! If a fault was found in the ASIC, it would take three months to get a new one, and thus delay the project by the same amount of time. There was no room for mistakes.

ORGANIZATION

The project was called *The Central Processor*. The project was huge, and consisted of seven major smaller projects, several of which also consisted of subprojects. The development of those smaller projects was dispersed according to the locations of the core competences. When activity was at its peak, about 300 people were involved in different projects located in seven different places across three countries and two continents (Sweden, Germany, and Australia). In addition, the project also used external consultative resources to a great extent.

The project was challenged by cultural differences, especially the corporate cultures. Many engineers in the project had been employed for 20 years or more in the same organizational setting. But at the time of the project, the company began to reorganize. Personnel from fixed telephone networks, mobile networks, and military networks each having different and quite embedded management styles and different approaches to system development were moved between the business units and would now have to cooperate with each other. The reorganization also implied that many managers came from other parts of EXT, a leading provider of telecommunication and data communication systems, into this project. This reorganization included a new CEO for the business unit where the project was located.

PROJECT RE-PLANNING

In the middle of the project, there was a change of personnel in key positions. The project had just passed Tollgate 2, which was the point in the EXT project model, where the decision is made to either go ahead or not. At this point, the organization realized that the project would be heavily delayed compared to the estimates in the prestudy. When the new project manager of the critical CPG integration project (one of the seven major projects) entered the team, it became clear that there was no control over the project status or when things should be done. As an effort to get to the bottom of the problem, the new project manager started to interview people involved in the project. At the same time, there were rumors about a new approach used in an earlier project at EXT: *The Japan Project*, when EXT became the first foreign supplier to get a foothold of the Japanese market for mobile systems. The company had painfully learned from earlier projects that: *"When managing complex projects, there is a need for a general picture."* The approach used in the Japan Project provided such a general picture: *the anatomy.*

Supported by the new management that had experience from the Japan Project, the CPG project manager got the go-ahead to try the same integration-driven approach as in the prior project in order to get things back on track. This approach was based on the anatomy and consisted of three phases: definition of the anatomy, dividing the development task into verifiable increments, and integration of those increments.

THE ANATOMY

An anatomy is created from the customers' point of view—in this case, the end customer, being the one using the final system as opposed to the internal customer ordering the project. The starting point in creating the anatomy begins by asking the questions: "What is the first thing customers do? The next? And the next? What do they expect out of this product? What is the end result?" The anatomy illustrates the dependencies between the capabilities the system must have; from switching the power on to getting out a stack of bills that can be sent out to customers.

CREATING THE ANATOMY

In the central processor project, creating the anatomy required about seven meetings. Each meeting included, at the most, 12 persons, but throughout the process more people were involved. The people who took part in creating the system anatomy were line managers, and people with knowledge regarding the system (system integrators, system leaders, product managers). Informal group leaders were included as well.

These people had a mix of hardware and software skills. Everyone involved was not active all of the time; some had knowledge of the start-up capabilities and some of other capabilities. There were many discussions over where the anatomy ended. However, the important issue was to create a common understanding of the final system solution, and to understand each others' "language."

INTEGRATION AND PROJECT PLANNING

CPG project manager: When we agreed what capabilities to deliver and how they depended on each other, we could take the next step to organize the work and create a plan for the work. First we did the anatomy, then the organic integration plan, and last, the integration and project planning. These latter two were not made within the same group of people that created the anatomy. In this process, the project manager was in charge. The focus of this process was to be able to test and integrate the system as early as possible and to also deliver to the customer as early as possible in order to avoid further delays.

The organic integration plan, as I call it, is when we try to organize the capabilities from the anatomy into suitable increments. We decide how long it will take, the number of test channels, etc. It is a lot to take into consideration, and personnel from system integration and verification are important to include in this task.

The integration plan is similar to the organic integration plan; the difference is that time is included in the integration plan.

CPG project manager: In this project, we flipped the integration plan so it went from left to right, like any other project plan. The reason was that it was difficult to get all the information into one picture, and then it worked better to flip it around. It is important that the images are explicit and intuitive to understand.

The anatomy focused on capabilities. However, when these capabilities were transferred to the integration plan, the increments were referred to as something directly related to what they should do. Sometimes, a larger system's name was put into the integration plan, even though only a small capability of the system was going to be used. This, however, helped people to recognize themselves in the integration plan.

When the integration plan was created, it served as the basis for a tollgate decision to go on with the project. If there was a change in the project, or the customer wanted to pay less, it was easy to cancel an increment or capability, and still know how this affected the rest of the system.

THE LEARNING

By restarting and implementing the anatomy approach, the new management felt they could take control of the project.

> **CPG project manager:** Before we drew up the integration plan, we thought that RPH (a subproject) was delayed. But then we saw that we did not need their output until a later date. In fact, they were perfectly on time.
>
> In other words, we had a different opinion about the dependencies in the project. It was the micro-programming which was wrong, totally wrong. They didn't even have the resources they needed in the project at that point in time. So they had to reschedule everything they had to do.
>
> This is an obvious effect of the anatomy work. It wasn't until then we saw how late we were. When doing the anatomy, we realized that we would be late anyhow. By implementing the approach, several other issues surfaced that needed to be addressed as well. Some of these are implicit in any project, but they became painstakingly clear when this method was applied.

RESISTANCE TO CHANGE

Change, however, does not come easy. The initiatives of the new management were interpreted as a devaluation of previous experience by senior engineers. On the other hand, new engineers and newcomers in the project organization approved it. And most importantly, the new CEO liked it because the status of the project became very clear. The resistance though led to frustration from both the new managers and the engineers that stemmed from the proud traditions of AXE development.

There were many who said: "We have been developing the APZ processor for 20 years. We know our work. People from other parishes should not come here and tell us how to do our job. It is possible that these methods have worked elsewhere, but our work is a specific case."

The anatomy enforces a certain mindset: "What is the first thing the customer does when he starts up the system? Push the start button!" This mindset is on the principle that a system should be developed in the same order as it "comes to life." This principle was somewhat hard to accept for some designers: "I'll design this function later; I'll start with the most difficult (and interesting) things first."

Some actors experienced the anatomy as a threat to their professional knowledge: "It can't be this bloody insignificant; I'm actually doing more difficult stuff" implying that their difficult everyday tasks were dwindled down to a mere functional box on the anatomy chart. Others claimed that they could not even see their particular contribution and context in the anatomy. Thus, in order for the anatomy to be accepted by the actors it is important that it is defined in such a way that each actor can see both his context and place in the overall picture.

CPG project manager: What we missed in the CPG project was the ability to split the anatomy into smaller parts so that individual designers could understand where they participated and where they didn't.

MAKING SENSE

CPG project manager: Some continued drawing up the anatomy chart for two months . . . In one case pertaining to a mature product; people could not agree (how the product worked) even though they have been working together and across the same corridor for 20 years.

These engineers had differing views of how the system actually worked. During the process of creating the anatomy, the crucial dependencies had to be articulated and agreed upon. For that specific system, it was the first time they discussed which were the crucial capabilities in the system and the interconnections between them.

CPG project manager: It is about creating a common image and not about having a 200-page document in English needed to be equally understood by all, regardless of being Swedish, German, or Japanese.

The most important quality of the anatomy is that it enables the actors to see their task in relation to other tasks. This is achieved by a visual picture which is preferably drawn on one page. Based on this picture, they can take the proper actions to achieve a common goal. No sophisticated tools are necessary. Mostly, PowerPoint is used since it is easy to learn and commonly available.

PROJECT PLANNING AND CONTROL

Hardware engineer: It is an extensive architectural change we are now approaching, involving a radically new approach. It is probably the largest step that has been taken when it comes to APZ processors.

The task of developing a central processor demanded a scheme where subtasks (i.e., the development of subsystems) needed to relate to each other as both input and output, implying a reciprocal interdependence. As such, the interrelated parties needed to communicate their requirements and respond to each others' needs. The central processor also had 29 different external dependencies for various communication systems that needed to be considered during the development effort, making it complex and difficult to manage.

> **Hardware engineer:** In this project, we have really been exposed to the general problem of development work: The fact that it has not been done before makes it impossible to know for sure how to do it and how long it will take.

In the project, no network plan or Gantt chart was used at the top level. The experience was that network diagrams were too time-consuming to update in projects where frequent changes were the order of the day. Thus, in order to plan and control the project, there was a need for two types of plans—the integration plan based on the anatomy, and the ordinary project plan (time and resource plan) for each increment which provided the necessary logistics for the execution of the project.

COMMITMENT AND RESPONSIBILITY

Commitments and responsibilities are important aspects when coordinating a complex development project. The anatomy provides a "push and pull" concept which makes these aspects clearly visible. There is a receiver for every delivery in the project, including the customer. If a subproject is delayed, this becomes very evident, something that wasn't always appreciated in the CPG project.

> **CPG project manager:** When you did progress control based on the anatomy it became very clear who was slacking, which was not very popular among many people. It became so evident on the map and that could be hell. But it was the truth.

Thus, the anatomy made commitments and responsibilities more transparent, which in turn made it easier to control the progress of the project. Concerning progress control, it is easy to show progress and problems with the help of the anatomy. A simple system of cues can be utilized for signifying progress, for example, traffic light signals.

CONFIGURATION MANAGEMENT

A drawback to the integration-driven method of working is that it is complicated for the system manager. Before, the PROPS model had milestones that were considered well defined. Now, some of these milestones could not be reached until every subproject had met its milestone, resulting in the passing of many subprojects' milestones at the same time. Even further, configuration management of the project became more complicated.

CPG project manager: It becomes a bit messy and requires a very competent configuration manager. That particular person acquired a higher status since he/she had to keep track of partial deliveries.

This, in turn, requires a tool supporting configuration management, and, if the project is very complex, a tool to keep track of all the increments and dependencies.

Discussion items

1. Explain the purpose(s) of project anatomy.
2. What are the disadvantages of project anatomy?
3. What are the differences between the project anatomy and the WBS? Is it any better? Why/why not?

Rapid Prototype for Fast Profits

Stevan Jovanovic

RAPID PROTOTYPING—NEW DISCIPLINE

Rapid prototyping has been a constant growing and evolving field since the late 1980s. As technology improved, so did the opportunities in new markets. The idea quickly evolved from its grassroots beginning to many small companies competing for a bigger share of the growing market.

Frank Billings was just another name in what was at that time a niche market. As a student in engineering school, he followed the development of the new prototyping techniques and realized their potential in the marketplace. His dream job was to work for a rapid prototype equipment manufacturer. There were only a few start-up companies in rapid prototype machine development, however, and none could pay the average engineer wage.

Like most engineering school graduates overloaded with school loans, he couldn't wait for his dream job to come along, so he went for a job at Cocable Company. Cocable designed and manufactured specialty cable and cable-related products. It had nothing to do with rapid prototyping, but it paid well.

He worked hard at Cocable and earned enough to pay down his debts. He proved to be an excellent engineer, earning a great reputation at Cocable and making many contacts along the way. In those three years at Cocable, however, he never stopped thinking about rapid prototype machines. He spent his free time coming up with a rapid prototype machine design, always dreaming of having his own company. Three years in, he was ready. He quit Cocable and started his own rapid prototype (RP) design business. He perfected his own RP machine design and was ready to prove himself in the growing field.

BUSINESS BOOMS

Like every start-up business in a new field, finding customers is tough. In the RP field, there are two types of work. The first includes owning an RP machine and doing prototypes per order. The second is selling RP machines to businesses that want the machine to do in-house RP. The latter option is far more profitable since the machines are more expensive than each prototype they produce. Frank would have been happy with either type of business, since at the time, he wasn't doing much business at all.

All those years making contacts at Cocable Company proved to be worth the time and effort. He had stayed in touch with these contacts and through them was happy to learn that Cocable had just been hired by GE to design and manufacture cable installations on their newest jet engine. Part of the wiring installation that Cocable had been hired to design included junction boxes and switch covers. The installation would be no simple task as these "boxes" are made of specialty materials with complex shapes and multiple designs, all needed for application. They had to be perfect from the start since airplane engines have no room for error. This was a huge job and the timeline was tight. Rapid prototypes were an absolute necessity for this job. Frank's knowledge of Cocable's needs made him perfect for the RP job. Cocable wanted full access to rapid prototyping so they decided to contract Frank to custom build four RP machines to their specifications. Frank could not be happier. The RP machine specs were given to Frank and he went to work.

WHO IS GOING TO PAY FOR THE CHANGES?

After three months of all-night work sessions, the machines were built to specification and ready for delivery to Cocable. Frank's daring steps into a new field were fully rewarded, he thought.

Everyone was ready for a test run, after the first machine was delivered to Cocable. The CAD model was loaded and it was time to hit the "Start" button.

Beep, beep, beep.

"That's not good," said Frank.

He felt embarrassed that the machine failed in front of everyone. He was sure the machine ran fine before it was delivered. He couldn't allow his first major deal to fail in any way.

The machine was checked over for shipping damages. The connections were double-checked. Everything appeared intact.

Frank sat down to review the CAD model and discovered the problem. The model was 62 inches long. This was an issue, considering the RP machines were designed for a maximum of 55 inches.

The original Cocable specs for Frank's RP machines were for a maximum length of 48 inches. Frank optimized his machines for a length of 48 inches, but to be on the safe side, the machines were capable of 55-inch designs. Sixty-two inches went outside that range. A machine that could make prototypes that long would require completely different processors, actuators, and adhesion processes. This would be a major redesign of the RP machines. This would take time and a lot of money.

Cocable claims that the original specs for a maximum of 48 inches came from GE. GE claims that it never gave Cocable a maximum length. The first design that GE requested from Cocable was 62 inches long and that had been weeks before. Cocable should have double-checked their RP specs.

Nobody wants to take the blame for specifying the prototype design sizes and Frank's first major product is now going nowhere. Everyone is dissatisfied and two things are for sure: (1) The entire project is running late, and (2) it will be way over budget.

Discussion items

1. What have you learned about the issue in the case?
2. Who do you think should pay for the changes?
3. What could have been done to make sure that the project scope was correct?

Chapter 6

PROJECT TIME MANAGEMENT

This chapter presents cases which address the processes required to manage timely completion of a project, which can be found in Chapter 6 of the *PMBOK®* *Guide*, Project Time Management. There are seven cases in this chapter—four critical incidents and three issue-based cases.

1. How Long Does It Take to Catch a Fish—TAD?
    The focus of this issue-based case study is scheduling, particularly the development of a time-scale arrow diagram (TAD) for a project. The TAD is a member of the network diagram family, where it requires a determination of each activity timeline. Similar to when using other scheduling tools, team members should be included in creating the schedule to ensure the accuracy of the project timeline.
2. Workshop: The Jogging Line in Action
    Workshop: The Jogging Line in Action is one of the critical incident cases in the Workshop series, which specifically demonstrates a jogging line. The joggling line exhibits the amount of time each project activity is ahead of or behind schedule, which can be viewed as a step in the proactive management of the schedule.
3. Sequencing
    Sequencing is an issue-based case discussing the basic concepts of schedule development. The case presents how a team defines some of their project activities and identifies the sequence of activities.

4. The Rolling Wave

    The Rolling Wave is an issue-based case study which concentrates on the development of the Rolling Wave schedule for a project under high uncertainty. The Rolling Wave is a form of progressing elaboration planning where the work to be accomplished in the near term is planned in detail and future work is planned at a higher level of the WBS.

5. Schedule Accuracy

    Schedule Accuracy is a critical incident that briefly talks about the PERT (Program Evaluation and Review Technique) approach (calls it "schedule accuracy") and explains major assumptions needed in identifying the probability of completing the project's critical path.

6. AtlasCom

    AtlasCom is a critical incident that talks about the Critical Path Method (CPM) for project scheduling. CPM is a technique used to predict duration of a project by analyzing which sequence of activities has the least amount of scheduling flexibility.

7. Workshop: The Milestone Chart

    Workshop: The Milestone Chart is one of the critical incident cases in the Workshop series, which specifically presents a milestone chart. The milestone chart shows milestones against the timescale in order to signify the key events and to draw management attention to them.

CHAPTER SUMMARY

Name of Case	Area Supported by Case	Case Type	Author of Case
How Long Does It Take to Catch a Fish—TAD?	Schedule Development (Time-scale Arrow Diagram)	Issue-based Case	Ferra Weyhuni
Workshop: The Jogging Line in Action	Control Schedule (Logging Line)	Critical Incident	Dragan Z. Milosevic, Peerasit Patanakul, and Sabin Srivannaboon
Sequencing	Activity Definition and Sequencing	Issue-based Case	Art Cabanban
The Rolling Wave	Activities Definition (Rolling Wave Planning)	Issue-based Case	Dan Itkes
Schedule Accuracy	Schedule Development (PERT)	Critical Incident	Dragan Z. Milosevic, Peerasit Patanakul, and Sabin Srivannaboon
AtlasCom	Schedule Development	Critical Incident	Dragan Z. Milosevic, Peerasit Patanakul, and Sabin Srivannaboon
Workshop: The Milestone Chart	Control Schedule (Milestone Chart)	Critical Incident	Dragan Z. Milosevic, Peerasit Patanakul, and Sabin Srivannaboon

How Long Does It Take to Catch a Fish—TAD?

Ferra Weyhuni

The meeting of the Project Review Board (PRB) was underway.

Shane: So, I've only got one custom order that needs to be reviewed, but I think it is quite a demanding order. I attached the customer requirement to the agenda of this meeting. Let me know if any of you did not get it. Markus, do you want to let everyone know what this order is?

Markus: Yeah, sure thing. This order was received at the end of last week from RedGate Technology. They ordered a customized wafer transfer system for our DL800 to process their 300 and 450 mm wafers. The tool they currently have can only handle the 300 mm wafers. They know that our tool can be used for 450 mm wafers, but they do not want to buy a new one. They proposed a dual 300/450 mm process capability.

Robbie: What is the timeline that they gave us?

Markus: Well, they are aware that we will have to start from our drawing board to design this new transfer system. They are asking for eight months delivery time from PO submission. They are planning to submit the PO within the next two weeks so if we start next week, then it'll be eight months and one week. Is that going to be achievable?

Robbie: I'm sure that we can modify our current design but I think hardware will be a long lead time item in our schedule. However, let's not jump to a conclusion of go/no-go. Let's give Maggie a couple of days to determine the schedule for this project.

Maggie: Yes. I'm going to call a meeting with them tomorrow morning to create the WBS and a schedule for the project. If you all can send me resources from your team for this project, that will be great.

Robbie: One thing to remember is that the hardware lead time for a wafer transfer system can be up to nine weeks. I suggest the prototype material be obtained from our in-house shop.

Maggie: Okay, thanks for your input. I will get together with the project team tomorrow morning and send you all the update by Monday morning. Should we get together on Monday afternoon to make a go/no-go decision on the project? Do we even have the option of saying no, Markus?

Markus: Yes, if that is justified. However, the customer is willing to put an NRE for this project. It has not been determined yet but I think our upper management will be pushing hard to get this done.

Maggie: Of course. Thanks for the warning.

Markus: No problem. I'm glad I can be of support for you. Okay, will we meet again on Monday afternoon? Let me recap action items from this meeting: Maggie is to call a meeting to create project's WBS and schedule, as well as to send the team the schedule by Monday morning. Everyone should review the schedule before the meeting so that we can get through it faster. Did I miss anything?

Maggie/Robbie: Nope. I think that's all. See you Monday.

BACKGROUND

The IEM Company is a high-tech company producing customized Ion and Electron Microscopes. The applications of their products can be used in a variety of fields, from academia to high-tech industries. Their customers are given the options of customizing the product to meet their specific needs in their processes. The company's financial profile shows that their sales revenue for last year exceeds $400 million.

PROJECT REVIEW BOARD

In order to fulfill their customers' custom orders, the IEM Company has the Project Review Board (PRB), which consists of representatives from the Project Management, Marketing, Manufacturing, Product Engineering, and R&D groups. The PRB team consists of the following representatives:

- PRB Lead: Ronan MacBride
- Project Management: Maggie Gerrard
- Marketing: Markus Riise
- Manufacturing: Steve Fowler
- Product Engineering: Rory Crouch
- R&D: Robbie Kewell

The group meets as custom orders arrive with main objectives of reviewing all customers' requirements for their products and, if necessary, creating new projects to deliver the requirements. If new projects are created, the group will assign a project manager to identify the Level of Effort (LOE), which consists of the resources, time, and budget required to complete the project.

PROJECT DEVELOPMENT

On Monday, the PRB team meets again to review the project schedule, the required budget and risks, and to decide whether the project should be done or not based on the time, cost, and profit.

In addition to the PRB, the project team consists of the following individuals:

- Project Manager: Maggie Gerrard
- Mechanical Engineer: Keith Bellamy
- System Engineer: Dylan Carragher
- Product Engineer: Emma Owen
- Supply Chain Lead: Paul Finnan

The meeting of the project review board is starting.

Ronan: The agenda for today is to review the project team's schedule and required budget. We need to make sure that the NRE can cover our required budget for this project. Maggie sent the Work Breakdown Structure (WBS) and the Time-Arrow Diagram (TAD) to everyone early this morning. I'm not sure if everyone has had a chance to look at it but Maggie, maybe you could tell us what you guys did and what you came up with.

Maggie: Yes, definitely. So the project team as you all know consists of Keith Bellamy as the mechanical engineer, Dylan Carragher as the system engineer, Emma Owen as the product engineer, and Paul Finnan as the supply chain and logistics lead. We met at the end of last week for the "Cage Day" event, which pretty much took all day, to identify an activity list, duration, and owners for the project. By the way, we came up with the project name of "The Combo Deal." We identified six major areas for the project: Software, Mechanical, System Test, Supply Chain, Work Instructions, and Deployment. We created the WBS to produce the "Combo Deal" wafer handling system (see Figure 6.1).

Maggie: From the WBS, we know that mechanical design does not need to start from zero. We can use the existing design and modify it for the new specifications. From there, we assigned times and owners for each activity. Then, we developed the TAD for the project along with all activity dependencies (see Figure 6.2).

Figure 6.1   The Combo Deal WBS

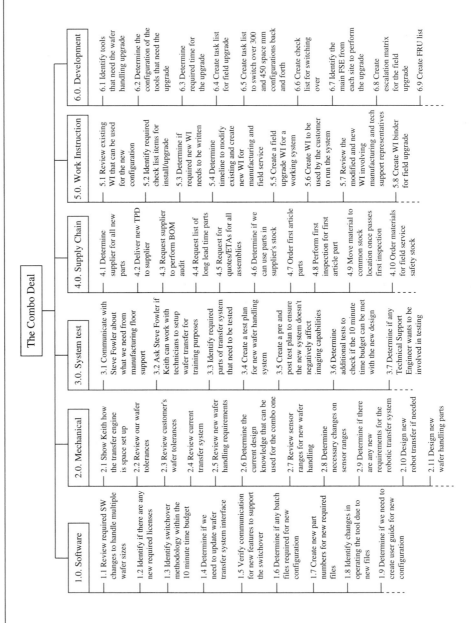

The Combo Deal

1.0. Software
- 1.1 Review required SW changes to handle multiple wafer sizes
- 1.2 Identify if there are any new required licenses
- 1.3 Identify switchover methodology within the 10 minute time budget
- 1.4 Determine if we need to update wafer transfer system interface
- 1.5 Verify communication for new features to support the switchover
- 1.6 Determine if any batch files required for new configuration
- 1.7 Create new part numbers for new required files
- 1.8 Identify changes in operating the tool due to new files
- 1.9 Determine if we need to create user guide for new configuration

2.0. Mechanical
- 2.1 Show Keith how the transfer engine is space set up
- 2.2 Review our wafer tolerances
- 2.3 Review customer's wafer tolerances
- 2.4 Review current transfer system
- 2.5 Review new wafer handling requirements
- 2.6 Determine the current design knowledge that can be used for the combo one
- 2.7 Review sensor ranges for new wafer handling
- 2.8 Determine necessary changes on sensor ranges
- 2.9 Determine if there are any new requirements for the robotic transfer system
- 2.10 Design new robot transfer if needed
- 2.11 Design new wafer handling parts

3.0. System test
- 3.1 Communicate with Steve Fowler about what we need from manufacturing floor support
- 3.2 Ask Steve Fowler if Keith can work with technicians to setup wafer transfer for training purposes
- 3.3 Identify required parts of transfer system that need to be tested
- 3.4 Create a test plan for new wafer handling system
- 3.5 Create a pre and post test plan to ensure the new system doesn't negatively affect imaging capabilities
- 3.6 Determine additional tests to check if the 10 minute time budget can be met with the new design
- 3.7 Determine if any Technical Support Engineer wants to be involved in testing

4.0. Supply Chain
- 4.1 Determine supplier for all new parts
- 4.2 Deliver new TPD to supplier
- 4.3 Request supplier to perform BOM audit
- 4.4 Request list of long lead time parts
- 4.5 Request for quotes/ETAs for all assemblies
- 4.6 Determine if we can use parts in supplier's stock
- 4.7 Order first article parts
- 4.8 Perform first inspection for first article part
- 4.9 Move material to common stock location once passes first inspection
- 4.10 Order materials for field service safety stock

5.0. Work Instruction
- 5.1 Review existing WI that can be used for the new configuration
- 5.2 Identify required check list items for install/upgrade
- 5.3 Determine if required new WI needs to be written
- 5.4 Determine timeline to modify existing and create new WI for manufacturing and field service
- 5.5 Create a field upgrade WI for a working system
- 5.6 Create WI to be used by the customer to run the system
- 5.7 Review the modified and new WI involving manufacturing and tech support representatives
- 5.8 Create WI binder for field upgrade

6.0. Development
- 6.1 Identify tools that need the wafer handling upgrade
- 6.2 Determine the configuration of the tools that need the upgrade
- 6.3 Determine required time for the upgrade
- 6.4 Create task list for field upgrade
- 6.5 Create task list to switch over 300 and 450 space mm configurations back and forth
- 6.6 Create check list for switching over
- 6.7 Identify the main FSE from each site to perform the upgrade
- 6.8 Create escalation matrix for the field upgrade
- 6.9 Create FRU list

Figure 6.2 The Combo Deal's TAD

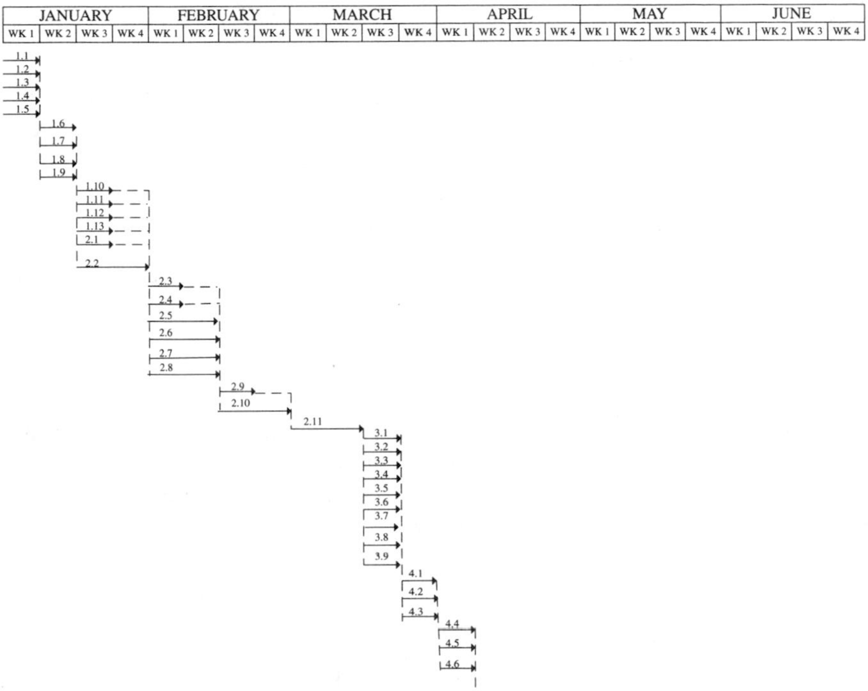

Maggie: So, from the TAD, our critical path is the supply chain of the parts. Since this is a very complex design, our supplier would not commit parts delivery sooner than eight weeks just by talking about potential changes from our current parts. That is after reprioritizing our orders and their WIP for the next few weeks. Also, the test plan shows that we need one week of tool time to test the first article before sending it out to the customer.

Steve: Does the supplier know how critical their delivery timeline is for this project?

Maggie: Yes, Paul has talked to them and offered some help to expedite the parts. They have not seen the updated drawings yet but they're very familiar with the piece parts. I'm sure the eight weeks prediction reflects reality. In addition to the mechanical design, the supplier is also concerned with the materials delivery timeline.

Markus: Okay, from your TAD I can see that you'll be testing the parts and then writing the upgrade work instructions, which requires about one week to perform. Is it possible to do that while waiting for the parts and testing them?

I'm not sure how difficult it is to write the work instructions before performing the installation in the tool, but I think we should be able to copy our existing work instructions for this project as well, shouldn't we?

Maggie: That's a very good point. Maybe we can ask Emma to audit the existing work instructions and see if we can modify it right away. We can use the draft for the first article testing and identify if the work instructions still need some work. I like that. Let me update the TAD to change the dependency from FS (Finish to Start) to SS (Start to Start). If we do that I think we can finish it in eight months with two weeks of buffer time.

Ronan: Great, I hope we won't have to use those two weeks. Does anybody have any more questions?

Discussion items

1. What else can the group do to shorten the project schedule if all activities are required to be performed to complete the project?
2. Does the activities' sequence in TAD reflect the WBS activities numbers?
3. What are the pros and cons of the TAD?

Workshop: The Jogging Line in Action

Dragan Z. Milosevic, Peerasit Patanakul, and Sabin Srivannaboon

Ball, Inc. is a local management consulting firm which is very well known in organizational change management assistance; development of coaching skills; and technology, project, and operations management improvements. The company's long success history has attracted a number of customers in different fields for many years. Certified as a Project Management Professional (PMP®) and Program Management Professional (PgMP®), Konrad Cerni is one of the renowned senior consultants in the company. Among his various skills, his proud specialization is project management tools and metrics.

At the time, Konrad was conducting a workshop at the client premises. The workshop was specifically designed for project management tools where 30+ project and program managers were expected to attend. The workshop was scheduled to run from eight to five, with a one-hour lunch break in between. Given the fact that more than 50 project management tools are available in practice and the amount of literature on project management, and considering the time he had, Konrad decided to pick only the important tools. A "Jogging Line" was one of them.

WHAT IS THE JOGGING LINE?

The jogging line is one of the widely used schedule control tools. It spells out the amount of time each project activity is ahead or behind schedule. The line indicates the fraction of completion for each activity to its left, and what remains to be completed to its right. Generally, it is viewed as a step in the proactive management of the schedule. In particular, the amount of time each activity is ahead or behind the baseline schedule is used to predict the project completion date and map corrective actions necessary to eradicate any potential delay.

Typically, to build a jogging line information is needed regarding the baseline schedule, performance reports or verbal information about actual performance, and any change requests. The first two components provide information on the planned progress (baseline) and the actual progress (report or verbal information) that the jogging line compares to establish where the project stands. Since the changes in the course of project implementation may occur such that they extend or accelerate the baseline, it is beneficial to take them into account when considering schedule progress and predicting completion date.

A PICTURE IS WORTH A THOUSAND WORDS

Konrad commented, "The jogging line is usually used as an informal tool. Its results are not recorded, and all you really need are a chart and marker. The jogging line has only three rules for using it:

1. Rule of the past
2. Rule of the present
3. Rule of proactive management

Rule of the past is everything to the left of the jogging line is done. Rule of the present means that everything on the jogging line is at this moment. Rule of proactive management, which is the most important, means that everything to the right of the jogging line remains to be done."

After Konrad explained the jogging line concept, he showed an example of the tool to the attendants (see Figure 6.3).

Figure 6.3 The Jogging Line Example

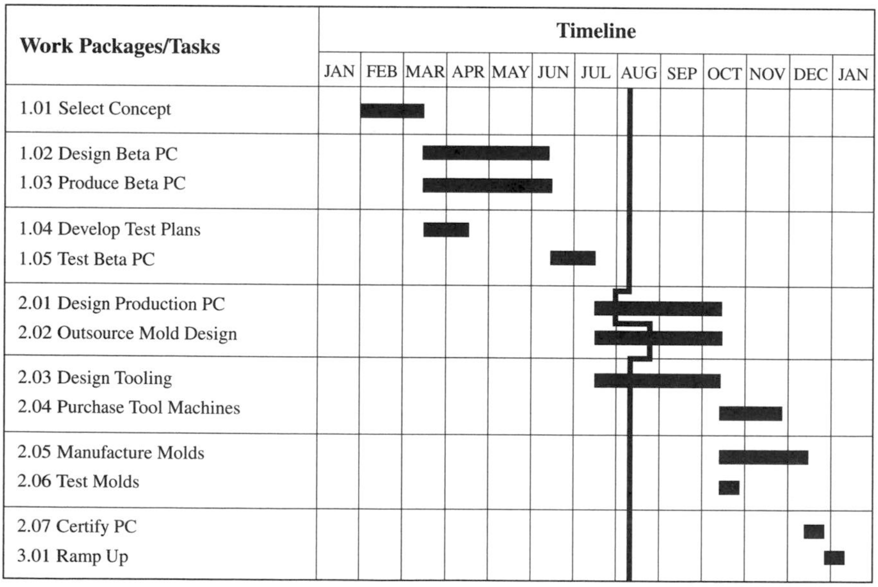

Work Packages/Tasks	Timeline												
	JAN	FEB	MAR	APR	MAY	JUN	JUL	AUG	SEP	OCT	NOV	DEC	JAN
1.01 Select Concept		▬											
1.02 Design Beta PC				▬▬▬									
1.03 Produce Beta PC				▬▬▬▬									
1.04 Develop Test Plans				▬									
1.05 Test Beta PC						▬							
2.01 Design Production PC								▬▬					
2.02 Outsource Mold Design								▬▬					
2.03 Design Tooling								▬▬▬					
2.04 Purchase Tool Machines										▬			
2.05 Manufacture Molds										▬▬			
2.06 Test Molds										▬			
2.07 Certify PC												▬	
3.01 Ramp Up													▬

"The three rules are clearly employed as the activities to the left of the jogging line are finished (e.g., 1.02 Design Beta PC and 1.03 Produce Beta PC), the activities on the jogging line are present (e.g., 2.03 Design Tooling), and the activities to the right of the jogging line remain to be completed (e.g., 2.04 Purchase Tool Machines)," summarized Konrad.

Discussion items

1. What are the pros and cons of the jogging line?
2. When should the jogging line be used?

Sequencing

Art Cabanban

Andy was nervous as the Leadership Team was going over the quarterly metrics. He would be updating the audience about the phototool conversion project that he and his small team had been working on over the past three weeks. The slides were ready to show how well the project went. The team worked furiously to plan, perform the various experiments, and come to conclusions with the selected film vendor. The presentation was just a formality to allow the factory's management to see the results of the testing and hear the recommendation from the team. But Andy continued to worry about the questions that may come. He had a transition plan, but was concerned that the directors and other managers wouldn't like it. He took a deep breath, closed his eyes, and remembered how the last few weeks had rushed by. This case focuses on the approach Andy used to develop his project schedule.

ABOUT ANDY

Andy was a new engineer at the Printed Circuit Board, Inc. (PCB). He had just finished his rotational training throughout the factory to understand how the business and process work. Andy was the sustaining engineer for the phototool area. Phototools are the images that help create the lines and spaces in the printed circuit board process. The PCB plant was very quality-oriented. All the areas used statistical process control to help monitor all the critical aspects of the process. Changes to equipment or procedures can cause a process to go out of control. New processes and tools go through a rigorous qualification process to ensure all the proper functions are controlled. The phototooling area was no exception. However, the phototooling vendor announced that they needed to change their equipment, chemicals, and film because of a business partnership they had entered. This required the plant to qualify the new supplies. The management decided that there was an opportunity to test and possibly qualify a new vendor with competing equipment, chemicals, and film. A small team was formed to identify all the activities required to test the two vendor tools, create the selection criteria, plan the transition, and finally implement the transition to the new phototooling.

PRINTED CIRCUIT BOARD, INC.

Printed Circuit Board, Inc. is a U.S. factory that manufactures printed circuit boards for companies that make computers, telecommunication devices, and other high-tech equipment. The company's annual revenue is about $300 million and it employs approximately 1,000 workers. The phototool vendor selection project had a minimal budget of $5,000. Much of the supplies and equipment would be supplied free of cost from the competing vendors.

IT'S NOT JUST A VENDOR SELECTION PROJECT

Vendor A was the current supplier of phototooling film and chemicals. The PCB plant had purchased phototooling development machines from Vendor A and had a maintenance contract with them. The plant was very satisfied with the current film process and the support they received. However, PCB management was perturbed when Vendor A announced that they would be changing their film and chemistry recipe. It was a surprise to PCB and the engineering team. It was decided that they would qualify Vendor A's new product and put it up against Vendor B's product. That would allow the plant to decide which product was the best for the company. The company would award the contract to one of the vendors based on a comparison of the technical aspects of the vendors' products, their support models, and overall price of materials.

GIVE ANDY A TRY

Vendor A's old product would start to be taken offline within one month. This required the PCB plant to have its transition solidly begun to ensure there would not be any supply issues if Vendor A were to remain the selected phototool supplier. Ron, the Manager of Inner Layer Engineering, met with the photo department manager and the directors of Engineering and Manufacturing to discuss the resources required to complete the project. The management group decided that Andy would lead the project with the support of two other engineers from Ron's group and a lead operator from the photo department. The other engineers would help with the measurements, statistics, and any other quality-related issues with the film. The photo department operator, who would assist in running the designed experiments, could delegate work as necessary to others in the department. There was some concern that the team was too inexperienced for such a critical and time-constrained project. Andy was new to the company. He and the other engineers on the team were fairly recent college graduates. There could also be some conflict with the veteran operators who had been working in the department for nearly 15 years.

ACTIVITY DEFINITION

The team has sat down for the first time after being selected. Andy was the project manager as he was the Phototooling Process Engineer. The area Andy supported was the one being affected by Vendor A's business decision. He was also the most familiar person with the technical aspects of the phototools. Ron has called a kick-off meeting with the team to begin their process. Andy was anxious to lead this project, as it would be his first big test at this company. Anthony was selected to help identify any mechanical needs and program the coordinate measuring machine (CMM). Sam would be helping with the statistics and running the statgraphics program. Jane was the selected operator to help with running the experiments.

>**Ron:** Thanks for meeting. As you know we've got a big task in running experiments on A's and B's tools and to help decide our vendor of choice. Today's agenda will be to figure out what we'll need to do. Andy will be running this project. With that I'll turn it over to Andy.

>**Andy:** Thanks Ron. We've got a big task ahead of us and don't have a lot of time. Let's brainstorm the issues at hand.

>**Anthony:** Well, you've got a lot of mechanical aspects to consider on the film.

>**Jane:** That's what the CMM is for. We'll be able to measure everything necessary. I'm sure Anthony can program anything on the machine.

>**Sam:** Statgraphics will help us with all sorts of statistical analyses.

>**Andy:** Great! So I'll start with a few things I think we need to do. First, we need to make sure we can get enough film and chemicals from both vendors.

>**Jane:** Are you going to use our current developers for Vendor A's new product?

>**Andy:** Yes, when Ron and I met with them, they said that we could do it.

>**Ron:** That's good. Vendor B said that we could borrow a developer. I think we can get a "no-cost" purchase order to be able to bring it in. I'll talk to Purchasing about that. But I know we've done that before with other vendor machines. I know Vendor B is hungry for our business, so that's why he's so agreeable.

>**Andy:** I believe both vendors will give us the film and development chemistry. Vendor A feels like they owe it to us. And like you said Ron, Vendor B wants our business.

>**Jane:** Since we have two production lines, we can set up one of the laser machines with the settings.

>**Anthony:** And then we can dial in settings based on the CMM measurements.

>**Andy:** Right, we could use the current SPC test. We will need to tune the laser exposure speed for the proper line widths. The developer speed has to be just right so that there isn't over- or underdevelopment. And we can set the laser dimensions to compensate for any overall size issues.

Sam: That sounds like a fairly simple design of experiment in order to get the proper line width, edge, and development.

Ron: Andy, did you capture all of that?

Andy: Yes, let me rattle off the activities: get film and chemistry from both vendors; talk to Purchasing to acquire a free developer from Vendor B; set up one of the laser machines; measure each film sample; and run a DOE to optimize both vendor films.

Anthony: How about running some development and etch samples on resist and copper? That'll help ensure that there are no quality issues that extend beyond the film's characteristics.

Andy: Excellent idea. Anyone else have any other ideas?

Ron: Just one more thing. Once you've made your schedule, run your experiments, and come up with some results, you'll have to pass that information to Purchasing as they'll help with the decision. Their input will be in regard to price for performance comparisons. Andy, you'll help with that aspect.

Jane: How long do we have?

Andy: We've got three weeks to get the results published. And we've got to present our plan to the Leadership team later this week.

Ron: That's right. When you have your plan, let's go over it with the rest of the Inner Layer Engineering team and the Photo department managers. We'll get their agreement and then we can present to the Leadership group for their final blessing.

Andy: Okay, let's meet later this afternoon to map out our activities. In the meantime Ron and I have to talk to Purchasing.

DISCUSSION OF DEPENDENCIES

Ron and Andy went to the Purchasing group to find out how they could get all their vendor supplies and machinery. They were pleased to find out that they just needed a few purchase orders specially written and then approved by Ron and his manager. Andy studied the SPC tool program to ensure it would operate properly in their experiments. He also talked to others in his Engineering team about testing new phototools with resist development and etching characteristics. There would have to be some copper samples available as well as machine time available on one of the resist laminators and on one of the developer-etch-strip (DES) machines.

The team met that afternoon to discuss and plan the activities.

Andy: Thanks for meeting again this afternoon. Take a look at the activities that we talked about this morning on the overhead:

Brainstorm List of Activities to Test New Phototools

- Get film and chemistry from both vendors
- Talk to Purchasing to acquire a free developer from Vendor B
- Set up one of the laser machines
- Measure each film sample
- Run a DOE to optimize both vendor films
- Run resist and etch samples for film

Andy: Did I capture everything?

Sam: It looks like you did.

Jane: Well, we're supposed to be done in about three weeks, right? So we need to put some dates around everything.

Andy: That's right. We've got a final date and we can work backward from there. Most of the activities have some dependency on a previous task. May 5 is when we've got to be done and submit our findings.

Anthony: So when will the vendors get us our test supplies?

Andy: Vendor A will be getting film and chemistry to us on Wednesday. Vendor B will get their film and chemistry tomorrow; and then their developer should be here next Monday. The no-cost purchase order should be approved by close of business tomorrow.

Jane: Andy, you can set up Laser Line 2 for the Vendor A film as soon as it arrives.

Sam: And we can run the first DOE for the Vendor A film.

Andy: Jane, I'll need some help taking Laser Line 2 offline to put the proper developer in the machine.

Anthony: Once you start running some Vendor A products, we'll be able to measure up and evaluate each DOE sample.

Sam: It shouldn't take more than a day to run everything, measure, and find the optimum laser and development settings. That's assuming the settings don't take too long to change and stabilize.

Jane: Nope, Andy does it when our SPC indicates we need to. He's got that down pretty well.

Andy: So we'll do the same thing for each vendor's product. And then we'll run our final quality samples on developed resist and copper etch samples. That may take a few days to schedule and monitor the runs, and then a day for us to analyze.

Sam: I've taken some notes and here's how our setup and testing will work. I'll put it up on the board.

Sam's List of Setup and Testing Sequences

- Receive Supplies
- Set Up Laser and Developer
- Perform DOE
- Measure Samples
- Refine Laser and Developer Settings Based on DOE Results
- Run Samples on New Settings
- Test DES samples

Andy: That's brilliant. We need to repeat all steps for each of the vendor samples. Based on Sam's notes, I'll come up with the final timeline and proposal for the Leadership team.

Jane: Sounds good. I can't wait to start.

PROPOSED ACTIVITY LIST

Andy took everyone's input and crafted a schedule into a simple Activities List. He decided that a fancy Gantt chart wasn't really required as it was going to be a quick and much focused project. All the resources had been allocated; it was just a matter of timing of all the activities. He also wanted something simple and easy to read for the Leadership team to review so that they could give their blessing to the project.

The next day, Andy showed the Activities List to the team for review. Andy's boss, Ron, was there to review as well.

Table 6.1 Activities List

Task	Date
Receive Vendor B Supplies	4/11
Receive Vendor A Supplies	4/12
Complete Vendor A DOE	4/14
Vendor B Developer On-line	4/18
Complete Vendor B DOE	4/19
Complete Resist and Etch test	4/26
Analyze Samples	5/1
Submit Findings and Recommendations	5/5

Andy: What do you think?

Sam: It looks great. It seems like you've built a bit of a buffer to Complete Resist and Etch Test and Analyze Samples.

Andy: Yes, I was concerned about the capacity of the DES machines.

Jane: What about our capacity?

Andy: Laser Line 1 will have plenty of capacity. After each vendor setup and run, I will put everything back to the default settings. But all the development will have to go to Laser Line 2's developer.

Jane: Oh, that's right. Laser Line 2's developer will have the new chemistry. And I know we'll have to feed Vendor B's film manually—take it off the laser and feed it into the developer.

Anthony: I've tuned in the CMM and it's ready to go. So we're ready when you are.

Ron: I like the plan and the enthusiasm. Andy, are you ready to present to the Leadership team?

Andy: Yes, let's go for it.

Andy presented the project to the Leadership team. There were some concerns about the plan's aggressiveness. The plant manager was already irked about being in this precarious situation with Vendor A's business decision, but he grilled Andy about the timeline and the methodology. One of the managers spoke up and reminded everyone that both customers and suppliers continually challenge the company. She also backed Andy up by saying that she was impressed with his performance when he did the rotational training in her department. After the discussions, everyone seemed pleased with the thoroughness of the evaluation. The project was given a "go."

Discussion items

1. Outline Work Breakdown Structure (WBS) and develop a project network.
2. What are some examples of external dependencies that could have been considered in Andy's project?
3. The Activities List was a very simple table of tasks and dates. Was that too simple of a list for the team? Was it the correct format for the Leadership team presentation? Is it a good medium for upper management to review?
4. What issues may arise as the team consisted of relatively inexperienced engineers along with a very seasoned operator? How could those issues be resolved?

The Rolling Wave

Dan Itkes

Doug's phone rang. It was Sam on the other end. Doug used to work with Sam on a couple of projects in the past, but it's been a few years since they've done anything together.

Sam: Doug, I have something for you. I know you're wrapping up another project, so I think we may be able to work on something together.

Doug: What do you have in mind?

Sam: Well, let's see. I need to build a software service to go together with the company's product. And since I know you've done it before, I am wondering if you'd like to do it again. We need to get traction in the market quick, or we will lose the momentum.

Doug: So do you know what's missing?

Sam: Well, it's clear that support is a big issue with customers not being able to set up their evaluation units. I think what we and they need is a remote management service that works over the Internet to help the tech set it up.

Doug: What kind of content do they use? (Doug remembers from his time working in digital signage space that content was a main issue).

Sam: Well, it's whatever they produce or what we give them.

Doug: Are they able to produce good content quickly?

Sam: Not really.

Doug: What kind of resources do you have for this project?

Sam: I've got one software engineer I can give you full-time, as well as access to several people who can work on it part-time.

Doug: And what kind of timeframe are we talking about?

Sam: If we don't fix it within three to six months, I'm afraid we're going to lose the momentum, and we worked real hard to get it.

Doug: Alright Sam, give me a couple of days to think it over.

Doug hung up the phone with a bit of an uneasy feeling. He liked working with Sam in their past experience together, but he wondered if this was the kind of project he wanted to get involved with at the present time. Doug thought Sam's pitch sounded like there were several issues involved: Sam thinks it's about the remote management, but Sam, being an excellent systems guy, is not putting on his marketing hat here. Just by asking a few questions, it seems like the problem might be something about being able to demo the product effectively with customers, especially when they're getting excited about building their own content and seeing results right away.

BACKGROUND

Caldera, a start-up in the Bay Area with ties to a media chip manufacturer, sports a line of digital media players designed to work in IPTV and Digital Signage applications. Caldera's business is half OEM (Original Equipment Manufacturer) and half direct. Caldera OEMs license its media players to some prominent display and network equipment manufacturers. Revenues are in the range of $5 to $10 million annually.

The IPTV market has been slow to take off, but with very high expectations by industry analysts. Caldera initially focused on the IPTV market, and while it had gained some traction, success was limited to some small-sized trials. Caldera found faster traction with commercial accounts looking to implement digital signage, which was accountable for most of the company's revenues.

Caldera has recently licensed its new player, DMS-5000, to a major Display OEM, Visualistic. While both companies were excited about this partnership, they were not getting the traction they needed in the marketplace.

MAKE IT HAPPEN

The next day, Doug called back Sam.

Doug: Sam, I'll take the project, but I'm not sure remote management is what we need here. Seems to me it should be about the automated sign generation.

Sam: Alright, let's get things rolling and then see. Oh, by the way, I already had the engineer work on the scheduling and remote management for several months. See if you can use what he's put together.

Doug called Bill, the engineer who started developing the system with little specifications from Caldera. "Well, what I have is a scheduler engine that will allow us to schedule content out." Doug talked with both Bill and his software engineer colleague Michael about the architecture. Michael said he could do the "Sign Builder" completely as a Windows application and render the interface as HTML on the digital media player. That would mean that the whole project would be dependent on Michael, who could dedicate 30 percent of his time to this at best. Doug also didn't like the idea of distributing a Windows application to the clients as it would mean having to do extensive testing for compatibilities, support, and robust upgrade procedures.

Bill, the dedicated software developer, was good at developing web applications in PHP and had his own framework developed, which would expedite the development but would also present a bit of a risk in the future as the framework was only known to him. Sam put a lot of stock in Michael's opinions and was a bit concerned about relying on Bill's framework alone from the future support standpoint.

The other question was what management features to implement. Sam felt strongly about content scheduling and box discovery. He also gave Doug access to another software developer, Jenny, who was skilled in in-network management systems, but she would also be able to dedicate only about 30 percent of her time, sporadically.

A few days later, Doug met with Visualistic, who confirmed that they indeed saw a problem with what they perceived as initial support for the trial customers and field salespeople. This conversation with Visualistic reinforced the feeling that if this project was to be successful, it would have to be about easy sign generation as much as anything else.

Up until then, most of the digital signage content was created by hiring creative agencies putting together movie clips for the venues. The process was long, expensive, and worst of all, should something change (pricing on the menu, new promo, etc.), the process would have to begin all over again.

"If we could only change the sign content generation around an online sign generation system," thought Doug, "we could cure all three problems with one fell swoop: (1) it would be easy to demo as there would be no content compatibility problems, no scheduling problems, and the customer would see exactly what *their* sign would look like; (2) the support problems would be greatly reduced as there would be a lot less configuration pieces on the client itself; and (3) content update becomes very simple for the customers to do themselves as they can make a textual change without disturbing the rest of the sign."

This was a pretty big departure from what both Sam and Visualistic viewed as a solution so Doug thought it would be best to (1) develop a prototype first, and (2) incorporate the system management aspects that Sam was talking about.

Given the lack of clarity on the requirements, resource, and time constraints, Doug decided to approach the project in the following way:

1. They would have to get everybody on board with the "hosted sign builder" solution changing the focus from alleviating the symptoms to trying to fix the root of the problem. This would take time. It would take prototypes,

functional demos, and some field results. It wouldn't be worth trying to get a consensus now. They would do it in conjunction with addressing the perceived problem.

2. Relying on undedicated resources would be problematic. They would be used for noncritical tasks and as consultants.

3. The development would have to be done in "agile" fashion—frequent and small demonstrable releases, but not show things too early to get rejected.

Doug then assigned Bill (dedicated software engineer) and Vlad (the designer) to work on what he thought was the critical piece in all of this—the hosted sign builder. The sign builder was architected so that the front end was loosely coupled with the back end allowing flexibility in swapping out both parts later down the road. He asked Michael to put together a prototype for his approach to the front end design. Finally he asked Jenny to work on the in-network management client (something she was familiar with) to address the device management objective.

CHANGE HAPPENS

Three months into the project Doug's phone rang. It was Sam.

Sam: Doug, you can congratulate us, we've been acquired.

Doug: That's great news, congratulations! How does it change the project though?

Sam: Nothing to worry about! If anything, it will be more strategic.

They chatted more about details and decided to keep everything on track. Three weeks later, however, Sam called again, this time with a different take on the situation. "You know," he said, "I didn't realize it at the time, but after discussing the project with the company that made the acquisition, they saw it as a competitive product to what they already have. We have to figure something out."

Doug: Ouch! How are they looking at it?

Sam: Well, they have a network management suite that they would like to use.

Doug: What about the sign generation piece?

Sam: That I'm not sure about. Let me see if I can get an exception for that.

A week or so later Sam delivered some good news: "Doug, no problem. I talked to the business unit and they didn't mind keeping the sign generation as an independent project, but they would like an option to roll it into the product." "Okay," Doug said. "Let's get this wrapped up then, see where it takes us."

Figure 6.4 shows the progression of the Rolling Wave Schedule used for this project.

Figure 6.4   Rolling Wave Schedule Progression

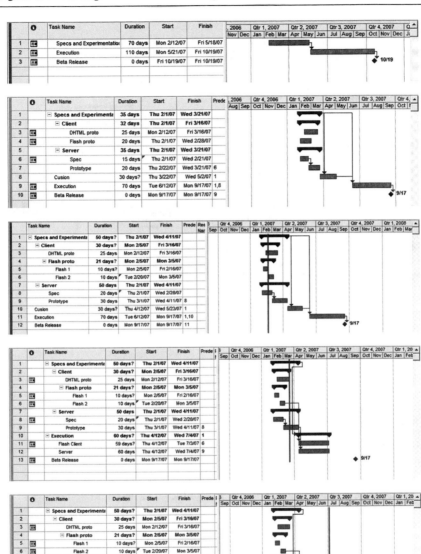

Figure 6.4   (Continued)

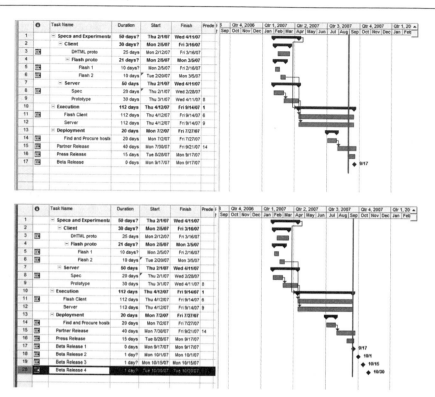

FINALE

Another three months have gone by. At the end of these three months, Doug had the following results:

- A functional prototype of the sign builder that he could demo
- The prototype that Michael had put together was clearly inferior to what Bill and Vlad had done
- The management piece was dropped due to the acquisition dynamic

Doug showed the prototype to both Sam and the Visualistic team. They were impressed with the ease of use and the look of the signs that were being built, but you could see they still had some doubts as to whether this was what would help improve the situation. They also wanted to see the system closer to the finish state.

Over the next three months, Doug was working close with Visualistic—their sales and product support team. During this time, Visualistic became fond of the new system, so much so that they changed most of their internal demos to it.

Fast-forward to InfoCom tradeshow later that year: At Visualistic's booth, visitors are impressed by how easy the product is to set up and maintain the content. System management questions rarely come up, and when they do customers see little need for the scheduler anymore, as the new way to construct the content has made playlist creation all but obsolete.

Visualistic's field embraced the new service and made it the staple of their DMS product sales allowing them to boost their sales by at least a factor of three.

Discussion items

1. Explain in your own words what rolling wave planning is.
2. What are the pros and cons of rolling wave planning?
3. What would have happened if Doug had waited until complete clarity was achieved?
4. What were the risks in this project?

Schedule Accuracy

Dragan Z. Milosevic, Peerasit Patanakul, and Sabin Srivannaboon

Being old friends, project managers Mark and Jerry came to a meeting organized by the engineering vice president, Barb. In addition to the three of them, there were twenty-one project managers sitting in the room. According to the invitation, the meeting was supposed to be a 45-minute session. So Mark and Jerry were hoping they could get back to work shortly after.

Barb: Thanks to everyone for coming. As already mentioned in the invitation, the purpose of this meeting is to define a strategy for our group. Last week we were trying to have a meeting about this, but many couldn't attend. So we moved it to this week, and this is the best time for most of us. To save time, let's cut to the chase.

Our business strategy is time-to-market. Therefore, accuracy of schedule is one of the key elements. But we never actually define what accuracy means. So people often claim that their project schedules are accurate. But to what extent? And based on what? We never know!

THE ACCURACY

Barb proceeded to give a theoretical explanation of "accuracy" as it relates to the business. She then continued with a specific explanation of what accuracy meant to her particular business unit.

Barb: The accuracy is viewed as a percentage to which our project schedules are likely to be accomplished. All our schedules will be in a network form. Specifically, we use the Time-Scaled Arrow Diagram, or TAD.

Barb continued her speech by showing the three steps in coping with schedule accuracy:

First step: The project team must develop a TAD of the project. We require team members to participate in the development as they are expected to discuss the length of each activity. In particular, they must give us three points of estimate of each activity in which they are involved. Typically, the three point estimates are the best case, the most likely case, and the worst case. Speaking mathematically, the range from optimistic to pessimistic should cover six standard deviations, and the estimates should be made at the 99 percent precision level in order to use the regular Beta formulas.

Second step: This three point estimate will be used to calculate the likelihood of the schedule. Generally, we assume the project activity durations fit a Beta distribution and are statistically independent, so we can calculate the activities' means and variances accordingly (based on the Beta formulas). Then, we find the longest path in time of the project, or the critical path, and assume that the overall project duration has a normal distribution. So, we can calculate the Z-value and look at the Z-table to find the probability of completing the project's "critical path" (note that this is not the probability of completing the "entire" project on time).

Third step: Generally speaking, those who are responsible for the schedules are project managers. So they have to make sure that all the resources are available when needed in the right quantity. After identifying the probability of accomplishing the project's critical path, they will meet with the clients to further discuss the project deadline.

After the brief explanation of the new approach to the schedule accuracy, everyone in the room looked at each other. They were thinking they needed a short statistics class to truly understand what Barb just said. But before they said anything, Barb continued, "Don't worry if you don't understand what I just said. I have arranged the tutorial session for you. And the instructor is already here; ready to start in five minutes." At that time, Mark and Jerry knew the meeting wouldn't end in just 45 minutes.

Discussion items

1. Define schedule accuracy. What are major assumptions?
2. Did the approach offer schedule accuracy or buffer against inaccuracy?
3. From your point of view, how important is the probability of completion to the customers?

AtlasCom

Dragan Z. Milosevic, Peerasit Patanakul, and Sabin Srivannaboon

In its 50+ year history, AtlasCom has excelled in manufacturing state-of-the-art specialized construction equipment. Recently, management has noticed that manufacturing productivity and quality has been falling behind the competitors. In response, several projects were launched, one of which was to focus on reengineering the factory layout. Management tasked a two-tier team to get the job done. The first tier was the core project team, a cross-functional group of middle managers responsible for managing the project effort. The second tier, the extended team of manufacturing specialists, was in charge of doing the project work.

Faced with the lack of a formal project management process and experience, the core team received basic project management training before hiring a consultant who helped develop a detailed CPM schedule. Gearing to launch the execution, the core team explained the CPM schedule to the extended team members, asking them to get the work started and to report progress in a week. The problem was, the extended team members commented, that because of CPM's complexity, they were not able to use it as a basis for planning, organizing, and reporting.

A few days later, a quality improvement project (QIP) including members from both teams was chartered to find a solution to the problem.

WHAT IS CPM?

Developed by the DuPont Corporation in the 1950s, Critical Path Method (CPM) is a network diagram technique for analyzing, planning, and scheduling projects. It provides a means of representing project activities as nodes or arrows, determining which of them are "critical" in their impact on project completion time and scheduling them in order to meet a target date at a minimum cost. CPM uses deterministic activity time estimates (the most likely), rather than probabilistic activity time estimates (e.g., the optimistic, pessimistic, and most likely) such as those used in Program Evaluation and Review Technique (PERT). The typical use of CPM was for construction projects, although it is nowadays applied in many types of projects across various industries.

Figure 6.5   Example of a CPM Diagram

Activity	Description	Immediate Predecessor	Duration (Days)
a	Start		0
b	Get Materials for a	a	10
c	Get Materials for b	a	20
d	Manufacture a	b, c	30
e	Manufacture b	b, c	20
f	Polish b	e	40
g	Assemble a and b	d, f	20
h	Finish	g	0

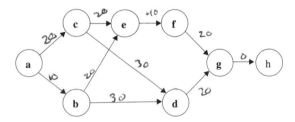

The process of constructing a CPM diagram requires the following inputs:

- A list of activities, from the Work Breakdown Structure (WBS), required to complete the project,
- The estimated duration of each activity, and
- The dependencies (predecessor/successor relationships) between the activities.

Normally, the diagram shows a number of different paths from Start to Finish, defined as sequences of dependent activities. To calculate the time to pass through a path, add up the times for all activities in the path. The "critical path" is the longest path in time from Start to Finish. It indicates the minimum time necessary to complete the entire project. Essentially, the critical path is the highest priority path to be managed.

Figure 6.5 shows an example of a CPM diagram.

WHAT WAS THE ATLASCOM SOLUTION?

In a week, QIP proposed the solution: *a short-term outlook schedule*. The solution suggested extracting activities from CPM that were owned by a team member and were to be done in the coming two weeks, rather than the entire schedule, presenting them in a Gantt chart format. Familiar with the Gantt chart, each member thus should have a user-friendly two-week plan that they could start using for organizing work and reporting and updating on a weekly basis.

Discussion items

1. Do you agree/disagree with the AtlasCom solution? Why?
2. What are the pros and cons of the CPM diagram?
3. From Figure 6.5, what is the project critical path? – *E·c*
4. Is it possible for a project to have more than one critical path? Justify your answer.

 → Yes,

Workshop: The Milestone Chart

Dragan Z. Milosevic, Peerasit Patanakul, and Sabin Srivannaboon

Konrad Cerni is one of the most well-known and well-respected people at Ball, Inc., a famous consulting firm in the area. He is 45 years old and has a very strong background in project management, especially in the field of project management tools and metrics. His fame comes from the fact that he was recently named the best employee of the year in 2009 for his outstanding performances and rave remarks from his customers. That's in addition to many other prestigious certifications he has received during his twenty-plus years of service, including the PMP® in 2002 and the PgMP® in 2008.

A company in the vicinity recently contacted Ball, Inc. and asked for training designed to focus on project management tools. Although Ball had done this kind of workshop a hundred times, this particular request was a little different and in fact a little challenging: the workshop was to be completed in one day, as opposed to a regular three-day workshop. Looking at a variety of project management tools (generally speaking 50+ tools are available in the practice and literature of project management), the company decided to send Konrad, the most experienced person, to arrange the workshop and give the training to the clients. Due to the time constraint, Konrad ignored quite a few project management tools and selected only the crucial ones; one of which was called a "Milestone Chart."

WHAT IS THE MILESTONE CHART?

The milestone chart is a chart that shows milestones against the timescale in order to signify the key events and to draw management attention to them. A milestone is defined as a point in time or event whose importance lies in it being the climax point for many converging dependencies. Hence, "complete requirements document" is a distinctive milestone for software applications development projects, and "complete market requirements document" is a characteristic milestone for product development projects. While these milestones relate to the completion of key deliverables, other types may include the start and finish of major project phases, major reviews, events external to the project (e.g., trade show date), and so forth.

A PICTURE IS WORTH A THOUSAND WORDS

Konrad commented, "The milestone has an advantage over other charts. The advantages are in terms of its visual characteristic making it simple and useful as both a planning and tracking tool. But it also shows some drawbacks. For example, when used separately from a detailed schedule with activity dependencies, it is difficult to understand how to reach a milestone, especially when there are many milestones within one project or one segment of a project. Also, as a number of milestones grows, the chart lines lose their appeal by being overcrowded."

Konrad added, "A chart with a few key milestones related to level one of WBS, for example, is in a solid position to capture and enjoy management attention and time with high-profile project events. Not so with charts including many milestones linked with work packages. The gain from such charts is the ability to beef up the emphasis on goal orientation (milestone accomplished or milestone not accomplished), while reducing focus on activity orientations ('I am working on it')."

"Now, let's see an example," Konrad switched to the picture on his slide (see Figure 6.6).

"In this case, there are four major milestones, including requirements walk-through, conceptual design walk-through, final implementation walk-through, and postmortem review, each of which signifies when it is expected to occur," Konrad said.

"Let me give you more examples. In the new product development environment, there are eight sacred cow milestones for a company's product development projects: product concept approval, requirements definitions, plan and specs review, design complete, product evaluations reviews, launch plan, launch complete, and product release. Spanning from one to two years and costing millions of dollars, these projects are the company's engine of growth. The milestones signify the end of key phases and require upper management review. They are displayed in the management milestone chart that is used to report progress to senior management.

Figure 6.6   Example of the Milestone Chart

Milestones	Half 1, 2001							Half 2, 2001		
	DEC	JAN	FEB	MAR	APR	MAY	JUN	JUL	AUG	SEP
Requirements Walk-through	◆ 12/28									
Conceptual Design Walk-through		◆ 1/31								
Final Implementation Walk-through								◆ 6/29		
Postmortem Review										◆ 8/31

He continued, "Allow me to give a few tips for milestone charts from my experience.

1. Don't cram milestones. Space them out.
2. Use both charts for key events and detailed milestones.
3. Use the charts in both large and small projects for both the plan and actual project progress.
4. Use the chart in conjunction with another schedule showing activity dependencies.
5. Team-developed milestone charts lead to higher quality, better buy-in, and stronger commitment of team members."

"Any questions?" Konrad asked the workshop attendants.

Discussion items

1. What is the purpose of the milestone chart? Describe it in your own words.
2. When should the milestone chart be used?

Chapter 7

PROJECT COST MANAGEMENT

This chapter presents cases that involve the processes in estimating, budgeting, and controlling costs in order for a project to be completed within the approved budget. This is, in other words, the project cost management, which is Chapter 7 of the *PMBOK® Guide*. There are six cases presented in this chapter.

1. The Court House Disaster
   The Court House Disaster is a critical incident that talks about one of the cost estimating techniques, called shadow estimating. The technique suggests the use of an estimating consultant, and was proposed for inclusion in the company's Best Known Method (BKM) for future projects.
2. Bad Metrics for Earned Value
   This issue-based case centers on the change of performance measurement baseline and consequent difficulties in maintaining the earned value metrics at desired levels. Different strategies are attempted to meet customer requirements. This case can also be discussed in the areas of monitoring and controlling process (e.g., scope management or time management).
3. The Museum Company
   The Museum Company is a critical incident that talks about the company's cash inflow issue. To respond to the issue, the company set up project management training that, they believed, could potentially help the company's leaders identify/solve the cash inflow problem.

4. Workshop: Parametric Estimate

   Workshop: Parametric Estimate is one of the critical incident cases in the Workshop series, which specifically explains a parametric estimate. The parametric estimate is an estimating technique that uses a statistical relationship between historical data and other variables to calculate an estimate for activity parameters, such as scope, cost, budget, and duration.

5. No Bottom-up Estimate, No Job!

   No Bottom-up Estimate, No Job! is a critical incident that exhibits the process of developing a bottom-up estimate. The technique suggests that the estimate be done on the individual work items first and then rolled up (totaled) to produce an estimate of the entire project.

6. Earned Tree Analysis

   This critical incident case presents the primitive use of Earned Value Management, techniques for project cost monitoring. The case presents how the cumulative percent complete is calculated and how the contractor gets paid accordingly. The case also challenges the reader to look at schedule and cost variances in the view of earned tree analysis.

CHAPTER SUMMARY

Name of Case	Area Supported by Case	Case Type	Author of Case
The Court House Disaster	Estimate Costs (Shadow Estimating)	Critical Incident	Dragan Z. Milosevic, Peerasit Patanakul, and Sabin Srivannaboon
Bad Metrics for Earned Value	Control Costs (Earned Value Analysis)	Issue-based Case	Don Hallum
The Museum Company	Estimate and Control Costs	Critical Incident	Jovana Riddle
Workshop: Parametric Estimate	Cost Estimation Technique	Critical Incident	Dragan Z. Milosevic, Peerasit Patanakul, and Sabin Srivannaboon
No Bottom-up Estimate, No Job!	Cost Estimation Technique	Critical Incident	Dragan Z. Milosevic, Peerasit Patanakul, and Sabin Srivannaboon
Earned Tree Analysis	Cost Control Technique	Critical Incident	Dragan Z. Milosevic

The Court House Disaster

Dragan Z. Milosevic, Peerasit Patanakul, and Sabin Srivannaboon

Mike Bitka, one of the three owners of BC Company, and George Slicker, the company CEO, met on a regular basis to discuss the company business. Usually very calm, Mike was nervous at this recent meeting.

Mike: I had a chance to talk to the other owners of the company last week about the Court House project, the one that was recently closed. They wanted me to discuss it with you so that we could improve our Best Known Methods (BKM) to prevent any similar disaster in the future. For the beginning, let me say that two months before the ending of the Court House project, we estimated to lose approximately $500,000 on a $1.5 million project. So from your point of view, what do you see as being the mistakes we should never repeat again?

George: I will be very brief. First, the major mistake was to hire Pete Tramp as the manager. He didn't have much experience in managing projects. Because the job market was very tight we hired him to estimate the Court House project. He did the bid and got the job. Apparently, the bid estimate was below the cost of that project, and I have to admit that Steve Johnson of E-contractors, another bidder, warned me that our bid was very low. But, I did not react.

Mike: Did anybody in the company check Pete's estimate?

George: No. According to our BKM, our estimating department should have verified Pete's estimate. But, they were too busy working on another two projects and did not have time to verify his estimate.

Mike: Couldn't our estimators rearrange their work schedule to make time to verify this important estimate?

George: I can't say for sure, I only can assume. But if our estimators wanted to have this project's estimate verified, they could've applied a method we learned in a project for Intel. It is called "shadow estimating." When you want to verify an estimate for a project, you hire an estimating consultant, ask him to develop an estimate for the project, but don't show him the estimate being verified. Of course, the consultant's estimate is assumed to be accurate. Then, the two estimates are compared, and if the difference is zero or close, the estimate being verified is considered good. The larger the difference, the weaker the estimate being verified. The downside is that you have to pay the consultant.

Mike: That sounds interesting. We should have done that with the Court House project. I also want to know something else. Could our earned-value based reporting system have helped to detect the problems with the Court House earlier?

George: Yes. But we did not have time to train Pete on our earned value system, so he couldn't use it to report. Instead he reported what he had completed, consumed labor hours, and installed cables and equipment—basically very reactive. So I believe it is a disaster that we must prevent from happening again. For this, we should add to our BKM that:

1. We only hire candidates that at least meet the job requirements.
2. We use the shadow estimating in all significant projects.
3. We train any new hire in our BKM before he/she starts the new job.

Do you have something to add?

Mike: Yes. I miss the $500,000 that Peter Tramp lost!

Discussion items

1. How would you rearrange the work of the BC Company's estimating department to make it possible for the department to verify Pete's bid estimate?
2. What are the pros and cons of the shadow estimating approach?
3. How could the earned value reporting have helped detect problems with the estimate of the Court House project earlier?
4. How could the periodic project audits, if the BC Company used them, have helped detect the estimate problem in the Court House project earlier?
5. If you were in charge in BC Company, would you consider hiring a professional estimating company to develop a bid estimate for this project?

Bad Metrics for Earned Value

Don Hallum

THE DEATH OF THE MANUFACTURING PROCESS

Mike Thompson, the Program Manager for the new development project, came into Howard Bono's cubicle. "Well, the new manufacturing process is dead for this project." Howard knew that this day might come, but it wasn't any easier to take; after all, everyone on the Mechanical Engineering team had spent a lot of hours trying to make it work. It had been the focus of much of Howard's first six months since accepting a new position as Technical Project Manager (TPM) at the beginning of the new calendar year. Having moved from the Mechanical Engineering department, the Director of Engineering thought it would be a good fit for Howard as a first assignment. While all knew there was a possibility that the customer might not accept the risks associated with the new process, it didn't make the news any easier to take. All of the assigned engineers had worked diligently toward making the process certifiable and knew that, given enough time, they could make it work. There just wasn't any more time left.

"What does that do to the financial numbers for the project?" Howard replied. "Is it still viable without the new process?" Howard knew the company had won this project from the competition based on, to a large extent, a lower recurring cost for the system, and attaining this level of recurring cost was dependent on using the new process. Any alternative process was estimated to cost at least 50 times more per structural part, and there were six structural parts per system, and two systems in every installation. Howard was wondering how he could have let this happen? What could he have done differently? It all seemed to have unraveled so quickly. Howard thought back to the previous December when he interviewed for the new TPM position.

BACKGROUND

Acme Avionics, Inc. is one of the leading companies for aviation electronics. The company has provided products to the military, commercial, and business aircraft market sectors since 1980s. Its annual sales are around $4 billion corporate-wide.

HIRING HOWARD BONO

Ward Robinson, the director of the Engineering department at a relatively small business unit, who had known Howard for years, seemed pleased that he made the effort to apply for the position.

Ward: Well, you're certainly ready for this.

Howard: I think it would be a good fit. I just finished my master's degree in Engineering Management; have worked in the business unit for 14 years now; and have worked with this customer many times before. I know they are demanding, but I've always been able to handle it as a lead engineer in the ME department.

Ward: You realize that you'll have to let go of most of the day-to-day technical involvement; you just won't have time to handle that and everything else that needs to be done.

Howard: Yes, I am going to have to let the lead engineers and their teams do their jobs.

Howard's family life wouldn't suffer too badly; the customer is located only an hour's flight from his location, and he wouldn't be gone from home more than one evening at a time when he had to travel. In fact, on most trips he could fly out in the morning and be home that same evening. But it was out of his control now. He had done the best he could during the interview; let the chips fall where they may.

A week later the voice message light on Howard's phone was blinking. Ward Robinson had left a message, "Hey, Howard, I want you to send me a short biography via e-mail by the end of the day today." Howard wondered, "Did I get the job? It must be; why else would he ask for something like that?"

The next morning it was in Howard's e-mail in box:

December 11, 2006

From: Ward Robinson

To: The business unit engineering department.

I am pleased to announce the assignment of Howard Bono as Technical Project Manager for the Boeing 787 Project. Howard will be replacing Greg Sanderson, who earlier this week accepted the position of Project Manager for Business and Regional airline projects. Howard has been a Mechanical Engineer with us for 14 years, and will be awarded a Master of Engineering in Engineering Management degree from Portland State University later this month. Howard has most recently been a lead engineer for Boeing Projects, and will be reporting directly to me. Please join me in welcoming Howard to his new position.

GETTING TO THE CORE OF THE PROBLEM

It wasn't too much later that Howard was called into the office of the ME manager, his now former supervisor. Ward Robinson was already there.

> **Ward:** I want you to spend the rest of this month in transition to your new position. The new manufacturing process is in need of attention, and your background as an ME will be very helpful. You need to get your arms around this as soon as possible.

Howard soon learned that the customer was responsible for certifying the system in the airplane; a brand new airplane, not a derivative of an existing certified airplane. If they were not comfortable with something, then most likely Howard's team had to fix it. Howard also learned that they weren't comfortable with the mechanical properties of the material produced with the new process. The strength of the material was thought to be extremely variable, and the material was very brittle. This made them wonder what the minimum margins of safety really were, and also made them think that the process was not yet "under control."

"Not a problem," Howard thought. "The load requirements that the customer had levied on us were fairly benign, and we had met all of them with a large margin of safety relative to the 'typical' strength properties of the new process," he further thought.

One of Howard's first tasks as the new TPM was to convince the customer's engineering representative and project manager that the process was viable—the margins were high enough—and that the material properties would only get better with time. While Howard could tell that they were skeptical, he managed to convince them that the potential for cost savings was worth the effort of sticking with the time-consuming process improvement steps. Their attitude was one of "We believe your analysis is accurate relative to the stresses in the assemblies, but after doing some research we've determined that the material strengths are extremely dependent on the part's geometry. We're going to require you to conduct some material property testing to prove that the margins of safety you have arrived at are also accurate."

While this manufacturing process was new to aviation "structural" applications, it wasn't exactly a new process. Parts had been made with this process for some time now, but none of them had been scrutinized for strength properties as yet. This was a high-volume process, and as such the piece part price was low relative to what Acme had been used to dealing with. In getting the system on board with this airplane, Acme's marketing department had accomplished something never before done: The system was standard on every airplane that left the factory, and there were two systems on every airplane. Up until this time, one system per airplane was the norm. Sales of the airplane exceeded expectations, and the high-volume process and low piece price seemed like a perfect fit.

PERFORMANCE MEASUREMENT BASELINE CHANGED

Howard met with Mike Thompson to briefly discuss the situation, and then they got on the phone with the vendor responsible for producing the structural parts with the new process. The vendor and Acme agreed to take parts from the last batch produced and cut them into tensile test specimens. The vendor would then subcontract a local test lab to conduct tensile testing of these parts; five specimens from each of the six structural parts. Howard then contacted the customer's project manager to inform him of their plans. He answered, "That sounds good, but we've been discussing this situation here and we want to get our material and process technology experts involved in this. The schedule and technical risks are too high, and we can't afford failure."

After they hung up, Mike and Howard started talking about the impact of the recent developments on the project's financial status.

Mike: The vendor isn't going to do this extra work for free you know, and I want our engineering people intimately involved. We can't afford to let the vendor be solely responsible for the success of these tests. The stakes are too high.

Howard's mind drifted back to a recent meeting the TPMs had with Ward Robinson:

Ward: I'm getting quite a bit of heat from corporate management to keep our earned value metrics near 100 percent. I want to remind you all that your primary responsibility is to keep your project's Schedule and Cost Performance Indices (SPI and CPI) in the 90 to 110 percent range. If you do that, it makes my life a lot easier, and we don't have to spend the time and effort explaining why we're deviating from these goals.

Howard continued with Mike:

Howard: You know this new effort is all out of scope relative to the project's baseline.

Mike: I know it might be, but what choice do we have? If we don't do this, the customer will never accept parts from the new process, and we don't have the time to develop an alternative in time for the first flight of the airplane in August. Besides, meeting our recurring cost targets depends on our being able to use the new process. I do know one thing, though. Developing the new process was definitely in the baselined project; maybe we just did a poor job of estimating the required extent of that effort.

ALL IS WELL THAT FINISHES WELL

Howard left Mike's office feeling a little uneasy, but consoled himself with the knowledge that their strength margins were high and that he was worrying for no good reason. A few days later Howard received a call from their mechanical lead engineer, Art Blake, who was at the test lab with the new process vendor.

Art: Well, the first round of tensile testing is done, and although we're not looking too bad overall, the numbers don't quite reach the typical values we use to determine the stress margins of safety. We did, however, have one specimen of the 30 that broke well under the typical value.

Howard: How bad was it?

Art: It was about 40 percent of what it should have been. But all the others were 75 percent or more of the typical strength, and our margins are high enough that this shouldn't present a problem. I will say, though, that the customer material and process engineer is very worried about the one weak specimen. He says that this proves that the process is not in control, and that we really don't know what the worst-case strength of the material is.

Howard: What if we use the strength from the one low test result, what does that do to our margins of safety?

Art: Well, for five of the six parts our margins would still be high enough to meet the certification regulations, but for one of them I don't think we could get there without a substantial redesign effort.

The next day a teleconference took place with the customer project manager and his boss. They stated that the one low specimen showed that the process was not adequately controlled, and that they were not sure how they could certify Acme's system on the new airplane. The time for first flight was fast approaching, and if Acme couldn't come up with a way to meet the certification regulations the system would have to be added at least a year after the airplane certified, providing enough time to develop an alternative process. This would be disastrous for the business unit's finances; Acme was counting on the revenue from two systems per airplane starting in the third quarter of the next fiscal year.

Monthly project and resource review meetings with the business unit's executives now became very uncomfortable. The project's estimate at completion (EAC) dollar value was steadily increasing. Dave Jansen, VP of the business unit, calmly (but firmly) stated that if the project didn't get a handle on the spending soon, it would eat into the unit's discretionary budget, and that this would hamper its ability to develop future products. This would have a horrible ripple effect; a reduction in new products meant that the revenue projections for future years would have to be scaled back. Layoffs were a definite possibility as a consequence. To add to an already uncomfortable position, Howard's project was now retaining engineering resources beyond the baseline completion dates. This was causing delays in starting new projects, and everyone wanted to know when they would be releasing them to work on those projects. The only alternative would be to go out and hire replacements, and finding candidates having experience with the product would be nearly impossible. The ramp-up time alone would be enough to jeopardize most new projects from the start.

By now Acme's vendor was getting nervous as well. They also were aware that the prospect of producing 12 structural parts per airplane was in jeopardy, and were willing to go to great lengths to prevent this from happening. Since the customer was concerned with the repeatability of the parts' strength, an effort was launched to understand how each of the process's control parameters affected the part strength. In Acme's weekly coordination meetings with the customer, their material and process engineers were saying that in order to prove that the new process was under control, hundreds—perhaps thousands—of test data points were necessary to establish statistical significance. Was everyone involved willing to commit to such an effort? "What choice did we have?" was the concurred response. Meanwhile, the project's EAC continually increased, to the dismay of the business unit's executives.

In the course of discussions with the vendor, Howard learned that there was one earlier application of the process for aviation use. How did it get certified? The answer was that each and every part was tested to its worst-case service load and inspected for deformation. If there was no deformation, the part was acceptable for installation in the airplane. Howard reviewed the regulations, and sure enough, if a material's strength could not be determined on a statistical basis, it was acceptable to test each part. While this gave him a glimmer of hope for the near term, each part could be tested until the independent statistical process control effort was complete, it meant more cost to the project. Test fixtures and methods would need to be established, and the part's recurring cost would be increased because of it. But it was a way to get the parts approved during the original airplane certification. A review of the analyzed stresses produced by the service load requirements given to Acme by the customer and the tensile test data gave Howard confidence that the process yield would be adequate, and that the scrap would be low enough to have an insignificant effect on recurring part cost. After discussions with the vendor and Mike Thompson, it was decided to pursue certifying by testing 100 percent of the parts produced.

Knowing that the airplane's first flight was imminent, Acme's engineers were working hard to stay on schedule, and the earned value metrics showed it. SPI was at or above 100 percent, but CPI had taken a nose dive. As the weeks progressed, the CPI had degraded to nearly 50 percent. Acme was staying on schedule, but it was costing them much, much more than they had originally planned. Howard felt it was time to talk to Ward Robinson about this, since he was the one taking the heat from corporate for the poor cost performance:

Howard: I've been concerned about our poor CPI metric, but I've been working under the assumption that this was okay as long as we stayed on schedule.

Ward: I think your assumptions are correct, Howard, and that's why I haven't intervened. When projects at corporate headquarters are in the same situation, schedule is king. It's a lot easier to justify a low CPI when you're on schedule than to justify both CPI and SPI being low, or having a CPI near 100 percent but being behind schedule. Given your situation, I think you're doing the right thing.

By now the customer felt that it was important that they and Acme gauge the reaction of the certification authority to Acme's plans to meet the regulations. After meeting with the customer to present Acme's plans for test fixtures and load determinations, a meeting was arranged at the regional Aircraft Certification Office of the Federal Aviation Administration. Since the customer was responsible for certifying the airplane, they presented the 100 percent test method as a means of complying with the strength regulations. While the FAA tentatively accepted the plan, they did express some reservations about the ability of the process to produce a consistently sound structure. Since there was no industry or military standard for which the new process could conform to, the FAA wanted to evaluate this risk. While it wasn't a slam dunk, Howard left the meeting feeling good about the prospects for the future.

What followed next was the biggest shock of Howard's young project management career. No, the FAA didn't reject the plan; the customer dramatically increased the 100 percent test load requirements. After reviewing the proposed load cases, the customer's stress engineers noticed that an important component of the requirements levied on Acme was missing. A critical certification test for the airframe was behind schedule, and so the customer would have to take their "best guess" based on past experience at the load produced on Acme's equipment by this test. When the new load was received from the customer, its duration seemed extremely long. Based on Howard's knowledge of similar tests on other airframes, he asked them if they had any analysis data to back up this load case, and they did not. However, they were willing to ask the subcontractor of the airframe section where Acme's equipment was mounted to produce a computer simulation of this test. But this meant more time would elapse before Acme knew the final load case for testing, and time was becoming something that was in short supply.

After two weeks the simulation results were in. The load duration had indeed reduced, but its magnitude was five times greater than any previous tests of its kind that Acme had experienced. Could the parts take a load of this magnitude without deforming? Howard questioned the load's magnitude with the customer, but their response was that it was the best they could do short of having actual test data. The new requirement stood.

The customer test data would not be available until a time that was near the planned date for submitting certification data to the FAA. That meant that if Acme felt their parts could not stand the new load without deformation, there were only two choices: (1) Acme could redesign the parts quickly using the new manufacturing process and hope that it could meet the loads within the space constraints imposed on them, or (2) Acme could take the existing design and move it to a more expensive, but certifiable manufacturing process. Tooling changes, more money, more time, more resources, lower profit margins: a very, very bad combination when considering earned value, project costs, engineering resources, business unit discretionary budget reduction, and a rapidly approaching certification date.

To assess Acme's technical risk for its existing design with the new load, another stress analysis had to be performed. The results were not good. The chances of performing the 100 percent test with this new load and not deforming the parts were slim to none. Howard passed the bad news on to the customer; Acme needed more time if this new load from the certification test simulation was a requirement.

Discussion items

1. Are earned value metrics always clear indicators of a project's true status? How might good earned value metrics be potentially deceiving?
2. When and how should customer-mandated changes in project scope be addressed?
3. At one time during the project, EAC is continually increasing. Which of the three values was growing the fastest: planned value, actual value, or performed (earned) value? What does it mean when EAC is continually increasing?
4. Interpret what CPI=50 percent and SPI=100 percent means. Why can we use CPI to correctly forecast EAC, but cannot use SPI to correctly forecast schedule at completion?
5. Should Acme's 787 project have been terminated? Why or why not? If so, at what point?

The Museum Company

Jovana Riddle

Oleg Zahar, the CEO of the Museum Company (MC), was an architect by training and had recently been appointment to the CEO role. His biggest challenge in this new position was getting his hands around the concept of baseline costs. Baseline costs are time-phased budgets used to measure and monitor the cost performance of projects. They ensure that each phase of a project is profitable and on time.

Despite being very good at what it did and having a large backlog of contracts, MC had no positive cash flow at the end of each month. A detailed analysis revealed that the lack of cash was a result of the cash inflow (money the company earned) being less than the cash outflow (money the company paid out) during the course of its projects. The only time the company's inflow and outflow were the same was after the completion of the projects. Thus, this lack of cash meant that the company was struggling to meet its payroll obligations and had to borrow money from the bank and incur high interest rates each month to pay its employees. The high interest fees the company was paying to its bank significantly ate away at its profitability and could eventually lead to the company's collapse.

BACKGROUND

The Museum Company was a reputable military contractor whose main line of work was building museum exhibits. The company was founded in 1977 and its annual sales were approximately $20 million. The differentiating factor in this industry was that the military, unlike most other entities, was more concerned with the quality of its exhibits than with the cost of each one. Thus, if they were working with a reputable contractor who they had experience with they were unlikely to question the price charged for the work being done.

MC was a reliable contractor that had the reputation of delivering on-time high quality exhibits among the industry players. Their reputation brought in a consistent stream of work and its pipeline was backfilled months in advance. However, despite its respectable external perceptions the internal practices of the Museum Company were questionable at best. Their biggest weakness, which was attributable to the lack of positive cash flow, was that they didn't negotiate payment deadlines in advance of performing/finishing work.

149

Thus, payments (cash inflows) would frequently be made long after project completion and were the root cause of the cash starvation the company was experiencing.

TRAINING

In order to address and fix the Museum Company's cash inflow issue, Oleg had some hard work ahead of him. First, he had to identify his most senior/experienced 15 project managers who would help him turn the company around. Once this team was identified he had to hire an experienced project management (PM) training company, which would set up a customized training program for him and his project managers to help them solve the cash flow problem. The company he hired helped him set up a four-day training that he and his core team would attend. The training consisted of four modules and would help the company's leaders identify/solve the cash inflow problem. The training would cover the following areas of project management:

1. Scope—During the first day of training the core team would identify a standard scope template to apply to each project. This would allow them to define what each project is supposed to accomplish and identify what the end result of each project should be. Furthermore, they would identify activities that would enable them to achieve the predetermined end result and the appropriate deliverables to accompany the project activities.
2. Time—The second day would be spent putting together a standard Gantt chart that the company would use for all its future projects. The chart would help identify start and finish dates of the core elements of each project as well as payment points that would ensure positive cash flows during the duration of each project.
3. Cost—The third day would allow the team to identify a standard cost estimate template to use for all upcoming projects. The cost estimates combined with cost contingencies and time-phased budgets would be used to establish cost baselines, which would ensure that all phases of a project would be profitable.
4. Integration—The last day the team would integrate the scope, time, and cost modules to establish an execution strategy/plan for all future projects. This would include identifying all changes the company would have to make in its daily operations in order to implement its new execution strategy.

During the last hour of training, the team would apply its execution strategy to its new and existing projects to see if the new approach would eliminate the cash inflow problem. This exercise would test their ability to work as a cohesive team as well as their ability to apply new knowledge to the problem and save their company.

LEARNING

The final step in fixing the Museum Company would take place after the training was complete. Implementing the execution strategy would require all PMs to have in-depth knowledge of military exhibit design. In addition to understanding design the team would have to understand how the architecture piece of each project fit in with the actual construction of each exhibit. Since this understanding could not be gained from a book, but rather from on-the-job experience, Oleg would have to hire a technical expert who would provide advice to the PMs and the young engineers until they gained the necessary knowledge and understanding of the coworkings of the two disciplines. Once the technical expert was in place and the execution strategy was implemented, the Museum Company would finally be on its way to being cash-flow positive during each phase of its projects.

Discussion items

1. How would each of the training areas (scope, time, cost, and integration) contribute to the solution of the company's cash inflow problem?
2. In your opinion, what is the most challenging area? Why?

Workshop: Parametric Estimate

Dragan Z. Milosevic, Peerasit Patanakul, and Sabin Srivannaboon

Ball, Inc. is a management consulting firm advising leading companies on issues of strategy, technology, projects, and operations. Focusing on local services, the company has more than 100 employees at three different locations around the country. Konrad Cerni is a senior consultant with expertise in project management. He is a very knowledgeable and resourceful person since he has worked in this field at a wide range of companies in different industries from traditional manufacturing to very complex aerospace.

At the time, Konrad was conducting a workshop exclusively designed for project management tools for about 30+ project and program managers. These managers work for the leading electronics manufacturing firm in the local area. Given that more than 50 project management tools are available in the practice and literature of project management, Konrad knew he could not cover them all, and had to carefully pick the tools that would be most useful to these managers. One of the tools he included was called a "Parametric Estimate."

WHAT IS A PARAMETRIC ESTIMATE?

A parametric estimate uses mathematical models to relate cost to one or more physical or performance characteristics (parameters) of a project that is being estimated. Typically, the model provides cost estimating relationship(s) that measures cost of the project being estimated to its physical or performance parameters, such as production capacity, size, volume, weight, power requirements, and so forth. Determining the estimate for a new power plant may be as simple as multiplying two parameters—the number of kilowatts of a new power plant by the anticipated dollars of kilowatt. Or it may be very complex, for instance, involving 32 parameters (also called factors or cost drivers) formulated into an equation to estimate the cost of a new software development project. Values of the parameters can be entered into the cost estimating relationship(s), and the results can be plotted on a graph or tabular format.

A PICTURE IS WORTH A THOUSAND WORDS

Konrad said, "To develop a proper parametric estimate, you may need to collect quality information inputs that include the following:

- Basic project scope
- Selected project parameters
- Historical information

Basic project scope description provides understanding of what is being estimated. Its parameters are identified on the basis of the nature of the cost estimating relationship model that will be used to collect and organize historical information, which will be related to the project being estimated.

Here is an example of a typical cost estimating relationship in a parametric estimate model where cost and area are expressed in a linear function." (See Figure 7.1.)

Konrad continued, "Many parametric software effort models are based on key software parameters such as cost drivers. They are usually based on the statistical analysis of the results of previous software development projects. These analyses include key parameters such as system size (e.g., line of code), complexity (e.g., degree of difficulty), type of application (e.g., real time), and

Figure 7.1   *Typical Cost Estimating Relationship in a Parametric Estimate Model*

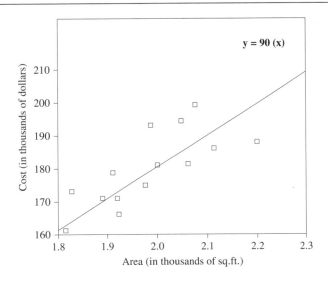

development productivity (e.g., productivity). Some experts suggest more than 59 parameters that can impact outcomes of these cost models. But a simple model can take the form of:

$$Z = CY^L$$

where

Z = estimated project effort (per months)
Y = Estimated project size (thousands of lines of code)
C = Regressions coefficient
L = Regression exponent

You can apply this model to estimate the effort for a new software development project by assuming the following values: C = 3.8, L = 1.4, Y = 2.

$$Z = CY^L = 3.8(2)^{1.4} = 10.03 \text{ person-months.}$$

"Now let's think about how you can apply the parametric estimate to your project," concluded Konrad.

Discussion items

1. What are the pros and cons of the parametric estimate?
2. When should the parametric estimate be used?

No Bottom-up Estimate, No Job!

Dragan Z. Milosevic, Peerasit Patanakul, and Sabin Srivannaboon

"We develop perfect quality software" was an informal motto of SP Group, a unit of a privately held company. Its clients, divisions of the same company, agreed to the very motto. SP Group was doing a great job of developing software applications that had almost no bugs. Happy with the quality, the clients didn't care much about the actual costs of the projects. For a project to be approved and paid for by the client, SP Group would simply submit an order-of-magnitude estimate ranging from 1,000 to 10,000 resource hours—and that's it. Then, a project would be easily approved and selected. Nevertheless, no one would know the exact cost of any project since the estimate was done based on an educational guess.

Then, the company went public and the trend of profit orientation and demonstrated cost efficiency took over. Unable to respond to the trends, all division managers were forced out and new, profit-oriented division executives were brought in. The change impacted the company in many ways. The game of cost estimating also changed. Now SP Group was requested to have better and more accurate cost estimates of its projects, and to provide proofs of such estimates. In particular, a Bottom-up Estimate was preferred.

WHAT IS A BOTTOM-UP ESTIMATE?

A bottom-up estimate is a cost estimation technique in which the individual costs are constructed based on the items identified in the Work Breakdown Structure (WBS). It relies on estimating the cost of individual work items, then adding them up to obtain a project total cost. Typically, an in-depth analysis of all project tasks, components, and processes is performed to estimate requirements for the items including labor and materials. The application of labor rates, material prices, and overhead to the requirements turns the estimate into monetary units. Figure 7.2 shows a generic version of the bottom-up estimate for simpler projects, with more complex projects having more details and documentation.

Figure 7.2   An Example of the Bottom-up Estimate

			COST ESTIMATE							
Project Name: Cablus			**Estimate #:** Cage - 010/1 **Compiled By:** E. Shaw					**Page #:** 1 of 1 **Estimate Date:** Aug. 5, 02		
1	2	3	4	5	6	7	8	9	10	11
Code	Item	Quantity	Labor				Overhead 25%	Materials		Total $ 7+8+10
			Unit Hours	Total Hours	Rate $/Hour	Amount $ 5×6		Unit Price	Amount $	
3210	1st Article	10	0.5	5	60	300	75	45	450	825
010	Project Total	1	291.5	291.5	65	18,947.5	4,737		900	24,584

The process of developing a bottom-up estimate and its accuracy heavily depend on the quality of the information inputs such as the following:

- Project scope
- Resource requirements
- Resource rates
- Historical information
- Project schedule

Project scope in the form of WBS provides a framework to organize an estimate and ensure that all work identified in the project is estimated. For this to happen, resource requirements that define types and quantities of resources necessary to complete the work are multiplied by resource rates to obtain a cost estimate. Typically, the rates come from historic records of previous project results, commercial databases, or personal experience of team members. Considering that some estimates contain an allowance for cost of financing such as interest charges, which are time-dependent, the durations of activities as defined in the project schedule are an important input.

WHAT ABOUT THE SP GROUP?

"Sharks," as project managers called the new division managers, flatly refused to even look at the order-of-magnitude estimates. Having profit-and-loss responsibility, sharks wanted to manage their cost and required bottom-up estimates to approve a project. Lacking the expertise to develop such estimates, the large majority of project managers were also forced out. Apparently, the time to learn how to develop a bottom-up estimate has come to SP Group.

Discussion items

1. What are the pros and cons of the bottom-up estimate?
2. When should the bottom-up estimate be used?

Earned Tree Analysis

Dragan Z. Milosevic

"Man, it is so funny," says Bruce Grinstein, the operations manager of PacCorp, a utility company catering to the Pacific Northwest. He goes on to say, "I've been using this method for 30+ years and all this time I haven't known that this is a real method and that it has name: Earned Value Analysis (EVA). Of course, I haven't used the full-scale and official terminology of EVA, but, rather, my own. And, I really have to laugh when I hear that EVA is a jewel in the crown. The funniest thing to me is when I heard that 0.1 percent of all projects in the United States use EVA. I think that number is way larger.

Why do I think so? I think that because a bunch of guys I know apply the method I use, so let's say it is EVA's primitive form, or only a small part of EVA that we apply. Don't think that I invented it. To tell the truth I learned it from my former boss. And all others who I know use it, probably learned it the same way.

Now, let me tell you about my method. As an operations manager for my region, which is Grants Pass, Oregon, I am responsible for guaranteeing that transmission lines work all the time. That also means that branches and twigs of the surrounding trees are not allowed to interfere with or to touch the lines. For that reason, I hire several tree contractors. With a contractor, I agree upon a number of trees he has to prune and a lump sum to pay him. I don't like to pay him per tree pruned. I am used to this way—lump sum—and contractors accept it. Trees mostly grow in the spring and I hire contractors for a year, from this spring season till next spring. I have the right to tell them which trees and when I need them pruned.

As the contractor prunes the specific number of trees, I count them, and divide the total number of trees pruned by the total number of tees to prune. What I obtain is a cumulative percentage complete. When I multiply that percent by the lump sum I arrive at the sum I cumulatively pay/owe the contractor. If the number of trees to prune increases, I increase the total lump sum I need to pay. And, I use the same formula to calculate the percent cumulative complete."

Discussion items

1. Is the method Bruce refers to EVA?
2. What is the official formula in EVA that Bruce is using to calculate "the sum I cumulatively pay/owe the contractor"?
3. Discuss how Bruce can calculate schedule variance and cost variance.

Chapter 8

PROJECT QUALITY MANAGEMENT

This chapter presents three case studies—one critical incident and two issue-based cases—relating to project quality management, Chapter 8 of the *PMBOK® Guide*. The cases discuss some quality control techniques and a quality management approach—Six Sigma.

1. Robots Fail Too

    Robots Fail Too is an issue-based case illustrating quality management practices of an organization in the high-tech industry. The case presents how the affinity and cause-and-effect diagrams are used. An example of a quality testing checklist is also presented.

2. The Peaceful Black Belt

    The Peaceful Black Belt is an issue-based case portraying the journey of a star employee in achieving a Six Sigma Black Belt certification. It is not an easy journey. The path to achieve a certification can involve some frustration along the way.

3. Workshop: Project Quality Program

    Workshop: Project Quality Program is one of the critical incident cases in the Workshop series, which talks about the process of developing a project quality program. The project quality program is an action plan that strives to ensure that the actual quality of a project will meet that which was planned.

CHAPTER SUMMARY

Name of Case	Area Supported by Case	Case Type	Author of Case
Robots Fail Too	Quality Control Techniques	Issue-based Case	Ferra Weyhuni
The Peaceful Black Belt	Quality Management Approach—Six Sigma	Issue-based Case	Marie Anne Lamb
Workshop: Project Quality Program	Quality Management	Critical Incident	Dragan Z. Milosevic, Peerasit Patanakul, and Sabin Srivannaboon

Robots Fail Too

Ferra Weyhuni

Within the recent month, there have been two sudden robot failures on two different tools during a build cycle. Lisa, the manufacturing engineer, has notified Nick, supplier quality engineer, about the failures, assuming that the two robots have some bad parts. She has requested that the two robots be sent back to the supplier for rework, even though no root cause has been identified. But, it seems that such a move has caused some to question where the blame should be placed. The focus of this case is related to project quality management.

OUR BUSINESS

The IEM Company is a high-tech company producing customized Ion and Electron Microscopes. The applications of their products can be used in a variety of fields, from academia to high-tech industries. Their customers are given the options of customizing the product to meet specific process needs. The company's financial profile shows that their sales revenue last year exceeds $400 million. The company is currently upgrading their tools for the improvement in the imaging and wafer transfer system. This is required to help expand the market size and to meet customers' satisfaction. This upgrading project was executed and is now in its operational stage.

WE HAVE A PROBLEM AND IT IS NOT OUR FAULT

Nick: How do you know it was the supplier's fault? Is there a chance that we damaged them during handling or installation?

Lisa: According to the Reject report, the technician said that the two robots were working fine for two weeks after installation. But then there were a few error lines such that the wafer transfer was stopped.

Nick: We don't really know if it's the supplier's fault or not. If it is their fault, those robots wouldn't have worked for two weeks, would they?

Lisa: True. However, anything is possible. I think we should send these machines back for them to check it out.

Nick: We can't just send them back without a well-documented "potential causes" report.

Lisa: We don't have time to do any tests or troubleshooting. They have the experts in their company who can test the robots to find out what's wrong with the machines. I suggest we send them back and save ourselves some time.

Nick agreed with Lisa's suggestion. The two robots were sent back to the supplier for investigation. One week later, similar problems occurred on several other machines. The problem became so big that the issue was elevated to Donnie, a manufacturing engineering manager. Donnie asked Lisa to form a team to identify the root cause of the problem. Lisa agreed to put together the team to brainstorm the root cause and the next course of action. She promised to follow the following steps: goal definition, root cause analysis, countermeasures identification, and standardization.

Lisa called a meeting with Nick and the other two manufacturing technicians, Joseph and Ryan. The team was working to get a list of possible causes for the problem. As a normal procedure in the team's analysis, the first thing to do was to create a fishbone diagram.

Joseph: As a starting point, can we capture what actually happened before the error message showed up on the screen?

Ryan: I don't really know what happened. I was just starting to teach the robot, following our procedure, but then the error message showed up.

Joseph: That doesn't make any sense. If nothing changed on the system itself, we shouldn't have gotten the error. There's got to be something changed on the system.

Lisa: Let's create a fishbone diagram for potential root causes of this problem.

The team brainstormed using the affinity diagram method. The purpose of this exercise was to ensure everyone's input was captured during the process. They determined the amount of time to be spent on brainstorming, and then went through each idea that each member came up with. When going through each idea, they also decided whether those ideas were candidates for root causes. If any of the ideas didn't make sense, they put them aside and noted them as "possible but not likely" causes. Some of the ideas are shown in Table 8.1.

Once the ideas of potential root causes were laid out, they started their fishbone diagram by grouping the potential causes into larger categories such as Software, Mechanical, etc. The fishbone diagram would be used as a tool to communicate with upper management as well as field personnel showing all possible items that needed to be checked if and when the errors occurred again. Figure 8.1 is an example of a fishbone diagram.

Table 8.1 Results from Brainstorming Session

Potential Causes	Possibility	To Be Tested (Y/N)
Robot's Firmware	High	Y
Robot's Controller	High	Y
Communication to Robot's Controller	Medium	Y
System's PC	Low	Y
Overall System's Communication	Low	Y
System's Software	Low	N (if overall system's communication passes the test)
Robot's Manual Controller	Medium	Y
Robot's Cables	Low	Y
Motion Controller	Low	Y
Motion Cables	Low	N (if motion controller passes the test)

Lisa: Here's the fishbone diagram you requested. We came up with a few things that need to be checked using our tools on the manufacturing floor.

Donnie: How much time do you need? Do you have a test plan for each item?

Lisa: I have not created the test plan yet but it should be straightforward.

Donnie: I think you should create a test plan to show us all what you're going to do and what the results would be. The customer does not know that we have this issue on the manufacturing floor and they don't know how severe it is. We should get to the root cause before it gets out of hand.

Lisa: I understand. However, I don't have the bandwidth to do all of this correctly.

Donnie: This is of the highest priority now.

Lisa: Okay. I will work on it.

Figure 8.1   Draft of the Fishbone Diagram for the Failures

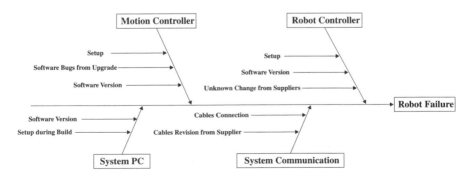

Figure 8.2   Quality Testing for the Robot to Be Used by Technicians

System used for testing:

Technician name:

Date:

Potential Cause	Activities	Results
Robot's firmware	Follow "Check Robot's Firmware" work instruction	
Robot's controller	(1) Check revision number for the controller (2) Match the controller version number with the Bills of Materials	
Communication to Robot's controller		
System's PC		
Overall system's communication		
System's software		
Robot's manual controller		
Robot's cables		
Motion controller		
Motion cables		

Lisa created a spreadsheet that could be used by technicians to test the tool for all possible causes (see Figure 8.2). This spreadsheet shows all activities to be performed to ensure there are no assumptions made by technicians. The results are recorded and anything worth noting during the test must be written down.

Discussion items

1. List the process that Lisa used to create the quality testing of the robot.
2. What can be done to improve the process that she's using?

The Peaceful Black Belt

Marie Ann Lamb

For the first time Milla Gold had to admit, at least to herself, that this might be beyond her capabilities; but within the field of program efficiency expertise, she felt she had little choice, unless she wanted to switch career paths. She felt lost in all the equations and multiple terms with the same statistical meaning. Milla could no longer rely on her tried, tested, and proven leadership and program management skills for her success. To her it seemed only years before that she was conquering the Pareto chart analysis with some confidence; now, if she was to become a six sigma black belt she also had to prove her other statistical skills.

A THREE-YEAR BUMPY ROAD TO TECHNICAL PROWESS

Milla works for a Fortune 500 consumer goods company with more than 50,000 employees. She was chosen from thousands of potential candidates to receive the expensive training because of her influencing abilities, leadership, and program management results within the company.

There were several issues, out of her control, that delayed Milla from gaining her six sigma black belt certification for another two and a half years. She had gone into the training thinking one was certified just by taking the training, as did 85 percent of her 30 fellow trainees representing seven different companies. In reality, this was just an initial training. Additional requirements for certification were: two six sigma black belt projects with at least $500,000 net savings each, proven six sigma leadership, and proven advanced statistical skills using less commonly known struggles encountered by six sigma black belt trainees.

In effect, the black belt's performance and path to certification could be frustrated and even curtailed by one of these struggles. One of these struggles is encountered at the onset in the training phase: Advanced statistical skills required for black belt work can be extremely difficult for nontechnical people.

As this lesser-known struggle indicates, becoming a certified six sigma black belt is not for those who want to take the easy path. Looking back, Milla started her black belt journey in 2004 with the misunderstanding that certification was achieved by simply sitting through a three-week black belt training requirement.

Two and a half years later, Milla finally achieved the much-sought-after six sigma black belt certification.

GUTS FROM THE START—OR IGNORANCE?

How naïve she was at the end of 2003, when Milla had created her development plan for the upcoming year; including a desire to receive six sigma black belt certification based on the belief that the process was simply a revised set of program management steps (called DMAIC); but with increased proven results from other program management philosophies. Also, Milla did not think it would be a negative to show her drive for self-improvement to upper-management levels. In April 2004, faced with an allotment of training dollars for six sigma black belt training for a handful of people, Milla's upper management immediately tapped her for the opportunity. Upper management had put a high priority on candidates with proven leadership and project results. What they did not know was that Milla had some inkling that it would be necessary to bone up on simple statistics and even suspected there might be more to the statistics than that; but thought she would have one or two years to do so before the opportunity arose. Milla uses one simple word to describe the scenario in which she found herself: ". . . Gulp."

The training program was to be provided by another Fortune 500 company, known to be one of the best, and most visible, black belt programs in the world. Milla decided she would improve her chances at doing well on the statistics by buddying with another program manager to work on the simple statistics pre-homework assignments. With some diligence they conquered the control and Pareto chart analysis sections with aplomb. She headed into the three-week training with renewed confidence.

ADVANCED STATISTICAL TRAINING NEARLY SANK HER

By the third day of training, Milla's brain felt shaken, and never righted itself throughout the remaining three weeks. She had taken on legions of challenges before this, but now found herself faced with what seemed an insurmountable number of new skills and approaches to becoming more technically savvy. For the first time, she could not rely only on her leadership and program management skills for her success. She spent those three weeks in a daze relying heavily on one set of skills she had honed over time: how to smile and nod even in the worst of circumstances. She reflects, "The one small moment of understanding I had in any of the statistical training was when we broke up into small teams, built paper planes, and conducted timed drop tests; which resulted in showing us the importance of various types of operator errors that can occur during a measurement." Milla further explained that she was lost in all the mathematical equations and numerous statistical terms. For her to succeed as a six sigma black belt, she had to prove her skills in:

1. Setting up a statistically valid data collection plan.
2. Setting up a measurement plan to ensure statistically valid repeatability, reproducibility, and part-to-part variation.
3. Determining minimum sample size.
4. Understanding variation, stability, and capability analysis.
5. Conducting hypothesis testing techniques.
6. Determining confidence intervals.
7. Doing 1- and 2-sample t-tests.
8. Conducting simple and multiple linear regressions.
9. Conducting 1-way and 2-way ANOVA.
10. Setting up and conducting a design for experiment.

AN UNEXPECTED REASON FOR HER PERSEVERANCE

The class moved fast, with each skill building onto the next. In the middle of this, Milla was able to glean enough high-level understanding to be able to say to all her relatives and co-workers that this training was important enough that "everyone should learn these skills." For several years Milla had been trying to improve her skills and program management methods to gain efficiencies; but she felt she had hit a wall as to the next level of efficiency gain. Within the second day of training, a lightbulb went off and Milla realized what was missing in her program management methods: a more technical and data-driven approach. From what she could glean at this high level, the six sigma black belt process and skills would close this gap for her. However, at the end of the training period, she estimates that she learned less than 10 percent of the knowledge and skills that could be realized from the training. She discovered that she'd been ill-prepared to the extent that she also could not comprehend what approach to take to correct the large gaps of understanding needed to learn from one day of training to the next. So, she continued to smile, nod, and sink further into a pool of what could only be called "statistical confusion."

She closed out her three-week training realizing the six sigma process and statistical skills would be paramount for her next level of development; but to do so would require re-training in approach and diligence to gaining technical skills. Prior to this, improvements in leadership and program management had built upon one another; whereas, the technical skills needed to be a black belt would require getting off one track she was already running on and getting onto another track at which she would be starting at a crawl. Then, if she could learn how to at least walk within this new track, she would have to quickly meld the two tracks together as this was expected in her leadership role at the company.

The path to technical skills was not an easy one for Milla. She returned from her initial training and rather than conquering this obstacle head-on, she fell back on comfortable leadership skills while using, as she is ashamed to admit, "big, technical words like 'measurement capability'" to keep the ill-informed from realizing she could not perform the technical black belt skills. This resulted in projects using some of the six sigma tools, but not all; therefore, not qualifying as full six sigma black belt projects. Milla continued to compensate by stepping up to be a voice/leader for six sigma and this later aided her in satisfying, in fact surpassing, the six sigma leadership requirements for certification. However, this experience left Milla feeling, for the first time, that she was a "talk-the-talk, not a walk-the-talk" type of person—unable to perform to what was promised. She was also perceptive enough to know statisticians, as well as the few six sigma black belts in the company, had a much higher standard of expectations.

After about a year and a half of being in the limelight for six sigma leadership, Milla, as part of her next development plan, took a basic to mid-level statistics course given within her company. The class was to be taught by statisticians over a four-month timeframe, consuming both company and personal time. As this class progressed at a relatively slow pace, Milla was not only able to put the big picture around how the statistical pieces fit together, but was also able to significantly increase her technical skills. The course, though, was taught by statisticians from her work group, so she felt hesitant about asking more questions to increase understanding—recall that she had already claimed to have been trained in this skill set!

Milla might have continued to progress at this pace, except that something happened to finally hone her concentration on the required statistical skills for certification. Up until this time, approximately mid-2006, training alone carried a relatively higher weight within her company than certification. Currently, however, Milla's company was planning a large layoff, and although not in a high-risk category, she realized that it would be next to impossible to receive black belt certification elsewhere as most of her previous work would be moot at other companies. The next six months progresses at a startlingly fast pace with focusing on six sigma project requirements; including "training while doing" statistical tasks and results for the project. Milla managed to swallow much of her pride and sign up for the internally taught advanced statistical course. She began to notice that as she conquered one statistical skill, it was easier to move onto the next. A doable list of technical skills to be checked off began to form in her mind. In this advanced class, although extremely difficult for her, Milla raised her hand and asked questions if she did not understand something—Gone were the days of the the empty nod to indicate understanding where there was none.

With her increased ability to say, "I don't understand and I need help" came respect and invaluable assistance from statisticians and black belts alike. This group became mentors who helped Milla to finally gain her six sigma black belt certification two and a half years after she had started down this road. For many reasons, two and a half years is not uncommon for gaining six sigma black belt certification; however, Milla often reflects on her experience. The 87-page document that was needed for certification, showcasing all her six sigma project accomplishments, leadership, and technical skills, remains as a great source of pride for Milla. What she often reflects on, though, is how did she manage to go so wrong from the start through to the middle of this certification process? "I'd rather take on a leadership challenge and speak in front of thousands of people; I had to do battle with myself and face many of my technical fears and this was one of the hardest tasks I have ever had to take on."

Discussion items

1. Where do you believe Milla Gold made some miscalculations in her approach to the technical aspects of six sigma black belt training?
2. What are some ideas for easing the path to understanding the advanced statistics necessary to become a six sigma black belt—particularly for nontechnical people? Include short-term and long-term actions.
3. Should six sigma black belt certification require advanced statistical skills?
4. Do you think the three criteria—influencing ability, leadership, and program management skills—used by Milla's management in choosing who to send to six sigma black belt training were the right ones?

Workshop: Project Quality Program

Dragan Z. Milosevic, Peerasit Patanakul, and Sabin Srivannaboon

Ball, Inc. is a management consulting firm with the objectives to locally provide its clients the best services in various fields such as organizational change management assistance; development of coaching skills; and technology, project, and operations management improvements. The company motto is: "Your one-stop management consulting for all your business." Konrad Cerni is one of not so many employees who have been working there since day one. He loves the company environment and enjoys working with his peers and clients. More importantly, he loves his job—project management training. Konrad is well known inside and outside the company. His current position is a senior consultant, and his specialization is in the areas of project management tools and metrics.

Recently, his local client requested Konrad conduct a workshop specifically designed for project management tools. Particularly, the workshop must be done within a very strict timeframe to accommodate the participants' schedules. With more than 50 tools that are available in the practice and literature of project management, Konrad knew he had no time to cover all of them. Thus, he had to pick only the important ones, and cover them in as many areas as possible. One of the tools he included is called the "project quality program."

WHAT IS A PROJECT QUALITY PROGRAM?

A project quality program is an action plan striving to ensure that the actual quality of a project will meet the planned one. Using WBS as a skeleton for integration, the program sets a quality level based entirely on customers' expectations and requirements. With such a strong customer focus, the project quality program translates the requirements into tangible quality standards, for whose accomplishment a set of tasks is defined. Explicitly defined responsibilities and timelines for the performance of the tasks add necessary elements to use the program as a project quality roadmap. In a nutshell, the project quality program states that this is what this project has to do to ensure that the quality of its deliverables is meeting our customers' requirements.

A PICTURE IS WORTH A THOUSAND WORDS

Konrad said, "The quality of the project quality program is heavily dependent on the quality of its inputs. In particular, the following inputs are known for their impact on the program:

- Quality policy and procedures,
- Voice of the customer, and
- Scope statement and WBS."

He continued, "The foundations of how quality is managed in an organization are described in its quality policy. Defined, documented, and supported by management, the policy is a statement of quality principles, beliefs, and key objectives for projects that set a general framework to carry out quality management actions in the organization. This framework is further detailed in quality procedures. Together, the policy and procedures set a direction for the program. For example, if the procedures mandate compliance with ISO 9000 standards, the program will have to comply.

Figure 8.3   An Example of the Project Quality Program

PROJECT QUALITY PROGRAM												
Project Name: SMP-1			Rev: 1									
Prepared By: Bob Maxwell			Sheet: 2 of 3					Date: May 10, 01				

1	2	3	4	5						6						
WBS Code	**WBS Element**	**Quality Standard**	**Quality Assurance Task**	**Responsibility Matrix**						**Schedule May/June 2001** **Week of**						
				Pete	Alan	Perry	Kim	DZM	Ian	5/7	5/14	5/21	5/28	6/4	6/14	6/21
		Flesh reading ease	Run test and rewrite				D					▰	▰			
		ISO 9000	Review	D	D	D								▰		
2.03	Project Mgml Manual	PMBOK	Review					D							▰	
		Brevity guidelines	Check and correct				D									▰
		Organization policy for writing manuals	Review	D					A							▰

Key: D-Do
A-Approve

Listening to the voice of the customer will not only help discover customers' needs; it will also help decipher customers' needs and translate them into the recognizable language of the project scope, establish units to measure customers' needs, and express them as quality standards, a crucial piece of the program. For this reason, the project quality program needs to be closely coordinated with the voice of the customer.

Finally, when you are setting project goals, the scope statement also sets a quality goal for the project. Along with this input goes the WBS which defines the project work for which a project quality program is developed. Therefore, both the scope statement and WBS are significant inputs to the quality program preparation."

Figure 8.3 is an example of the project quality program which also shows the responsible parties and the timelines of the WBS elements.

Discussion items

1. What are the pros and cons of the project quality program?
2. When should the project quality program be used?

Chapter 9

PROJECT HUMAN RESOURCE MANAGEMENT

This chapter contains five case studies—one critical incident and four issue-based cases. These cases relate to Human Resource Management, Chapter 9 of the *PMBOK® Guide*. Several topics are discussed such as conflicts, culture, virtual team, and performance appraisal.

1. The Bully, Subversive, Prima Donna, Etc.

   The Bully, Subversive, Prima Donna, etc. is an issue-based case discussing conflict management. In particular, the case discusses how personality can be a source of conflict. The case provides some specific situations where conflicts related to personality took place and how they were solved.

2. Startups Born with Conflicts

   Startups Born with Conflicts is an issue-based case discussing a specific situation where there is conflict between departments. The case details how the organization resolves the conflict and establishes a new approach to prevent similar future problems and conflicts.

3. We Do Not Speak the Same Language

   We Do Not Speak the Same Language is an issue-based case dealing with a virtual team. The case discusses a specific situation where different cultures and different working styles can be a source of misunderstandings and conflicts.

4. My Job Was to Integrate Two Cultures
    My Job Was to Integrate Two Cultures is a critical incident. With the prevalent practice of outsourcing, cross-cultural integration is a challenging task of several project managers. The case portrays an approach a project manager used to integrate two cultures.
5. Rate and Rank
    Rate and Rank is an issue-based case. It details an approach that one company uses for performance evaluation, called Rating and Ranking. Based on the information provided by the case, the readers should be able to identify pros and cons of such an approach.

CHAPTER SUMMARY

Name of Case	Area Supported by Case	Case Type	Author of Case
The Bully, Subversive, Prima Donna, Etc.	Personality as a Source of Conflict	Issue-based Case	Diane Yates
Startups Born with Conflicts	Conflict Resolution	Issue-based Case	Priya Venugopal
We Do Not Speak the Same Language	Virtual Team	Issue-based Case	Diane Yates
My Job Was to Integrate Two Cultures	Cultural Issues	Critical Incident	Dragan Z. Milosevic, Russ J. Martinelli, and James M. Waddell
Rate and Rank	Performance Evaluation Process	Issue-based Case	Rhaba Khamis

The Bully, Subversive, Prima Donna, Etc.

Diane Yates

One of the biggest issues managers have to deal with is conflict in the workplace. Dealing with conflict constructively in an environment with various personalities is as much a work of art as it is good managerial style. Read any number of articles that list "five ways to deal with workplace conflict" and you might be tempted to think that solving it is as easy as finishing this sentence. Despite all the bullet points enumerating methods of resolution and books written about managers as negotiators, workplace conflict continues to be a major problem in many organizations.

No one knows this better than Janet Miller. Janet is the Human Resources (HR) manager for Customer Support in a mid-sized software company located in the Midwest. A petite woman with an easy smile, her charm immediately wins people over. The company she works for has around 4,000 employees, and has posted annual revenues of about $800 million in 2007. Janet is one of about 25 HR managers worldwide. She has been with the company for 22 years.

Janet has seen just about every type of conflict in the workplace. She knows firsthand how demoralizing it can be, and the importance of reigning it in. She estimates that employee turnover from unresolved conflict costs the company millions of dollars in lost productivity and revenue every year.

"We require all of our managers to take conflict resolution training every year," she said. "This is on top of the cultural sensitivity training and harassment in the workforce training. We have some of the best conflict resolution training in the industry, where our managers and employees engage in scenarios and role playing. Still, no workshop or training session can prepare our people for every contingency."

Janet agreed to share some of her more memorable experiences in dealing with conflict. She started by relating a story about an incident she was presently dealing with. She thoughtfully straightened a stack of papers on her desk, and then folded her hands in her lap as she began.

THE BULLY

"By the time this problem came to my attention, it was already out of control," she said. "The problem exists with a current employee, whom I will refer to as Matt.

"You see, Matt is a bully. He screams and yells at people in meetings. He is subordinate to his boss. He comes to work when and if he feels like it. Other employees resent him because they feel he gets preferential treatment, which, of course, he does. Why does Matt get away with it? Because he makes the company a lot of money."

Janet paused for a moment and sighed. "Matt was our top salesman for three years straight. He brings in at least $50 million a year. Our customers absolutely love him. The big bosses love the money he makes. His managers were told to let him do what he wants—obviously he's got the Midas touch. What no one figured on was how difficult this man was going to turn out to be."

"Now even the company president and CEO realize he is out of control. Someone from corporate engineering complained all the way to the top. Apparently, Matt stepped on some toes there. Allegedly, he went behind someone's back and made a deal that took the sale away from another corporate engineer. He's done this not once, but many times in the past. People have quit their jobs because of Matt."

"The division has created a monster and now do not know how to contain him. His boss has talked to him, but to no avail. His attitude is, 'I'm making the company a lot of money, and if you have a problem with the way I do it, go ahead and fire me.' Now it is my job to see if I can get through to him. The company wants him to take sensitivity training so he can learn to get along with others. I am not hopeful he will agree to it, but I have permission to offer him an ultimatum—take the training, learn how to act around your fellow employees and superiors, or leave the company."

THE SUBVERSIVE

"A lot of times, we have to deal with people who are not open about their hostilities, but their attitudes still permeate through workplace relationships, and the work that they do.

"Most of the time, people conduct themselves professionally. We require rank-and-file employees to take harassment and cultural sensitivity training every two years. Our company prides itself on being diverse. We benefit from the richness different cultures bring to the corporate world, and since we do business worldwide, having members from different cultures helps facilitate that.

"The main purpose of training is to inform our employees of our standards of conduct, and what will happen if they cross the line. We realize that taking a class is not going to erase years and years of ingrained thinking, but they need to be aware that they are held to a certain level of conduct within the company, and they imperil themselves if they do not adhere to it.

"Our company employs a lot of women in leadership and managerial roles. This has worked very well, for the most part. Still, every so often you come across an employee who has a difficult time working with women.

"We had one employee who hated working with women. He later said that he felt a woman's role was to be at home, raising children. He was on a project where there was a woman leading it. He did not voice his malcontent to her, but would badmouth her behind her back. If he was given an assignment from her to do, he would do it poorly or not at all. Often, he would take a long time completing it in order to make her look bad to her superiors. However, if a man gave him something to do, he would complete it on time and it was perfect. His attitude was dismissive toward her, if he acknowledged her at all. Most of the time, he ignored her.

"He would try to get other employees to say negative things about her to her superiors in order to get her fired. Needless to say, this was causing problems with the team. He complained bitterly about his 'female boss' and how 'stupid' she was and how 'she doesn't know anything.' Often, he would complain about how she was 'taking a job away from a man.'

"Pretty soon, the gossip got back to her. She confronted the man, who told her that in his culture, 'men do not take orders from women.' The woman did not immediately complain to her superiors, but tried reasoning with him. After another unfruitful week, she went to her manager and complained. He just would not cooperate or complete the tasks she assigned to him.

"A meeting was called between the managers, the team lead, and the disgruntled employee. By this time, other employees were complaining about him. They said his continual carping was causing morale to drop, and his lack of cooperation was sabotaging the project. In short, he was ineffective and was causing problems for other employees as well as the project.

"When confronted, the man simply was unrepentant. Management knew what he said about the woman was untrue; she was an excellent employee and others enjoyed working with her. And although he was a very talented technologist, he was fired on the spot.

"You see, there is no room for that kind of thinking in the company," Janet said. "We can accommodate cultural differences to a certain extent, but that sort of behavior—being boorish and sexist—is also against the law."

THE PRIMA DONNA

Janet stopped for a moment to sip some coffee from her cup. She smiled over the cup. "You would think that in an organization made up of professionals that people would behave accordingly, but often this is not the case. I deal with the 'human factor' all the time. Conflict in the workplace is pretty normal.

"I remember another case dealing with an employee who made a bunch of unreasonable demands. She demanded that we give her an office with a locking door, although we only supply doors that lock to the managerial staff. She complained that we treated other employees better than her, although the examples she cited were unsubstantiated. She insisted on making her own work schedule, and insisted that her manager work around it, which he did.

"Most of her demands were met, because it was simply easier to meet them, if possible, and avoid trouble. She made it very clear that she would not hesitate to sue the company if there was a single misstep in how she perceived we treated her.

"She was very, very high maintenance. One time, she picked a fight with the manager of the Operations (Ops) team. She was sure this woman was disrespecting her in some fashion, and there was a blowup over a perceived slight regarding a softball team (employees often would play sports in the spring to facilitate 'team building'). The employee went straight to HR with her complaint. She didn't even try to work it out with her manager and the manager of the Operations team.

"She claimed that the Ops manager had 'something' against her, and was trying to humiliate her in front of fellow employees. When we probed further, all we discovered was that the Ops manager had asked her why she didn't try out for her softball team."

Janet waited for me to react to this. Incredulous, I asked, "But why would she be upset over being asked to join a softball team?"

Janet set the coffee cup down carefully on her desk. "Well, apparently, the Ops manager 'confronted' this woman in the hallway, while she and her team were going to lunch. The other woman felt like she was being cornered and put on the spot."

Janet continued, "However, things are rarely as simple as they seem. When we dug a little further, we found out that this woman had felt for a long time that the Ops team was snubbing her. She said that most of the time, they wouldn't even talk to her, so why now were they showing interest in her joining their softball team?

"When we talked to the Ops manager and her team, we discovered that this woman made them feel uncomfortable and resentful because of her unreasonable demands. Around the hallway she was known as a troublemaker. There were a couple of instances where they felt she had been abrasive toward them. However, the gesture from the Ops manager was meant to be a goodwill gesture, intended to mend differences, not create problems.

"The situation didn't resolve itself that year, but eventually the two sides made an effort to get along, although the Ops team still felt like they had to tiptoe around this woman.

"One thing I can tell you," Janet continued, "when you are always running to HR to solve your problem it is not looked upon favorably. We do expect people to try to resolve their problems first before coming to HR. Of course, not everyone can handle conflict effectively, so we get involved a lot. I think conflict resolution is a very difficult subject in the workplace. Managers call me all the time, asking for help with resolving workplace conflict."

WHEN THE MANAGER IS THE PROBLEM

"We try to address conflict as it comes up, before resentments and problems have had time to fester," Janet said. "As I've said previously, dealing with conflict can be really tricky. For example, we had a situation where the boss was causing problems and her employees were having a difficult time dealing with her. We didn't find out about it for a couple of years because employees were afraid to open their mouths."

She continued, "When the problem is a manager, this presents a unique set of problems for employees. Fear of reprisal is the number one reason employees cite for not coming forward with employee/manager workplace issues. And sadly, they may be correct in assuming that their managers will take it out on them if they find out that employees have been saying things behind their backs. Retaliation can be difficult to prove if it doesn't involve whistle blowing or harassment, and no one else comes forward with complaints.

"In the case of this manager, we later learned that she had terrible people skills. Often engineers or other technologists are promoted to the level of manager, but these individuals often lack any kind of people skills. Often they are promoted because they are the 'star pupil,' but it sometimes is not a good fit, and can be a disaster.

"This manager used retaliatory practices to get back at employees who complained about her. People would complain that she would berate them in front of other employees during staff meetings. She played favorites among the employees, granting special privileges to those individuals whom she liked. Other employees called these people 'suck ups.' Her team suffered from low morale, and would transfer out of her area as soon as another opportunity presented itself.

"The high turnover rate was noticed by HR, but she always had a plausible reason why someone did not work out. Finally, an incident occurred that broke the issue wide open and other employees decided to come forward with their complaints.

"There was a very capable engineer on her team who was well liked and respected by other team members. He was responsible for resolving all the difficult cases. (The team was in the software support division where troubleshooting customers' issues was part of the job.) The job involved talking to customers on the phone and resolving software problems.

"Call times and problem resolution issues were logged on a monthly basis. Because this employee took on complicated cases, his resolution times were longer than the other employees', as you might imagine. This had never been a problem in the past, because everyone knew he fixed the really difficult problems.

He also was the team lead and other engineers came to him when they couldn't fix something, and he always resolved their issues as well.

"When this manager was hired on to this team, there was immediate conflict between the two. And to make a long story short, the manager decided to get rid of this engineer, citing his call times and issue resolution times were too long. He wasn't even placed on an employee improvement plan—standard protocol in the company for underperforming employees. However, everyone knew that she fired him because she didn't like him. This caused more fear, yes, but it also caused resentment. Employees no longer had an experienced mentor they could go to for help with difficult cases. Call times went up for everyone, and so did the amount of time it took to resolve issues. Calls stayed longer in the queue as well. The effect backfired on her, because she now had the highest turnover rate in the Support Division. We started hearing from other employees about her lack of people skills and other issues.

"Finally, upper management decided to move her to a position where she could use her technical skills, but she wouldn't have employees reporting to her. The company can't afford to keep training new employees, so this was deemed a workable solution, since she had proven skills that could benefit the company."

Discussion item

1. In each situation presented in the case, discuss whether or not you agree with the conflict resolution approach.

Startups Born with Conflicts

Priya Venugopal

Northwest FCV is a new startup in an automobile assembly business. The company's assembly lines include, for example, brake rotor assembly, battery system assembly, and engine assembly. As a startup company, Northwest FCV's annual sales are around $2 million with a projection of 500 percent growth in the coming years. The company employs 50 employees. Each department has around five people. The management is made up of the President, the Vice President, and the Chief Financial Officer. Other departmental heads include, supply chain manager, purchasing manager, engineering manager and design manager. As conflict is inevitable in any company, Northwest FCV has developed a unique way to deal with conflicts. They organize a task force meeting so that responsible engineers and managers can discuss conflicts and their resolution. A typical meeting occurs in the following fashion.

AT A TASK FORCE MEETING

Jim: I wonder why we have this meeting today. The engineering team is doing a great job in designing the components required.

Judy: The purchasing department ordered a few wrong parts last week and that affected the shop floor and our company was unable to deliver the assembly to the customer at the right time.

Jim: But why did they order the wrong parts?

Judy: I have no clue. I remember we did a few revisions to those parts a couple of times.

Jim: I assume the purchasing department got the final revision. Correct?

Judy: I believe so. The person who made the changes should have contacted them before they decided to order and have the parts shipped from the vendor.

Mark: Just before coming here, I sent everybody the escalation report explaining what exactly happened in this whole process. You may not have had a chance to read it. What happened was that the purchasing department had the old version of the parts and they ordered the same parts. They did not receive any no-go signal from your department either. And when the parts arrived they did not fit properly and hence they had to be reworked again in-house. We have an engineering BOM (Bill of Material) created for all the designed parts. There you can always find the updated parts. The purchasing person should have pulled up the latest revision that needed to be ordered.

(Steve, the Purchasing Manager, arrives furious to the conference room.)

Steve: I think all the engineers are busy creating new parts whenever they want.

Jim: Steve, did you receive all the updated design parts?

Steve: If that were the case, I would not have purchased the wrong parts.

Jim: Sorry about that. I think we all should meet more often and discuss what is happening in our departments, if there are any new changes taking place or any other deviations from what was originally planned. This information can help all of us in knowing the current status of the organization.

Mark: Yes, also we can review the latest BOM and all the related queries from every department can be answered.

Judy: There should also be a standard process which all the engineers follow for updating the BOM and informing the purchasing department as to what to buy from the external suppliers.

Steve: Sounds like a plan. This will help us to acquire what we need rather than paying for the parts that need to rework again or are of no use to us.

Mark: Can we have this meeting every other week then?

Jim: Of course. Let's create a group called the Project Communications Management team, comprising of all the functional group leaders. Judy, please make sure you schedule this meeting accordingly.

Judy: Sure, I can do that Jim.

Discussion items

1. Out of different conflict resolution modes—withdrawing, accommodating, compromising, forcing, problem solving, etc.—which mode was used in the case?
2. Do you agree that startups are born with conflict? Why or why not?
3. Would the implementation of a Project Communications Management team help prevent similar problems in the future?
4. Besides a Project Communications Management team, should Northwest FCV revise its procurement process to prevent future mistakes?

We Do Not Speak the Same Language

Diane Yates

Dale Rodriguez thought she had it all figured out. Having the team in Egypt create the web training would free her instructional designer to do other, more urgent, projects. Like many managers, Dale is forced to do more with less—less people, less money, less time.

Even so, her instructional designers are not able to update all of the training. Each instructional designer is able to create one or two new courses a year, and update two to three courses a year, depending on the length of the class and difficulty of the course. With all of her team working full time, they create or maintain 25 to 30 courses a year. This means that Dale must look for other ways to update courses. Sometimes she has the customer support teams update classes. Sometimes product teams or documentation teams update classes.

For this particular web-based training, she decided to let the Egyptian team create the training. She would use Aidan Quinn, one of her on-site instructional designers, to review the training the Egyptian team created, checking it for technical and grammatical accuracy. The two teams would work in tandem, with the bulk of the work being done by the team in Egypt. Aidan liked not being over-involved with the project; because he was busy creating a brand new course that was taking up much of his time. Dale was proud of the fact that she could get things done in a creative and cost-effective manner. Another big plus to having the Egyptian team create the training was that they worked more cheaply than the Americans did. Dale had used virtual teams before, with much success, so she was certain the arrangement would work well for this project.

BACKGROUND

Dale has five full-time instructional designers that work at the company headquarters in the United States. They are part of a mid-sized software company that creates electronic design automation (EDA) tools for chip makers, PCB board designers, electric harnesses, and other high-tech disciplines. The company posted $920 million in sales in 2007, and employs 4,000 people worldwide. Dale manages a group of employees whose job is to create tool training classes for customers. The company owns the licenses to about 400 software tools, and is adding more to its portfolio every year. Out of these 400 tools, training is created for only 57 of its best-selling software packages. Well, she planned to involve them more in customer training projects.

THE WEB-BASED TRAINING PROJECT

The web-based training would consist of lectures and labs. The lectures would be of various topics, such as simulation and debugging techniques, followed by hands-on labs. The class would be self-paced, and would not have an instructor, but students would have the opportunity to contact a subject matter expert by email if they needed help with a topic. The turnaround time for questions was guaranteed to be within a 24-hour period.

The Egyptian team would build the web-based class using existing training materials based on current instructor-led classes. They would take existing PowerPoint modules, update the material with new screen shots and new tool features, and convert these into self-paced online tutorials. Once each module was built, they would be sent to Aidan for approval. After all modules were built, text files would be created to narrate each page of the lecture. These would be sent to Aidan to look over for technical accuracy. Aidan would send the vetted materials back, and the Egyptian team would then record the sound files for the lecture materials.

Massoud Ahmet was the team's manager in Egypt, and the person who Aidan would interface with. Individuals on Massoud's team would send their completed work directly to Aidan, but Dale instructed him not to contact them directly; instead, if he had concerns he was to go to Massoud.

THE VIRTUAL WORLD BRINGS PROBLEMS

Pretty soon completed work began to trickle in for Aidan's approval. Aidan began to notice discrepancies between the lecture topic and the screen captures, in the form of GIFs, placed onto the web pages. The screen capture often did not match the lecture topic, rendering the picture useless—or worse—potentially confusing to students. Also, many of the pictures were old and outdated, as if the team had not bothered to update the existing ones. Aidan was sure that the Egyptian team had the lab data needed to run the labs with which to create new pictures, but he decided to send a copy along again anyway, just in case. Since Egypt was half a world away, Aidan decided it would not be practical to call Massoud, since he most likely would not be at work, due to the time difference. He emailed Massoud, outlying his concerns, with unambiguous directions on how to fix the problems. He included the lab data so that Massoud's team would be able to use it in case they did not have it.

He expected to hear back from Massoud within a few days. Instead, more work came from the Egyptian team, still containing inaccuracies and mismatches with regard to lecture topics and screen captures. It was as if Aidan hadn't made contact at all. Not only that, but Aidan was spending considerable amounts of time poring through the "finished" materials they sent him. He was finding considerable amounts of grammatical errors in the online text as well. Often it is difficult for non-native English speakers to master the finer points of the language. 'So, this maybe to organizational barriers or is not,' is an example of what Aidan was getting. The team was using some colloquialisms as well—something that technical writers know not to do, but engineers and laypeople often do not think about.

Aidan wanted to give Massoud's team the benefit of the doubt. Perhaps Massoud was busy with other duties and did not have the time to check his email everyday. Aidan decided to call him. He left a message on Massoud's voicemail, explaining the situation he had outlined in the original email. He asked Massoud to call him back and let him know he had received the message.

Several days passed, and still no word from Massoud. The lecture material continued to come in with technical inaccuracies. Aidan was growing frustrated and concerned, trying to understand two cultures' differences of time. He was spending too much time supervising a project in which he was supposed to have a peripheral role. Every time he had to touch the work of one of the other team members, it added extra cost and time to the project, not to mention his own project started to fall behind.

He decided to email one of the other team members with his concerns and directions on how to fix the problems. This action was met with some success, but some of the screen shots were still not correct. Aidan decided he would take the screen shots himself, and send them back directly to the other team members with explicit directions on what they should do with them.

By this time, three weeks had passed. Aidan had given up working on the new project altogether and spent all of his time working on the web-based training project. Worried about his own project falling behind, he finally raised his concerns with Dale.

Dale was surprised that Aidan was so involved with the project. Secretly, she wondered if Aidan wasn't being a little dramatic about the quality of the work he was getting from the Egyptian team. She agreed to contact Massoud herself to see what was going on. She called Massoud and left him a voicemail.

Massoud emailed Dale the following day. He said that his team was making good progress, and everything looked good. They were nearly finished with building the web pages and soon would start recording the sound files.

Aidan continued reviewing the web pages, providing the screen captures needed in order for the pages to be correct. He sent the GIFs to the Egyptian team with detailed instructions on how to place the pictures on the web page, and what the content should be. He had the best luck working directly with the team itself, and so he decided not to go through Massoud. For reasons he did not understand, Massoud would not respond to him, despite repeated attempts at communication.

However, he simply did not have the time to find out what the problem was. He didn't think it would help much if he involved Dale, so he decided to forge ahead with the project himself.

What Aidan and Dale did not realize was that Massoud thought Aidan was disrespectful. To him, it sounded like Aidan was trying to tell him and his workers what to do. Massoud did not feel that he had the authority to do that, and was breaching etiquette. Personally, he found the American to be a bit abrasive and pushy, not giving his workers enough time to figure out what to do on their own. He felt that he was trying to make his team look bad.

Almost four weeks later, the web pages were finished and they looked good. While the Egyptian team was creating the sound file to go with the web pages, Aidan was able to catch up a little on his own project. By this time he was working nights and weekends trying not to fall behind.

The Egyptian team sent the sound text file to Aidan for review. Upon opening the file, his heart sank. Each page was filled with minor errors and technical inaccuracies. He would have to go through each page line by line, checking for inconsistencies and mistakes. It was evident that whoever created the file was not a technical person, and did not understand the subject. He would have to fill in the technical information so that the web page made sense. It took Aidan a week and a half to correct the file and send it back.

Finally, the project was completed and a CD of the product was sent to Aidan for a final review. He placed the CD in the disc drive and waited for the program to load. He clicked on the menu and loaded Module 1. All the graphics—the screen captures Aidan had made of the software performing a task—looked good. He clicked a button to turn on the sound recording to narrate the page.

A woman's voice with a thick Egyptian accent came on, enunciating the narration script perfectly. Aidan winced. "This will never do," he said to himself. "We need someone to narrate the text who has a neutral accent, if possible."

Aidan did not burden Dale with the details. When he found someone to narrate the training, he went to Dale for a purchase order. He let her listen to the original recording. She agreed that it had to be changed. She let Aidan hire the narrator. Since the team in Egypt had built the training, the recording had to be sent back to them so they could incorporate it into the training.

Finally the product was finished. Everything looked and sounded good. Aidan couldn't wait until he saw the final product on the internal website. He was proud of the work he did, and felt that without him, the project would have been a disaster. He was planning on placing the project in his portfolio of completed projects for the year.

ACCOLADES, ETC.

Later that month, at the Training Services quarterly meeting, Dale spoke about the successes her team had. She presented numbers showing an increase in classes created and maintained. She posted numbers that showed an increase in training revenue. She thanked her team and others who had helped to keep training up to date.

"I especially want to thank our team in Egypt for the fine job they did with the web-based training," she said. "They worked quickly and efficiently, and the quality of their work was superb. My experience with them is that once you give them a task to do, they put their heads down and get it done." She smiled, and continued, "The product group was impressed with the quality of their work. As a result, they are going to work closely with them to develop several other online training classes." She nodded to the Director of Training Services, John Bigelow, who was seated at the table. "John was so impressed with the work Massoud's team has done, that he is talking about creating a separate, permanent web-based training division. Massoud's team will head the division, and Massoud will be promoted to manager of web-based training services. They will continue to work with our people here to ensure the success of converting instructor-led training classes to self-paced web-based training classes.

"I have always promoted the collaboration between our team members all over the world. It has been my experience that when you bring a group of talented individuals together, no matter where they are, the benefit to our organization and our customers is nothing short of amazing. We will continue to use whatever resources are available in order to deliver the quality training that our customers have come to expect from us. And, it is people like our team in Egypt that help establish a world presence. As we move forward in the years to come, I am sure that we will rely more and more on team members that are scattered in offices across the globe to represent the face of our company. Therefore, it is important to work together to make our team the global leader in the EDA industry throughout the world."

She turned and faced Aidan. "The team in Egypt would not have been able to work as quickly or smoothly without the help of Aidan Quinn. Aidan overcame difficulties of time and distance to help facilitate the project. He gave considerable amounts of his time and talents, despite the fact that he had his own deadline to deal with. He really stepped up to the task and went beyond his duty to make this project a success."

Discussion items

1. Is Dale's final statement justified? Why or why not?
2. What are the key problems in the case? Would you do anything differently to solve these problems? What would you do differently?
3. What are the key factors for promoting team effectiveness when implementing virtual team?

My Job Was to Integrate Two Cultures

Dragan Z. Milosevic, Russ J. Martinelli, and James M. Waddell

"All my professional life I have dealt with software development—banging out code," began Jerry Dorsey, now one of several project managers for the geographically and culturally dispersed Dacia project. "I have always managed local, co-located software development teams, so I was stunned when my boss summoned me and asked me to manage a project with an outsourced development team in Romania. I think I asked the same question three times—'Romania'?" This case shows an approach for solving a cross-culture problem by integrating the Romanian team with the U.S. team.

UNDERSTAND THEIR CULTURE

Jerry continued, "At first I was shocked, since I didn't know the first thing about cross-cultural integration, but I began communicating with members of the Romanian team to learn about how they worked and what they valued." Jerry soon discovered that corporate culture and national culture often collide. "The Romanians were used to being tasked," said Jerry. "They had an attitude toward me that 'he's the boss,' and, therefore, I should have all the answers. The concept of brainstorming solutions, which is a common part of our company culture, was completely unknown to them."

Jerry continued, "They would also never say no. I could just give them more and more to do and they'd try to get it all completed. So, I had to learn how much I could actually give them by monitoring the progress of the deliverables. As long as they met their deliverables on time, I figured they weren't being overtasked."

HOW I DEAL WITH IT

The biggest lesson for Jerry, however, had little to do with managing the development of the software. As he explained, "Building strong personal relationships

190

was the most critical element in integrating the Romanian team into our company and program culture. We were able to bring the key technical leaders from Romania to the United States early in the planning phase to meet and interface directly with their U.S. counterparts. There's no better way to build mutual trust! I also made a point to travel to Romania once every two to three months to get to know the Romanians and make myself directly accessible to them.

"At the end of the day," concluded Jerry, "this was a great experience for me personally and for my career. I got firsthand experience on what it means to be a program manager, and it's definitely an avenue I'd like to continue to pursue."

Discussion items

1. Do you agree with Jerry's approach to integrate teams of different cultures? Why or why not?
2. Suggest an approach for a cross-cultural integration?

Rate and Rank

Rabah Khamis

It is that time of the year when you can feel the tension in the air at SEMITech; signs of stress are apparent in managers' faces, employees try harder than usual to focus, some conference rooms' little glass windows have been blanketed with presentation paper to prevent passersby from finding out who is meeting inside, and email responses are slower than usual. In one conference room, the tension is high between peer managers who normally collaborate to help each other. The issue at hand is each manager is trying to polish his/her employees' accomplishments to protect some of them from being put on "performance improvement needed" or "below expectation" rating. Managers want to rank their employees' performance as high as possible as it is a reflection of their own ability to lead. An example of a typical meeting of managers follows.

Herb: Arkay has joined my team in the middle of the year. Since he joined the team, he really matured and influenced the team in a positive way. He jumped in and volunteered to work nights and weekends tirelessly to help the team with their commitments. The team delivered very nicely since he joined; his team really likes him; and more importantly, he took on leadership roles where he worked with other groups to solve or facilitate the resolution of several bugs. The high level validation committee speaks very highly of him.

Mark: I like your enthusiasm Herb but I am not sure what technical competencies Arkay has that distinguish him from other team members.

Harry: (second-level manager): Yes Herb, what distinguishes Arkay technically over his peers?

Herb: For one, he has excellent planning skills. First, the guy comes from a different group with no chip architecture background and planned the Chip Power ON activity for our latest product, and we all know how critical that was for the company. He planned in-depth testing which got the thumbs-up from the group's technical leaders.

Harry: Okay, I buy that. That is a great skill and he did prove himself there very well. But what other technical skills set him apart?

Herb: He managed the offsite test house to supplement our testing. With his leadership, he managed to create test content, escalated high-priority sightings the test house found, and removed all blocks to get this test house busy testing our software and hardware. He used them as a resource to get the team commitments met on time every time.

Harry: Yes, he gets credit for that.

Herb: If you look at his accomplishments with his old team, he did work directly with big customers to enable them to go to market with their platforms that adopt our technology. He created test contents and resolved Kerberos bugs for IT, he created Network setup BKMs for our company and customers, and he still gets contacted by his old team for consultation on technical issues he is experienced with. He is currently building his technical experience in chip architecture. How I see it, given some time, he will definitely surpass his peers.

Mark: Well, I am sure he is a good guy but I do not feel comfortable promoting him, considering my employees did very visible and important work for the company.

Herb: All I am saying is Arkay has exceeded his grade level and he proved it to me month after month. I know several of his peers with the same grade level do not have his interpersonal or technical skills. Look, I feel I can throw anything at this guy and I am sure he will come through. I cannot say the same for Sharma, Sonny, Juviani, or Aaron. He modeled several company values such as "Discipline," "Risk Taking," "Great Place to Work," and "Results Focus."

Harry: Good point. When I . . .

Mark: Wait Herb, Aaron's grade level is lower than Arkay's. You can't rank him against Arkay.

Herb: Yes, but you definitely agree that Arkay is more productive than Aaron and he definitely ranks higher than the rest I just mentioned.

Mark: Well that is debatable. . . .

And the Ranking and Rating meeting continued for hours but the final ranking and rating was not finalized for another few meetings.

SEMITECH

SEMITech is a multibillion dollar semiconductor company that designs and manufactures electronic components for computers, cell phones, ASIC designs, embedded processors, and software. The company employs more than 70,000 employees and has annual sales of $30 billion. The company's mission statement is to "tirelessly pursue excellence in delivering technologies that become essential to the way we live and work." The company has several values which it expects its employees to model in their work: "Customer Satisfaction," "Results Focus," "Discipline," "Unmatched Quality," "Collaboration," and "Risk Taking."

The company considers its employee base a major part of its competitive advantage. To realize its mission, SEMITech's HR strategy is to recruit and maintain the best human talent in their fields. The company recruits the best graduates from the best accredited schools; recruits the most experienced people from competitors, and compensates its employees generously based on their performance. It also has created a review process to reward the best employees, identify rooms for improvement and development of all employees, and build on employees' strengths to advance their careers.

THE ANNUAL PERFORMANCE REVIEW PROCESS

SEMITech does yearly reviews on all employees. Employees are evaluated based on their managers' expectations, which are drawn from the organization expectations, as well as being evaluated against their peers' performance. The performance is measured in line with departmental and corporate business objectives, the impact made by the employee on the team (or business group and company), their performance relative to their peers' performance, performance against set expectations according to the job level of the employee, and the completion of their deliverables. At the end of the review process, rewards are allocated based on merit; low performance is addressed; and expectations and development areas are set for the next year. The review/evaluation process consists of:

1. **Initial employee performance assessment:** The employee turns in a self-assessment of his/her accomplishments, strengths, and areas of development. The manager measures the employee performance against agreed upon MBOs (Management by Objective Deliverables) and the manager's expectations which were set during the year. The manager also solicits feedback on the employee's performance from his peers and customers and integrates that with the self-assessment.
2. **Rating the employee's performance by the direct manager:** The employee performance is graded by the manager as: Outstanding, Overachieve, Meet Expectation, Below Expectation, or Need Improvement. Those grades are measured against a manager's expectations relative to a rating scale. See Table 9.1.
3. **Allocation of financial reward based on performance:** After the manager assesses the employee performance and rates it, he/she decides on the financial reward based on HR's set guidelines. The rewards can include a salary raise, stock options, promotion to a higher level, or a combination of all these. An employee could also be penalized for poor performance by not getting rewards, being demoted, putting on an improvement plan, or possibly being terminated. Rewards vary from group to group and may vary from individual to individual, depending on performance. A manager uses his/her discretion to allocate rewards but it is based on very well-defined guidelines.

4. **Ranking employee against other employees of the same organization:** In this step, employees from different teams are put in the same pool and ranked against each other according to the impact each one of them made during the year. Usually, high-level managers attend this ranking meeting along with direct managers and make their decisions on the ranking process. In this meeting, an employee could have been very successful according to a manager's expectations but ranked with lower performance than an employee of another team who worked on a critical and highly visible project. An employee who solves a customer issue may impact the company's financial performance more effectively than an employee who patents a new invention and, as such, be ranked higher. This part of the process is normally the most stressful for managers, and requires a lot of negotiations and compromise. A final ranking for all employees of the organization is the final outcome of these meetings.

5. **Review results communication to the employee:** This is the last step of the review/evaluation process in which the employee is communicated to by his manager the result of his/her performance last year and the next steps for him/her to advance to the next level. This part of the process is the most stressful for employees and managers. A manager has to explain to his employees what they have done right, what they need to improve on, and convince them this is in their best interest. Managers also have to explain very clearly to their employees what they need to do to improve. Employees have to prepare for all possible results.

Table 9.1 Example of Employee Rating for the First Four Levels at SEMITech

	Level 1	Level 2	Level 3 Levels 1, 2 Items Plus . . .	Level 4 Level 3 Plus
Product Impact & Scope of Innovation	Accomplishes technical tasks specific to sub-project.	Accomplishes technical tasks specific to sub-project.	Clearly documents and communicates about the project.	Strong technical influence and contribution to important aspects of the project.
Organizational Impact & Influence	Clearly communicates work and ideas to team members and supervisor.	Clearly communicates work and ideas to team members and supervisor. Contacts are primarily with immediate supervisor and others on the team. Sought by project peer to supply specific information on current project responsibility.	Uses relationships and analysis of data to gain support for proposals. Decisions frequently affect the performance and success of the project.	Influences tactical business issues that impact the entire team. Interacts with senior internal and external personnel to get updated information, answers, or advice to shorten own learning curve. Strategies are influenced by his/her recommendations.

(Continued)

	Level 1	Level 2	Level 3 Levels 1, 2 Items Plus . . .	Level 4 Level 3 Plus
Technical Expertise	Demonstrates basic engineering skills. Work is of good quality. Works on tasks of low to medium complexity, based on specifications developed by others.	Working technical knowledge in one technical area. Work employs company-wide BKMs. Shows consistent growth/ improvement in areas of contribution. Works on tasks of low to medium complexity, based on specifications developed by others.	Regarded as an expert in area of contribution. Work has high quality consistent with the complexity and risk of assignment. Work either extends or employs company-wide BKMs. Shows consistent growth/ improvement in area of expertise and its application. Needs minimal assistance on medium to complex tasks.	Regarded as an expert in several technical areas. Understands risks and manages them effectively. Exhibits initiative/ independence in finding ways to ramp up in new technical areas. Comfortable with medium to high complexity projects.
Teamwork and Leadership	Viewed by peers as a positive team player; demonstrates a professional attitude. Gives credit where credit is due.	Viewed by peers as a positive team player; demonstrates a professional attitude. Displays a willingness to volunteer for projects outside job scope. Gives credit where credit is due.	Viewed by peers as an excellent role model for communication, teamwork, and leadership skills. Shifts between leader and follower as needed. Sets clear expectations and requirements for team. Networks effectively to share methods and information to uncover/solve issues. Openly shares and accepts ideas.	Viewed by manager as more of a peer than a subordinate. Builds credibility and consensus both within team and external to team. Identifies problems and solves them.
Business Understanding	Understands high level of how his work impacts the whole project.	Understands how his/her work and that of his/her immediate work group fits business goals.	Independently adapts his/her work and that of his/her immediate work group based on a solid understanding of business goals.	Identifies, quantifies, and flags problems at a team level. Proactively generates possible solutions to problems including cost/ benefit analyses; drives solutions across the project.

Problem Solving	Actively seeks appropriate guidance to overcome roadblocks/issues.	Identifies, quantifies, and flags problems. Proposes viable solutions to problems and analyzes options with stakeholders. Investigates and overcomes challenges through creative methods, principals, and practices.	Learnings from other projects are incorporated such that problems encountered previously are avoided on current project. May contribute to project-level productivity enhancements. Correctly implements technical solutions of a project/task involving a small team of engineers.	Crisply identifies problem statement and develops a phase solution plan. Contributes to productivity enhancements. Responsible for developing a function, reusable by other projects.
Planning & Scheduling	Executes to schedule on assigned work with attention to detail. Clearly communicates work/schedule to supervisor.	Tracks progress against schedule. Detects and promptly flags schedule risks. Clearly communicates work/schedule to supervisor.	Develops own plans/schedule; can organize and schedule group tasks. Recognizes the importance of setting, tracking, and meeting schedules. Identifies schedule-limiting tasks and proactively searches for improvements. Deals effectively with dynamically changing circumstances and minimizes negative impact/consequences.	Performs proper scoping of tasks and risk assessment. Detects schedule risks and communicates and addresses them quickly. Proactively provides options for controlling schedule change.
Coach, Train & Mentor	Freely and proactively shares knowledge with others.	Freely and proactively shares knowledge with others.	Provides guidance to the project/team in area of expertise. May lead or act as primary reviewer during design/project reviews.	Considered the primary reviewer for high-level product documents. Clearly presents concepts to outside groups and upper-level management. Produces clear technical documents and training materials.
Minimum Qualification Guidelines	BS & 0+ years experience.	MS & 0+ years experience.	BS & 3+ years experience or MS & 2+ years experience.	BS & 4+ years experience or MS & 3+ years experience or PhD & 0+ years experience.

RATING AND RANKING

The outcomes of the evaluation process are a measurement of employees' performances, rewarding those who performed to expectations, developing all employees, and adjusting their compensations. Some people think that the review and evaluation is done at a certain time, such as the beginning of the year, but the truth is that the review process is a year-long process. Managers' observations of employee performance are continuous. Managers set the expectations for employee performance at the beginning of the year and communicate them clearly to employees. Table 9.1 shows the level of performance set by the organization for the first four job levels at SEMITech. Each employee should know these expectations by heart and perform according to their job level or higher. After all, employees are going to be measured against these expectations which means they are going to be "Rated and Ranked" with these performance metrics in mind.

Direct Manager Rating (DMR)

Managers evaluate annual employees' performances and determine their ratings. The "rating" is just an indication of the employee's performance compared to the manager's expectation and compared also to peers' performance. Simply put, it is a score of "how each employee performed." The "measuring stick" is a matrix, such as the one in Table 9.1, preset by the company or the specific department. The manager's expectation is this measuring stick. It should not be set higher or lower than the expectation set for the employee's job level. For example, if an employee is level 1, the manager's expectation should not be that the employee should manage his/her own tasks or perform on par. At the same time, a level 4 employee should not be expected to be given low-level instructions on how to perform their tasks.

For SEMITech, an employee gets one of five ratings:

1. Outstanding (O): This rating indicates the employee consistently outperforms peers with similar job scope and responsibilities.
2. Overachieve (OA): This indicates that the employee achieved results that go beyond the requirements of the job in all key areas. The employee often outperforms others with similar job descriptions and level.
3. Meet Expectation (ME): The employee makes a solid contribution in key areas of responsibility with some guidance and supervision. The employee performs on par with peers with the same level and similar responsibilities.
4. Below Expectation (BE): The employee successfully meets some but not all of the responsibilities and expectations outlined in Table 9.1. He/she requires substantial supervision for the level of experience at which he/she was hired. The employee performs below peers with similar responsibilities.
5. Need Improvement (NI): Frequently does not meet job requirements and needs substantial supervision and more guidance than is justified to carry

out responsibilities. He/she consistently underperforms job requirements compared to peers with a similar level of experience.

The DMR process is as follows:

1. The employee provides a self-assessment sheet describing the employee's view of his/her performance. The self-assessment contains the employee's current position and area of responsibilities, main achievements of the previous year, main strengths, and areas of improvements.
2. The manager and the employee discuss the self-assessment and feedback is provided by the employee's peers and customers about his/her performance. They also discuss a development plan. The manager integrates the feedback with the self-assessment and has an initial review for the employee.
3. Based on the discussion, the performance matrix is created showing the employee's self-assessment, the feedback solicited from peers, the manager's observations, and a comparison of employee's performance to his peers in the same team. The manager then assigns a "Rating" to the employee
4. The manager discusses his decision with his manager to ensure proper rating distribution. Then the manager creates a matrix with all employees' names and their ratings. The matrix also includes employees' levels and positions. See Table 9.2.

Organization Rating and Ranking Session

After the manager decides on a rating for each of his subordinates, an objective rating and ranking calibration session is assembled on the organizational level

Table 9.2 Direct Manager's Rating and Ranking Matrix

Rank	Name	Level	Position
1	John Smith I	2	SW Engineer
2	Jane Doe	4	HW Engineer
3	Samantha Jay	1	SW Engineer
4	Dennis McDonald	3	SW Engineer
5	Vijay Krishna	5	SW Engineer
6	Hou Meng	4	SW Engineer
7	Sasha Tee	3	SW Engineer
8	Matt Pen	1	SW Engineer
9	Jacob Gauge	1	SW Engineer
10	Kyle Mist	1	HW Engineer

and attended by all peer managers of the same organization and the second- or third-level managers. The purpose of this staff level review is to ensure each employee's performance is evaluated relative to the expectations of the job level set by the organization and the peer performance in other teams. In other words, the employee performance is not only measured against his immediate peers of the same team, but also against peers with the same job level from other teams in the organization. This is to ensure that peers from different teams in the same big organization are treated and compensated equally. This is an attempt to neutralize a manager's bias.

The following steps are SEMITech's organization level rating and ranking process which spans across multiple sessions over one to three months.

1. Each manager provides his own list of employees with their performance ratings, positions, and job levels (see Table 9.2).
2. A matrix is created with all employees in the organization of the same job level and function. They are grouped together to rank them against one another.
3. The high-level managers review the matrix in depth and analyze the achievements of all employees. These managers mark questionable ratings and make notes to question managers for details.
4. An R&R meeting session is scheduled. The meeting starts with each manager justifying to the second- or third-level managers' questionable ratings, promotions, and demotions of his/her employees. This step is very emotional, as managers try to defend their employees' achievements as they reflect on their own performance eventually. It is very likely that high-level managers will ask their subordinate managers to reconsider some ratings for some employees and maybe re-rate them and submit a modified matrix.
5. It may take several sessions to reach a final Organization Rating and Ranking. The final outcome of these sessions is a calibrated rating and ranking matrix which means some ratings for some employees may change. An employee may not get promoted this year as his/her manager would have wished. Some employees may get promoted and some may get demoted. Historically speaking, about 90 to 95 percent of submitted employees' ratings are approved by the second- or third-level managers from the first time because all managers go through review training beforehand and they realize they have to justify every decision they make on employees' ratings.
6. The second- or third-level manager approves the final rating and ranking matrix after he/she considers SEMITech's own performance rating distribution guideline. The guideline is as follows:

- 15 to 20 percent for Outstanding/Overachievers
- 65 to 75 percent Meet Expectation
- 5 to 15 percent Below Expectation/Need Improvement

7. After finalizing the organization R&R matrix, each manager takes the ratings and integrates them in the employees' reviews with compensation changes made where applicable.

Many of SEMITech's employees wonder if their annual review and rating and ranking processes are the best method to measure employee performance. Employees point to the fact that the average employee worked for SEMITech for only five years. Senior management thinks the process provides the best and most fair method for employee evaluation. They point to the evidence that the company has been very successful in the industry for more than 30 years.

Discussion items

1. Discuss the advantages and disadvantages of SEMITech's rating and ranking approach.
2. Does the employee's promotion depend heavily on the marketing skill of their manager to represent them well during the R&R session?
3. How do the government's diversity quota and affirmative action impact SEMITech's R&R approach?

Chapter 10

PROJECT COMMUNICATIONS MANAGEMENT

This chapter contains six cases—one critical incident and five issue-based cases. The cases relate to Project Communications Management, Chapter 10 of the *PMBOK® Guide*. The cases illustrate different approaches to promote communication among project stakeholders both at the beginning of a project and during the mainstream activities.

1. The Russians Join Us Late at Night

    The Russians Join Us Late at Night is a critical incident, discussing an approach to promote communication among team members of different time zones. The case shows that the team members, especially the project manager, must be flexible in communication. It sounds easy. Is it also easy in practice?

2. Quest for Clear

    Quest for Clear is an issue-based case. It details an implementation of new change management software in one organization replacing the existing software. It is typical that such an initiative must involve managing changes. This case portrays the importance of communication to such an initiative and how the project manager must practice strong communication management.

3. Electronic Medical Record

    Electronic Medical Record is an issue-base case. It focuses on project communication in the early phases of a project life cycle. Such communication helps initiate conversations among project stakeholders, especially during the requirement gathering process.

4. Improving Public Health Informatics

Improving Public Health Informatics is an issue-based case. It provides an example of a project communication plan used by a project team. Such a communication plan is necessary for successful communication, especially for a project involving multiple groups of stakeholders.

5. A Simple Metric Goes a Long Way

A Simple Metric Goes a Long Way is an issue-based case, discussing the development of a simple metric to report the status of projects. The metric is expected to promote cross-project coordination and executive oversight.

6. Executive Project Metrics

As an issue-based case, Executive Project Metrics discusses the issue of how to communicate project status to senior executives. The case suggests some parameters that should be of executives' interest. It also provides an example of how such metrics work.

CHAPTER SUMMARY

Name of Case	Area Supported by Case	Case Type	Author of Case
The Russians Join Us Late at Night	Communication in Different Time Zones	Critical Incident	Dragan Z. Milosevic, Russ J. Martinelli, and James M. Waddell
Quest for Clear	Communication Management	Issue-based Case	Mathias Sunardi
Electronic Medical Record	Communication Among Stakeholders	Issue-based Case	Mathius Sunardi and Abdi Mousar
Improving Public Health Informatics	Communication Plan	Issue-based Case	Abdi Mousar
A Simple Metric Goes a Long Way	Project Status Report	Issue-based Case	Art Cabanban
Executive Project Metrics	Project Status Report	Issue-based Case	Dragan Z. Milosevic, Peerasit Patanakul, and Sabin Srivannaboon

The Russians Join Us Late at Night

Dragan Z. Milosevic, Russ J. Martinelli, and James M. Waddell

"Communication is the key," says Sri Rastogi. Sri is a project manager on a project that is geographically dispersed, with part of the team in Portland, Oregon; part in Houston, Texas; and part in Moscow, Russia. The following shows how Sri views communication when dealing with geographically dispersed teams.

HOW TO DEAL WITH A DISPERSED TEAM

"Someone said that communicating with the team from Houston is tough. I said, 'Let me tell you the story,'" says Sri. "There is an eleven-hour time difference between Portland and Russia, so finding a good time to communicate in person is tough. One of the things the Moscow team has done is to shift their workday. They now come in about 10 or 11 o'clock in the morning, then go home anywhere between 8 and 10 o'clock in the evening. We now have overlap at the end of the day where we can usually find people in the office.

"Instant messaging technologies have also helped a lot," says Sri. "I log in from home for an hour each night and turn on my instant messenger. If anyone in Russia needs to contact me during their morning, they can do so, and I'll respond immediately," he says.

MY WAY

"I don't know how it is for the rest of the company, but the fact that I make myself available at 11 o'clock at night on a daily basis during the development cycle is a necessity to help communication channels stay open on a geo-dispersed team. It's not rocket science, but it works!"

Discussion items

1. Do you agree with Sri's approach in dealing with team members in different time zones? Why or why not?
2. Would Sri's approach work if he leads multiple projects, say if he also leads another project where the team members are in New Zealand?
3. Suggest a better way to promote communication among team members of different time zones.

Quest for Clear

Mathias Sunardi

Jim Nasyum was vey furious and close to exploding. No wonder. It was because Jim just learned that Copernicus, a major project he was overseeing, might be two months late. It was supposed to be a six-month project. Jim was a senior software engineer of Jtronics. He was responsible for the development of software for extracting information from the measurement equipment developed by the company.

Jim heard this bad news from the man who ran the project, Ide Home. In fact, Ide is Jim's star project manager. The sheer consequences of this delay terrified Jim and he already visualized the face of the easy-go-ballistic vice-president when she hears about Copernicus' delay. It is going to halt the release of four new product lines that are planned to use the Copernicus software, which Jtronics nationally advertised.

More than anything else, Jim was furious because of Ide's explanations about the causes of the delay. Ide explained that they had problems with change management software which caused several losses of data and information. "This is not the first time that we have had problems with this software," Jim thought. Jim had to deal with the delay before the news reached the vice-president.

JTRONICS

Jtronics is a U.S. company, ranking as a number two in market share in the electronic testing and measurement equipment industries. The company has buyers in both conventional and technology industries; and enjoys sales of more than $1 billion. More than half of that amount comes from foreign markets. Almost all products are software rich, causing the company to pay serious attention to software development, treating the software component on par with the hardware. Each of its new product development projects, organized by means of programs, has software and hardware managers all reporting directly to the program manager.

CHANGES IN THE SOFTWARE DEVELOPMENT PROJECT

As in all software development projects, there are always changes that occur during development, and many revisions are made to the software. To manage these

Figure 10.1   The Flowchart of the Process of Using Shadow

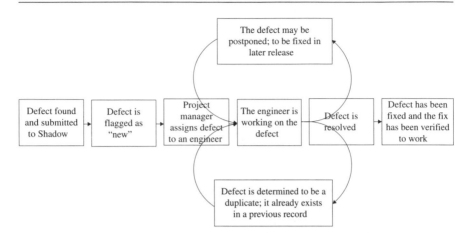

changes, Jtronics uses change request management software called Shadow (see Figure 10.1). Shadow allows users to submit defect or bug reports through emails using an array of report templates and metrics, and any changes or resolutions made are reported back to the team. The Shadow is coupled with software called ClearCase, which is versioning software that maintains the versions of the software under development.

Shadow and ClearCase have been used for several years by Jtronics, and since then users of the tools from different departments throughout the company have filed many complaints and recommendations to improve the software, as they feel that it does not fit their needs. The users prefer a system that can expedite the whole process of bug submission and change request. Most users complain that the tool is too complex; that the process is too tedious and takes a lot of time to complete the request. No immediate actions are taken to these inputs. Jim has noticed that there has been a backlog of complaints on the current change management software. According to Jim, if every engineer can save 30 seconds off of the process, the total savings for the company would be $4 million each year. Now, with the visibility of Copernicus problems, Jim hoped it was going to be easier to build the case—to improve change management software. However, he also knew that he had to do something to prevent fiasco related to the Copernicus delay.

CHANGE MANAGEMENT AT JTRONICS

Jtronics has some formal guidelines for managing changes in the organization, specifically for implementing new systems. For example, when considering implementing a new system, a business case must be presented and discussed with a committee; a person is appointed in charge of the change; the details of the change (reason, description, etc.) are documented in a change report form, and the document is signed off by the committee; then the change is approved. The process or procedure may differ a little, depending on the type and/or scope of the change. However, only the formal process is provided with a guideline, while the informal processes (such as getting buy-in from the stakeholders)—which sometimes are important to the change's success—are not described. Jim realized the importance of marketing the change, and getting stakeholders buy-in.

GETTING BUY-IN: COMMUNICATION

Jim recognized that it was necessary to get everybody to buy-in—both management and technicians on the floor. In order to do that, Jim created a sort of network of the stakeholders. Actually, he drew a kind of interrelationship diagraph, as it is known in quality business (see Figure 10.2). He started with a node that represents him, and then he created a node for every person he has direct connection to and drew a line that connects his node with that person's node. On the line, he wrote down what is his contribution to that person, and vice versa. Then, if that person had a direct connection to another person, either the ones he had direct connection to, or the ones he did not have, and did the same with the edges and contributions. This way he can see who and where the stakeholders are, and what their relationship is with him and to others. He can also see who he should contact to reach a person whom he does not have a direct connection to. He then was able to strategically communicate his idea to get buy-in from the stakeholders.

Figure 10.2   Jim's Interrelationship Diagraph

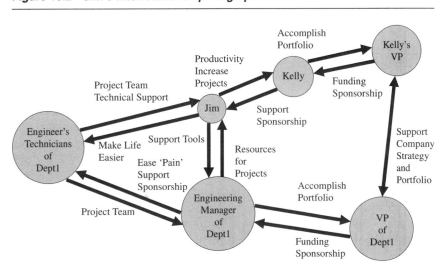

Jim started off by going back to his camp—people in favor of improving the change management software. He had the direct contact with them. From them, he found out what they liked about the current system; what they didn't like about it; and what they would like the change management software to be. For instance, he found that Shadow uses 127 different forms to report different kinds of bugs and defects, and the users find it cumbersome making the whole process an additional burden to their already over-swamped workday.

Since Jim understood that Shadow was used company-wide, and any changes to the system would affect the whole company, he shared his concern with his manager, Kelly Braid, with whom he also had direct contact. Kelly understood the issue with Shadow and supported Jim's initiative to take action on the matter. They worked together to create a business case and identified the alternative solutions. Mainly, they considered two options—fixing the system or buying new software from a third-party vendor. Based on the requirements gathered from the users so far, they found that a third-party vendor software called ClearQuest met their needs. Kelly and Jim started introducing ClearQuest to the technicians and engineers. All agreed that it was a better alternative.

Buying ClearQuest will cost about $500,000. At Jtronics, a project with such value requires approval from the executive management at the CEO level. Jim realized that in order to get funding for the project, he had to have buy-in from the vice-president to get approval. Unfortunately, Jim lacked direct connection to the VP of his department. But Kelly had this direct connection. So Jim and Kelly worked on developing a business case that was clear and understandable for the VP.

With regard to this, in one of the meetings, Jim commented, "It was cool—everybody already understood the issue, and the engineering managers were able to convince their VPs, while my VP already understood the issue because my boss and I already explained it to him beforehand." Afterward, it was just a quick and easy discussion between the VPs to decide whether or not this project was worth funding.

SPEED BUMPS

Finally, they agreed that the project had a valid business case, and deserved a "go." However, Jtronics was going through some changes at the time including shifting of management and other financial challenges. The new management was more interested in projects that are aimed toward saving money, and a half-million-dollar project such as changing Shadow to ClearQuest was not exactly attractive. Also, with Jtronics' financial situation at the time, Jim's project was placed under low priority and it went under the radar for years.

Later, Rational Software, the company that developed ClearQuest, was bought by IBM. After the acquisition, IBM reviewed all the projects done by Rational and noticed the project with Jtronics, which had been delayed for quite some time. They decided to follow-up, and contacted Jim. They offered help to proceed with the project by giving Jtronics a discount of $150,000. Jim was ecstatic and proposed the project back to his new manager. And sure enough, the project finally received the green light it had been waiting for.

IMPLEMENTATION

Although most of the users understood and accepted the change, Jim was aware that the switch was not going to be simple. First of all, the users who were receiving the change are technicians and engineers. Since Jtronics is a company that relies on the productivity of its crews, a system change like this is likely to cause a lot of down time while getting used to the new system. To overcome this, Jim and his team customized the "look and feel" of the ClearQuest interface so it closely resembled Shadow. Besides the common functions, they also included some of the new features of ClearQuest to the interface. This way, the users were able to quickly grasp the new tool because it was familiar to them, and it only required one or two hours for training. In the meantime, users gave feedback (comments, recommendations, bug reports, etc.) to Jim's team to improve the functionality of the tool.

Secondly, not everybody was ready for a change. There were some users who were eager for the change. They surprised Jim by stopping by his office and telling him directly that they wanted to use the tool immediately. There were also users who were not ready for the tool; either they were busy with their project at the time and could not spare enough time for learning the new tool, or they simply did not see the value of it. Jim described this situation as analogous to the technology adoption life cycle, where there are users who are enthusiastic about the new technology and cannot wait to start using it (early adopters), and there are users who do not adopt the technology until late in the life cycle (late adopters/laggards). Jim devised a queuing strategy for adoption. Users who were ready (and eager) to change, were placed in the front of the queue, while the late adopters were later. This way, the training was done in several sequences, one group at a time (each group consisted of one or more different departments).

COPERNICUS' FATE

By the way, as mentioned earlier, Jim had to do something to prevent a possible fiasco related to the Copernicus delay. And, he did. He took two major actions. First, he approved Copernicus to outsource their quality testing, do it overnight, and save two months' time. Second, he added to the team three full-time programmers to speed up the coding work. Jim knew the dangers of Brook's law (adding people to a late project will make it later). With all these, the Copernicus was well on its way, and Jtronics was more prepared than ever to proceed with their release of the next four new product lines, and their future projects.

Discussion items

1. How important is communication to a successful change management project? Propose a strategy for effective communication.
2. What should be a strategy to address the organization's members who are reluctant to change?

Electronic Medical Record

Mathius Sunardi and Abdi Mousar

In the summer of 2008 the County Health Department's director, Peter Beckham, realized that the department had a serious problem. Its clients' health information medical chart system was aging and becoming increasingly ineffective. Worse yet, the system was not integrated and was extremely time-consuming for healthcare providers to continuously update clients' medical charts by having to manually shuffle through sometimes more than 50 pages, depending on the number of years of each client's visits. Results, in terms of time clients would spend in the waiting room or any other time measure, were way below mediocre. (Table 10.1 is a sample chart depicting minutes and dates waited by patients to be seen.)

Table 10.1 Time Spent During Visitation

Service Provider A	Waiting Room	Exam Room to Checkout	Total Actual Cycle Time
06/2008	22.1	113.5	35.6
09/2008	30.8	106.9	137.7

Not that it was new. For a few years it kept popping up every time as a burning issue. Peter knew that he had to do something, or he might lose his job. He immediately consulted people who enjoyed expert status among employees—Vanessa Harley, director of the integrated clinical services; Jacob Marsh, project manager (PM) clinical services; and Dr. Adam Smith, a member of a leadership team. They looked at the results sheets, and tried to understand what could be done to solve the problem.

Vanessa: It is part of our strategic plan to effectively use information technology to improve the delivery of our healthcare and reduce costs.

Dr. Adam: That is correct and we need to come up with an alternative technology system that meets our all stakeholders' needs.

Jacob: Hundreds of healthcare systems around the country use an Electronic Medical Record (EMR) system to improve the care they provide to their clients. With EMR, computers are used to store medical information that would otherwise exist on a paper chart. And when you visit a clinic, your provider will use a computer in the exam room to look at and update your medical record. The EMR will also allow the clinic to connect your Laboratory, Radiology, and Pharmacy information with our healthcare provider's notes. This means your providers will have instant access to all of your healthcare information, without having to shuffle through paper charts. If you go to a different clinic, the provider will have access to all of your healthcare information.

Peter: Let's organize a joint meeting with our stakeholders, and Jacob, you take the lead of this project. I will make an announcement to our employees to get their support and involvement.

SELLING THE EMR PROJECT TO THE EMPLOYEES

To ensure a successful implementation, Peter's job was to sell the EMR system to his employees. He enthusiastically delivered the following announcement:

"Installing the EMR system in the health department is a major undertaking and investment. Numerous experts have identified the enormous potential of information technology to improve the delivery of healthcare while reducing costs. The core of this potential is a secure, patient-centered system. In order to do that our department has decided to implement an Electronic Medical Record (EMR) system. In the project we have to collaborate with the State Community Health Information Network, SCHIN. This collaboration allowed the county to spend less money on a high-quality EMR system as it cannot even afford the license alone. The three-year project was estimated at $4.2 million and will be funded through SCHIN (which is a nonprofit organization financed by federal grants and fees collected from their licenses that they purchase on behalf of their healthcare providers' partnerships); Care Willamette; and state, federal, and clinical resources. The EMR will also provide regional and statewide data, addressing the problem of access to healthcare for uninsured and under-insured -state residents.

"It is worth noting that EMR will complete the third phase of an integrated information technology initiative for the health department. The integrated EMR system will have the capacity to handle the medical records for 250,000 visits a year. The integration of our SAP system of financial and human resources and our EPIC system for patient scheduling and billing along with EMR supports the Health Department's goals and strategies for reducing costs and improving care while facing a future limited resource. Remember that SAP and EPIC were also made possible by SCHIN."

UNDERSTANDING SCOPE: STAKEHOLDERS' INVOLVEMENT

After Peter's great announcement, Jacob did all he could to get participation of the stakeholders. The hope was to increase stakeholders' commitment to the project. He started off by including the stakeholders in the development of a project plan.

Prior to developing the project plan, Jacob organized a joint stakeholders meeting. The meeting also included a walkthrough of the department's clinics to check the possibility of the equipment placements and exam room arrangements. The meeting and walkthrough facilitated conversations among stakeholders. The concern of State Community Health Information Network (SCHIN), one of the financiers and service providers, was the integration of EMR to SAP and EPIC, the existing applications. The health department was concerned about the continuation of its client services during the equipment installation. In addition to the interoperability of the overall system, IT people wanted to make sure that they placed network data ports and PCs in the right places in the exam rooms. The facility guys wanted to make sure that all hardware and network equipment conformed to the safety requirement of the building. The design team wanted to make sure that the EMR configuration fit any clinic's day-to-day operations. Stakeholders wanted their needs to be addressed as well.

Jacob recalled: "EMR was such an important project. Luckily, it started off well. I could see the power of bringing all stakeholders together to facilitate the conversation among them. At the end of the day, we got a rough idea of what the scope of this project should be and stakeholders were pretty much buying into it.

AFTER EMR IMPLEMENTATION

In the trial phase, the team implemented EMR to some of the providers. The doctors, nurses, and patients had experienced a significant improvement in terms of the patients' time spent during each visit. The leaders of the health department were very pleased with the results. Table 10.2 depicts the improved wait times following the implementation of the EMR system.

Table 10.2 Time Spent During Visitation After Implementing EMR

	Service Provider A	Waiting Room	Exam Room to Checkout	Total Actual Cycle Time
Before	06/2008	22.1	113.5	135.6
	09/2008	30.8	106.9	137.7
After	02/2009	19.2	56.6	75.8
	05/2009	28.9	59.0	87.9

Dr. Adam: Now that we have more providers on the system we are beginning to see the power of EMR. It has been great seeing the synergy between providers and, more importantly, between providers at different clinics! This is the beginning of the network we have always wanted and needed.

Peter: There are strong indicators and trends that show the EMR brings about positive changes and it was well worth the expense and effort of implementation.

Discussion items

1. What prompted the County Health Department's strategy in implementing the EMR system?
2. How significant was Peter's announcement of the EMR implementation project?
3. What are the benefits of a joint stakeholder meeting?

Improving Public Health Informatics

Abdi Mousar

In the event of a major disease outbreak, senior managers need to respond in a timely fashion. To be able to do so, one needs to have all the latest information and data to make an appropriate decision. The County Health Department (CHD)'s strategic plan specifies increasing capacities for the program evaluation and response to major communicable disease outbreaks. Senior managers have identified the need to improve informatics capacity as an important step toward meeting these goals.

Currently, each community health services program of CHD has information systems that effectively track specific information for their assigned work. However, the systems were designed reactively without considering future expansions or an updated public health data plan. Each data base is almost a standalone system. While the state data bases are linked to CHD's data bases, the environmental data bases are not linked with communicable disease data bases. Representatives from the Tuberculosis (TB) program, for example, had to enter identical client information in four different places. Moreover, the state mandates the public health department report the health conditions of its citizens and control infectious diseases. In order for this to happen, one needs informatics systems that are capable of tracking infectious diseases and reporting all investigations in a timely manner.

While the benefits of an integrated data base stand out, most employees in CHD are reluctant to change, particularly with regard to adopting and using a new technology as evidenced by the implementations of the EMR systems and other specifically designed data bases. To ensure success of the implementation, the project manager must provide relevant information to the stakeholders on a regular basis. This case focuses on project communication management, especially within the communication plan.

THE BIG PROJECT

It all began when a new epidemiologist, hired by CHD, articulated the need to have an integrated public health informatics system. With the new system, the reporting of infectious diseases as well as possible future outbreaks could be done in a timely fashion. The decision makers can then make an appropriate decision.

The communicable disease section of the county health department has taken the lead in championing this project. The team includes two IT engineers and the epidemiologist, who will be overseeing the project. The team is responsible for gathering business processes and information system requirements, working closely with stakeholders, and at the same time, training IT engineers using the best practices developed by the Public Health Informatics Institute (PHII). The project duration is expected to be from six to twelve months and the team will provide monthly updates as well as ad hoc updates whenever specific information is needed by the stakeholders.

The team has extensive experience in working together on previous successful projects such as the EMR implementation, EPIC implementation, and the communicable disease data bases' statewide expansion. It is the team's firm belief that this project, to integrate a variety of data bases into a system that can communicate to one another, is doable.

PM: The initial idea of this project was formulated several years ago but it did not get the support it needed. Now we are bringing it back to life. And I know that you guys (IT engineers) are in favor of it and are ready to take the lead in this, with interoperability and data sharing in mind, while at the same time conforming to Public Health Informatics Institute's best practices.

Dan: Correct. However, I am wondering who is going to fund this project as the health department is struggling and currently lacks the necessary resources.

PM: There is a grant from the Robert Wood Johnson Foundation (RWJF) which is affiliated with the Public Health Informatics Institute who specifically funds the struggling local health department's public health information systems to meet their daily operational needs. I am going to put together the grant proposal and I need inputs from you guys.

Tim: Let me know what you need.

Dan: Do we have full support of the department leadership this time?

PM: Yes, this time we have the absolute support of the entire leadership team. However, we have to make sure that we keep them updated with project status. In fact, learning from past projects, communication is very important for system implementation. We have to make sure that we communicate well with all project stakeholders.

Tim: I agree. But *all* stakeholders? That will be tough.

COLLABORATIVE PROCESS AND COMMUNICATION PLAN

To engage project stakeholders, the team followed a collaborative process (Figure 10.3). By following the process, external and internal stakeholders understood the information system's support role as well as how it worked.

The team also developed a communication plan and used it to avoid communication breakdown and, more importantly, to ensure that appropriate correspondence existed between all stakeholders in the project. An example of the communication plan is shown in Table 10.3.

Figure 10.3 *The Collaborative Methodology Process*

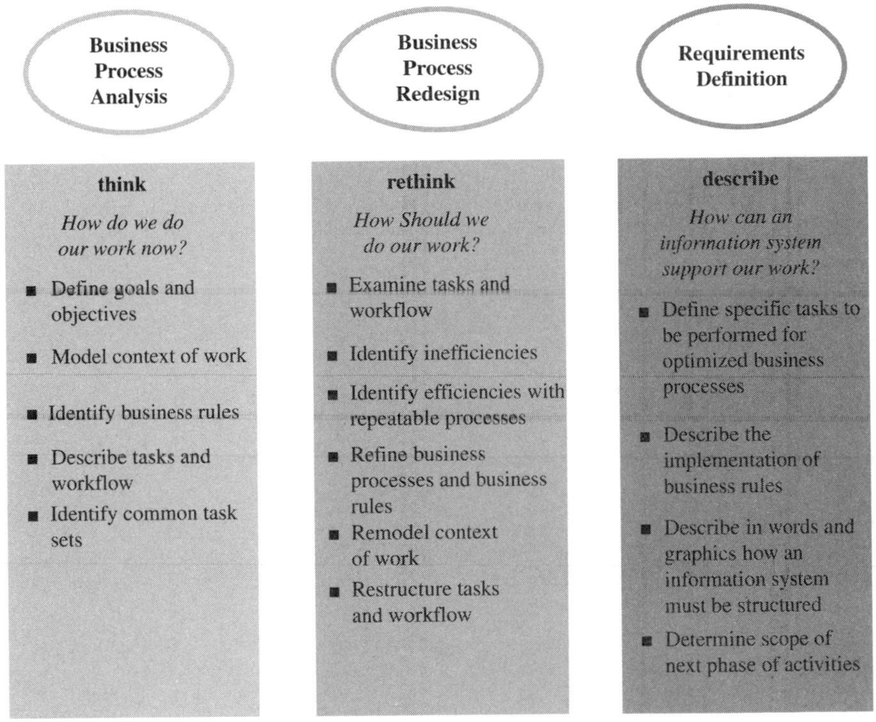

Business Process Analysis	Business Process Redesign	Requirements Definition
think	**rethink**	**describe**
How do we do our work now?	*How Should we do our work?*	*How can an information system support our work?*
■ Define goals and objectives	■ Examine tasks and workflow	■ Define specific tasks to be performed for optimized business processes
■ Model context of work	■ Identify inefficiencies	
■ Identify business rules	■ Identify efficiencies with repeatable processes	■ Describe the implementation of business rules
■ Describe tasks and workflow	■ Refine business processes and business rules	
■ Identify common task sets	■ Remodel context of work	■ Describe in words and graphics how an information system must be structured
	■ Restructure tasks and workflow	■ Determine scope of next phase of activities

Table 10.3 Communication Plan

Project Name: Improving County Public Health Informatics Systems

Prepared by: Project Manager

Date: 12/13/08

Key Stakeholders (Distribution Schedule)	Stakeholder Issues	Key Messages to Communicate	Communication Methods to Be Used (Written, One-on-One, Electronic, Meetings, Etc.)	Description of Specific Communications (Content, Format, Level of Detail, Etc.)	Timing Issues (See Also *Bar Chart, Project Schedule)*	Other
Client	Providing continuous information on the project status	Project status and key milestones and any relevant issues relating to the project	Email, meeting, telephoning, and teleconferencing	Highly detailed, formal communication	Based on project schedule	N/A
Senior Management	Providing continuous information on the project status	Project status and key milestones and any relevant issues relating to the project	Email, meeting, telephoning, and teleconferencing	Highly detailed, formal communication	Based on project schedule	N/A
Sponsor	Providing continuous information on the project status	Project status and key milestones and any relevant issues relating to the project	Email, meeting, telephoning, and teleconferencing	Highly detailed, formal communication	Based on project schedule	N/A
Project Team Members	Providing continuous information on the project status	Project status and key milestones and any relevant issues relating to the project	Email, meeting, telephoning, and teleconferencing	Highly detailed, formal communication	Based on project schedule	N/A

Key Stakeholders (Distribution Schedule)	Stakeholder Issues	Key Messages to Communicate	Communication Methods to Be Used (Written, One-on-One, Electronic, Meetings, Etc.)	Description of Specific Communications (Content, Format, Level of Detail, Etc.)	Timing Issues (See Also *Bar Chart, Project Schedule*)	Other
Employees	Providing continuous information on the project status	Project status and key milestones and any relevant issues relating to the project	Email, meeting, telephoning, and teleconferencing	Highly detailed, formal communication	Based on project schedule	N/A
Subcontractors	No subcontractors of this project	N/A	N/A	N/A	N/A	N/A
Suppliers	Most products available off the shelf	N/A	N/A	N/A	N/A	N/A
Unions	Communicated that the project will cause no layoff	Benefits that the project will bring to employees	Email, meeting, telephoning, and teleconferencing	Highly detailed, formal communication	Based on project schedule	N/A
Government Agencies	Providing information on the project status whenever requested	Benefits that the project will bring to public safety and cost savings	Email, meeting, telephoning, and teleconferencing	Highly detailed, formal communication	Based on project schedule	N/A
News Media	Providing information on the project status whenever requested	Benefits that the project will bring to public safety and cost savings	Email, meeting, telephoning, and teleconferencing	Medium-level detailed, formal communication	Based on project schedule	Press release
Community	Providing information on the project status whenever requested	Benefits that the project will bring to public safety and cost savings	Town Hall Meeting and conferences	Highly detailed, formal communication	Based on project schedule	N/A
Other	N/A	N/A	N/A	N/A	N/A	N/A

Discussion items

1. Is the communication plan adopted in this project realistic enough in terms of communicating to all stakeholders of the project?

2. Publicly run projects are quite different from privately run projects. Can you specify different communications that would be needed in a project involving, say, unions as opposed to a private project?
3. Are the communications plans compatible with *PMBOK's Guide*'s Project Communication Management? Why?

A Simple Metric Goes a Long Way

Art Cabanban

Al has been a bit scattered in his thoughts lately. The company is in the middle of a new product introduction and the production engineers are having difficulties communicating among themselves. The production engineers have a wide range of responsibilities at High-Tech. The team meets daily and seems to have a grip on action items, projects, and support. But Al wants the engineers to work more cohesively. The focus of this case is on project communication, especially the communication across projects.

A LITTLE PIECE OF A BIG PIE

The Production Engineering (PE) department is a part of the entire Engineering Support organization within Global Manufacturing and Operations of High-Tech. Its responsibility is to create processes, define material flow, and assist in continuous process improvement. Although good portions of the PE tasks are related to providing support (e.g., updating incorrect procedures, addressing corrective action, implementing changes to processes, etc.), much of the work can be categorized as projects that support corrective actions or process improvements. PE represents a small population within High-Tech. Doing business globally, High-Tech has over 30,000 employees. The annual sales of its "complex electronic system" are approximately $12 billion.

STARTING OFF WITH A SIMPLE PROJECT DASHBOARD

To make sure that all projects in his organization have executive oversight, Al gathered a team of his engineers to brainstorm about a project dashboard (the PE Dashboard). The team agreed that while the dashboard shows the status of a project, it must not be time-consuming when inputting information. It must be straightforward and easy to use. "The last thing I want to do is fill out a nonsense project report," commented Sanjay, one of the production engineers.

Bob, another production engineer, proposed an idea for the dashboard (see Table 10.4). It is a simple metric that shows the status of the project and important

Table 10.4 Production Engineering Web-based Dashboard

Project	Owner	Status	Issues
Project BIG Expansion - Increased capacity	Al	Y	- Power upgrade 11/1; PO submitted - Cables 11/1; PO submitted; vendor behind schedule
Project BIG Metrics - Test Time Measurement	Tim	G	- Team agreement on goals 9/14 – DONE - Weekly reports 10/30 – DONE - Monthly site reviews to begin 11/07
Blocks Line Material Handling - Decreased assembly time	Joe	G	- Procedures written and released 9/17 – DONE - Area layout 9/18 – DONE - Training and implementation 9/25 – in progress
New Project	?	N/A	- Xxxxx

project issues. Bob suggested that this simple metric can be used to communicate with various managers. He can develop an online feature so that everyone can access it 24/7.

The PE Dashboard was soon put into use. In prior project status meetings, each project team reviewed the status and made necessary updates. With this new process, the team seems to be more engaged during the meeting as they can see the goals, projects, status, and give suggestions and feedback to the activities.

The team continues to use the PE Dashboard. It can sometimes be cumbersome as it requires regular maintenance and effort, but it keeps the team diligent and honest on their status. It also allows management to simultaneously check status on many different projects and actions.

Discussion items

1. What other communication means could be used to improve project communication?
2. Describe advantages and disadvantages of this newly implemented online dashboard. Is it too simple?

Executive Project Metrics

Dragan Z. Milosevic, Peerasit Patanakul, and Sabin Srivannaboon

Mick Dobroff, CEO of Interconnecting Cable, Inc. (ICC), knew roughly what he wanted of vice-presidents of engineering, Ian Plachy, and marketing, Rod Stewart. He saw it in a management conference, where they called it the Project Dashboard. The Dashboard is a tool that highlights and briefly describes the status of the project by reporting on progress toward achievement of the major business goals of the initiative. Project Dashboard is used to predict the future state of the project based on past and current performance. Mick wanted the Dashboard to show the status of all company projects. More precisely, Mick's intent was to have the vice-presidents create such a report for him. In order for Mick to have these reports resemble exactly what he wanted, he needed to issue detailed specifications to Ian and Rod.

OUR CABLE BUSINESS

ICC makes custom-made interconnecting cables for the health, computer, and other industries, with annual sales of $140 million. It is a market leader and business unit of a large conglomerate. Typically, ICC would develop platform cables (big projects) per customer requirement, which could be changed to a minor degree—color, slight changes of material—(called small projects), or to a great degree—color, major changes of materials, length, etc.—(called medium projects).

EXECUTIVE DASHBOARD

Mick decided to use one free Saturday, which was usually a day off, to come to the office to tackle many miscellaneous tasks, among them the Dashboard specification. "What sort of Dashboard do I need?" Mick asked himself. He wrote out a stream of thoughts: "First, I need it to include four to six project metrics that would help me follow the health of the engineering department on a regular, periodic basis, what engineering folks do each month and what kind of performance they put out, as well as how they fare against engineering departments throughout the company. No, I do not want to have only the financial metrics. On the contrary, I want and need as much information as possible—a group of well-balanced financial, process, customer, and learning metrics.

"In addition, the Dashboard must be a one-pager and should take no more than three minutes to interpret it. No words in the Dashboard. . ." Mick kept on writing the specification for the Dashboard. "Only graphical signs, colors—green, yellow, and red—should be used to indicate the progress toward reaching the project goals. Green indicates the project is on track to meeting the goals, yellow indicates a goal is in jeopardy, and red indicates a goal has been compromised and immediate action is needed to recover. The use of colors to indicate project status reminds many users of the automobile dashboard, where color lamps indicate the status of all major car systems, hence the name of this tool. Also shown may be a project summary including status of work completed, significant accomplishments, and risks and issues currently being addressed by the project team."

"Frequency? Once a month, that's it," thought Mick. "Let me not forget that at times the Project Dashboard should predict trend, and state actions required to overcome issues and reverse a negative trend. The Dashboard will be used for multiple projects, actually for all company projects—240 projects per year, i.e., roughly 20 a month. And, yes, I want both types, driving metrics (also called lagging) to show and warn me the intermediate results and trend, and outcome metrics (also called leading metrics) to indicate the final results."

IMPLEMENTATION

To respond to the request of CEO Dobroff, an interdepartmental project team of engineering, finance, and marketing was formed that proposed the following metrics:

- EBIT/FTE Engineer Number of EBIT dollars earned per an engineer; also a measure of an engineer's productivity expressed in dollars; the larger the metric, the better.
- Average project profitability index: A sum of profits from all projects divided by the sum of all projects' cost; also a measure of a project's profitability; the larger the profitability index, the better—the larger a project's profits.
- Customer satisfaction with product quality: Customers' assessment of product quality on a scale of 1 to 5, 1 being the lowest, 5 the highest measure of product quality; measure of perception of customer satisfaction with product quality.
- Customer satisfaction with milestones' accomplishment: Customers' assessment of project milestones' punctuality on a scale of 1 to 5, 1 being the lowest, 5 the highest measure of timeliness; measure of perception of customer satisfaction with accomplishing project milestones timely.

- Percent of milestones accomplished: Percent of all project milestones that are met in a certain time period expressed as percentage of all project milestones planned to be met in the same time period; also, a measure of punctuality of project delivery.
- Percent of milestone budgets met: Percent of all project milestone budgets that are met in a certain time period expressed as percentage of all project milestones planned to be met in the same time period; also, a measure of adherence to project budget.

CEO Mick Dobroff was content with the proposed metrics but wanted to see the reaction of internal and external customers, those who provide metrics data. Mick's intent was to see if the data obtained could be processed in a way that benefited ICC and for engineering to be able to hear the customers' voice. Following this line thought, the project team developed a survey to get data for the above metrics. They sent out the survey and the response obtained produced this information:

- EBIT/FTE better engineer (not available)
- Average project profitability index = 1.8
- Customer satisfaction with product quality = 4.3/5.0
- Customer satisfaction with milestones' accomplishment = 1.9/5.0
- Percent of milestones accomplished = 38%
- Percent of milestone budgets met = 19%

These metrics really surprised Mick. Now, he had something to seriously think about with his engineers.

Discussion items

1. Do individual metrics that the project team propose meet Mick's specification to include four to six project metrics that are well-balanced financial, process, customer, and learning metrics? Which metric is of what kind?
2. Can Mick use the metrics to follow the health of the engineering department?
3. What do you think of ICC's average project profitability index of 1.8? Explain.
4. Should ICC be satisfied with the customer satisfaction with product quality of 4.3/5.0? Explain.
5. What should ICC do about customer satisfaction with the milestones' accomplishment of 1.9/5.0? Elaborate.
6. What should ICC do about the 38 percent of milestones accomplished statistic? Elaborate.
7. What should ICC do about the 19 percent of milestones budgets met metric? Elaborate.

Chapter 11

PROJECT RISK MANAGEMENT

Project risk management typically involves the processes of risk management planning, risk identification, risk analysis, risk response planning, and risk monitoring and control. Project risk management is one of the critical activities impacting the success of a project. This chapter contains four issue-based cases relating to Project Risk Management, Chapter 11 of the *PMBOK® Guide*.

1. Risk Policies in Project Russia

   Risk Policies in Project Russia is a comprehensive case. It brings the readers back in time to the war between France and Russia in 1812. The case details project risk management in that famous war project.

2. Risk under the Microscope

   As an issue-based case, Risk under the Microscope shows how a project team practices project risk management. The case also illustrates how communication plays an important role in successful risk management.

3. Monte Carlo in Italy

   Monte Carlo in Italy is an issue-base cased. It portrays a risk management practice of a company. The case discusses the use of Monte Carlo Analysis, a quantitative risk analysis tool.

4. Probability and Impact

   Probability and Impact is an issue-based case. It presents the use of probability and impact as a risk analysis procedure. The case also discusses the development of appropriate risk thresholds for the nature of risk events.

CHAPTER SUMMARY

Name of Case	Area Supported by Case	Case Type	Author of Case
Risk Policies in Project Russia	Risk Management in a War Project	Issue-based Case	Dragan Z. Milosevic
Risk under the Microscope	Risk Management Process	Issue-based Case	Ferra Weyahuni
Monte Carlo in Italy	Risk Management Process	Issue-based Case	Meghana Rao
Probability and Impact	Qualitative Risk Analysis	Issue-based Case	Jovana Riddle

Risk Policies in Project Russia

Dragan Z. Milosevic

This case study reviews some of the major risk management tactics used in a typical war project—Napoleon's war with Russia in 1812. The war outcome had a stunning end and caused turbulent ramifications for the European map. A lot of ink was poured to explain the destruction of Napoleon's forces, known as the Grand Army, and experts only agreed on the fact that the Russian winter had a major impact on the war outcome. In the study, we take a risk view of the war conflict.

MISERY AND DEATH WAITED THE GRAND ARMY

For Napoleon, many dilemmas stayed unresolved even after entering Russia. He looked amazed by the glory awaiting conquerors of Russia but at the same time he was painfully aware that he might make the same error Charles XII, Swedish military genius, committed one century earlier—attacking the Russians in the winter. Listen to what Count de Segur, who was with him in Russia, has to say about that.

"The last days of July and the first ones of August in 1812 were stiflingly hot in Vitebsk. In the old city palace Napoleon Bonaparte, Emperor of the French, prowled restlessly from room to room in his undergarments. His mind, brilliant author of 12 years of triumphs and 20 famous victories, was torn between prudent counsel to encamp now against the coming winter and bold counsel to march straight on to Moscow. So he paced . . . in this state of perplexity he spoke a few disconnected words . . . 'Well, what are we going to do?' 'Shall we stay here?' 'Shall we advance?' 'How can we stop on the road to glory?'

For 15 torrid days, Napoleon groaned under the weight of his thoughts. By night he tossed his coat, arising frequently from his biography of Charles XII—he would never . . . He shouted, 'I'll never repeat the folly of Charles.'"

Instead of the glory, misery and death were awaiting the Grand Army by the time it exited the "sacred soil of Russia", as the Russians had the habit of saying in Kovno, December 13, 1812. Segur was a witness: "Instead of the four hundred thousand soldiers who fought so many successful battles with them, who had rushed so valiantly into Russia, they saw issuing from the white, ice-bound desert only one thousand soldiers and troopers still armed and twenty thousand being clothed in rags, with bowed head, dull eyes, ashy, cadaverous faces, and long ice-stiffened beards . . . And, this was the Grand Army!" So, 380,000 soldiers of the Grand Army perished. How was this possible? Do the risk planning practices of Napoleon have anything to do with this? Let's review major events and practices in turn.

BACKGROUND

The underlying cause of almost each war tends to be of an economic nature. This war, in Europe from the beginning of the 19th century, is no exception. The three strategic players to the conflict are the great powers of England, France, and Russia. France and England are major rivals and contenders for the central place in Europe. One more power is involved—Russia. The main hero of this conflict who was heavily favored to win this war was France. How was the strategic triangle formed?

The French wanted to execute their economic blockade of England, reduce their goods from European markets, and thus stifle the economy of the country. The French threatened to attack any country that would violate the blockade. Candidates for such a violation risked war with Napoleon's Grand Army, the most famous army of its time.

To beat Russia was a big feat for each army; a trophy for every general. Napoleon had already defeated all armies he could dream of. But nobody had beat Russia! To be truthful, many had famous warriors had tried but the Russian winter proved to be an invincible opponent and a major ally of Russia. Some lived to send the message about what they learned to potential invaders. Swedish King Charles XII, known as Charles the Madman, the warrior with the pedigree, in 1709 attacked Russia in the winter, lost the whole army, and proclaimed in writing: Don't attack Russia in the winter! But future invaders did not listen.

Napoleon saw Russia as a great trophy. Yes, alleged Russian violation of the blockade was used for the proxy cause of his attack, but this had not been proven. Some rumors circulated that said that Napoleon, the second most successful general of all time (Alexander the Great was considered the first), dreamed of overtaking Alexander the Great, and becoming number 1. Possible, but such were ambitions of Napoleon, who was often called the Anti-Christ and the tormentor of Europe. In truth, the Grand Army was made up of all European nations. There were, for example, 79,000 Bavarian, Italian, and French soldiers and 34,000 Austrian soldiers. The Grand Army was a microcosm of European armies. But they did not volunteer in the army; they had to serve because their country was subdued by France. In case of Napoleon's serious defeat, his army could face the rebellion of the foreign soldiers which meant that war carried the potential for freeing Europe from the domination of Napoleon.

Lastly, winds of bourgeois revolution were felt throughout Europe. Napoleon's expectations were that Russian farmers would accept their own revolution and that they would take his side. He was incorrect, although he hoped to export the revolution and expand its ideas.

IT STARTED LONG BEFORE THE WAR BEGAN

Paris balls and parties as social gatherings were very much appreciated and among the best of their kind in Europe. The winter balls were especially good. People who typically frequented the balls were nobility and the politically elite. If you had gone to a classy Paris ball you would have had a chance to see some very well-known and powerful people. There was a hierarchy of balls, that depended on the power of or historic strength (tradition) of the host(ess) or the list of invitees. The higher those were, the higher significance of the ball. If Napoleon was in attendance, the ball would be of the first level of hierarchy. Such are the reasons that balls and parties of a higher rank were used to host the people with high intelligence knowledge, and so were automatically suitable for gathering intelligence data.

People of Russian nobility and the politically elite used to be frequent guests of these balls and parties. And they fit very well. Be reminded that that the higher class in Russia had spoken French as a first language. So, Russians felt at home in Paris. As people with strong social capital they liked to mingle with the French crowd of high social standing, and were constantly invited to the balls.

Napoleon liked to frequent the balls. He had a specific question for the high military visitors from Russia who were there: "What would you do if you commandeered the Russian Army and I attacked you (with the Grand Army)?" Different visitors gave him different answers, often made up. But some replied extremely truthfully like general Beniggsen, who happened to be the commanding officer of the Russian army at the time of Napoleon's attack. He was asked by Napoleon and his answer was exactly what he would do a few years later: "I would never fight your army back, it is too strong; I would retreat and retreat, waiting for the winter to finish you." It is not clear how Napoleon processed this information nor whether he believed it. But it is a fact that Napoleon had a sparrow in his hand, whether consciously knowing it or not.

LESSONS OF THE PAST

During the times of Charles XII, one of the greatest secrets was the size of the population of Sweden. Why? The King did not want anyone outside the country to know just how many people lived in his country who were available to fight in a war. That number would actually show a small country that does not have the population that supports its war policies. Despite a ferocious reputation, Swedish soldiers, whose ancestors the Vikings also enjoyed standing of exceptional fighters, would be defeated more frequently should their enemies know how many Swedish opponents they faced in the long term.

In the beginning of the 18th century, Charles found himself in the long war with Peter the Great, the Russian tsar. Charles dominated the Baltics, where Peter wanted to build the Russian fleet which would become the influential force. After several smaller victories of Charles' army, the major battle took place in 1709 near the river Poltava. Charles' army was not only defeated but it was destroyed. The ruler Charles managed to flee southward to Russian arch nemesis Turkey. He published a book (actually his aid authored the book) whose major message to future invaders of Russia was very simple: "Don't attack Russia in the winter . . ."

Napoleon not only read the book carefully but liked to be seen reading it over and over. He tried to serve as an example to his generals and made sure they read the book and understood the experiences of Charles. In fact, the book had immeasurable value for the Grand Army. In terms of risk planning processes, Napoleon used the book in a proper way, at least initially.

RUSSIAN WINTER

In a simplified manner, first, judging by the campaigns before Russia, e.g., Egypt, as usual, Napoleon did not have but the slightest of sketches about how to direct the war against Russia. Troops were told that Napoleon was burning from desire to have the decisive battle against Russians as soon as possible, defeat them, and make them surrender. Over! And all of that was to happen before the infamous Russian winter came. The Russian strategy was diametrically opposed. Exactly like general Beniggsen—German by origin at the time of the vision—the commander of the Russian army predicted they would retreat, retreat, and retreat (surprisingly Segur observed that "there was more order in their victory than in our victory.") and told his troops to avoid a decisive battle as much as possible, and wait for the Russian winter to come and help finish off French forces. Well, two strategies look very mutually exclusive, if one happens, that one excludes the other. Let's see how the two strategies unfolded in several risk events.

On June 20, 1812, unknown to the Russians, the multinational Grand Army entered Russian territory. To their surprise they were able to set foot on Russian soil without meeting with any resistance. They found peace there: they had left war on their side. However, a single Russian officer commanding a night patrol soon appeared. He asked the intruders who they were. 'Frenchmen,' they told him. 'What do you want?' he questioned further. 'And why have you come to Russia?' One of the sappers answered bluntly, 'To make war on you!' While a stealthy entry was favorable to the Grand Army from the aspect of having the opposing army surrender, it was not so. Namely, for an army to surrender, it has to be formed, which the Russian Army was not.

The battle of Borodino lasted one day and was the only battle in the war of greater interest, but was not the decisive battle. That occurred on September 7, 1812, 79 days after entry of the Grand Army into Russia. The battle had an enormous number of casualties—43 generals of the Grand Army were wounded or killed; 20,000 killed or wounded troops—but failed to produce a clear winner, although the Russians went back to their retreating strategy and disappeared for a time. In strategy terms, Napoleon's officers believed that their army made a big mistake, not keeping in contact, chasing the opponent and trying to destroy them. Instead, they regrouped allowing the Russians to take off. So Napoleon had a chance to finish the war early enough to avoid the trouble of winter. Appatently Napoleon had no great desire to accelerate his army and force a decisive battle. So the Russians continued to buy time and kept waiting for winter to do its job, increasing Napoleon's war risk.

Napoleon entered a burned and deserted Moscow on September 14, 1812. The Russians destroyed the city in order to prevent the Grand Army from using Moscow supplies. At this season of the year, Russia is fully aware of his advantage. From there they continued to negotiate the Russian surrender who did not intend to surrender but again, buy time. As Segur says, "Thus far Napoleon had conquered only space." The retreating Russian armies were in front of him and Moscow was but 20 days away.

In a situation, when every date meant a lot for survival of the Grand Army, the Russians outsmarted Napoleon and opened the possibility of winning the war. Again, Napoleon did not show a willingness to change strategy and catch the Russians, thus reducing their risk. Amazing was the French lack of attention to details and no contingency plann.

RISK TREATMENT

It is interesting to observe how the best of the best, for instance, the Grand Army, follow the normal risk policy which, in this case, would be among one of the widely accepted policies such as *PMBOK*'s. It has six processes: *risk management planning, risk identification, qualitative risk analysis, quantitative risk analysis, risk response planning,* and *risk monitoring and control.*

Grand parties and balls, as a place for top intelligence, probably covered the processes of risk management planning and risk identification, more so than, say, qualitative risk analysis. There were no indications that risk events like a retreating strategy and the Russian winter were subjected to risk analysis, risk response planning, or risk monitoring and control. No details in Segur's book hint to a mitigation or adaptation of the military strategy to account for Russian continuous retreating as a valid military strategy. Hence, at best, the French heard about Russian intentions in terms of a Russian approach, but did not take the action to prevent that or learn to move the soldiers faster.

As for the writings of Charles XII, they had significant influence among the top officers of the Grand Army. How significant? He did some sort of risk management planning and identified risks such as those offered in the book and he assessed risks qualitatively by describing the heavy impact of the season. However, we don't see that Napoleon did any risk analysis, let alone response planning or taking any risk monitoring and control steps. Napoleon didn't make any adaptation in his military strategy to not be facing the Russians during winter. Nor did he quit after studying the message of Charles to not attack Russians during winter. If we take Napoleon's approach as insufficient, we conclude that he didn't really listen to Charles' advice. Probably, he saw the quality of his Grand Army as incomparably higher than the one of the Swedish. The fact is, then, that the analysts considered the French advantage to be a better army, but the Russians had familiarity of terrain and climate. Maybe Napoleon was right, maybe not, in leading his solders to death.

Speaking of the Borodino, the battle there had enormous importance. Technically viewed, it is not known whether any steps in a risk analysis were even considered. This means that risk management planning, risk identification, qualitative risk analysis, quantitative risk analysis, risk response planning, and risk monitoring and control were not considered relevant. But wait a minute, Napoleon's decision at Borodino to allow the Russians to run away made some of his generals angry and some spoke of treason. Maybe French nationalism played a role, or maybe Napoleon thought there had to be one more battle to settle the account, but the mistake to let the Russians go and not make it the central piece of their risk strategy were blunders. The French didn't have the luxury of seeing such a chance again.

The importance of an empty and burned Moscow, aside from public relations, had one more cause of importance. This was a time for diplomatic moves, to have a Russian surrender. Napoleon thought that the Russians did not want to surrender but only pretended to. He believed, and some French generals as well, that at this point Russians had the advantage. "Napoleon entered Moscow with only 90,000 troops." Russians played this negotiation game, just for one reason—to buy time and prolong the French stay on Russian soil until the winter would finish them off. In such conditions, *PMBOK*'s six risk policies did not have their usual significance. More accurately, the French ship had already sunk enough by then, and it was time for the Russians to secure the win.

Discussion items

1. Identify major risk events, perform risk analyses, and develop risk response plans.
2. In your opinion, how did Napoleon control each of the major risk events on your list?
3. Do you think the way in which Napoleon controlled risks related to major risks influenced the war outcome?

Risk under the Microscope

Ferra Wayahuni

The Field Service Engineer (FSE) at RedGate Technology had called an escalation Level 4 meeting to resolve a computer failure issue during the project—tool upgrade for imaging and wafer transfer improvement. The company's escalation process requires that the FSE escalate the problem, if it has not been resolved after six hours. The escalation is now closed; however, the tool was down for three days, which means daily schedule of the upgrade may need to be revisited. The tool at the RedGate Technology site is the first tool to be upgraded so the team wants to capture as much learning experience as possible to make the next upgrades run more smoothly. This is a case about the unknown project problem that suddenly occurred and how risk planning prepared the team for mitigation of damage, even though a root cause was not known.

THE MEETING

The project update meeting with the core team is underway:

> Product Engineer: Adam McAllister
> Technical Support Engineer: Donna Nolan
> Systems Engineer: Calvin James
> Project Manager: Jason Orange
> Program Manager: Julia Gallagher

Jason: Hi everyone. Thanks for coming. Julia suggested we meet and review our upgrade schedule. We have another tool to be upgraded in two months so we should make quick changes on any design or work instructions that need to be improved. Donna sent out the escalation report describing what happened and the root cause of the issue.

Donna: Yes, I sent it to the team as well as the Field Service Engineers (FSE) at the site. Calvin and Adam are working on retesting the computer and identifying the root cause, and the fix, so that we know what to do if it happens again. I hope it won't happen again, however.

Jason: Good. It sounds like everything is under control now in terms of troubleshooting the tool.

Calvin: Well, actually, I haven't had any luck on reproducing the error that they saw at the site. I talked to Nick Filan, the lead FSE, about what happened. He told me that it was the "blue screen" phenomena—the computer worked fine and then the screen was blue all of the sudden. They were not able to recover the error at all.

Jason: Have you seen this happening before?

Calvin: I don't have much experience on this tool so I don't have any historical issues that I've seen with my own eyes. I couldn't reproduce what they said was happening prior to the "blue screen" phenomena, but I couldn't get the blue screen on my simulator.

Jason: Yes, I realized Steve Huggins did not give you any briefing before he left the company. It would be nice if he had documented everything that he'd seen when designing this system.

Adam: I haven't heard of it happening before. Calvin and I can ask other engineers to get their input on this issue. The bad thing is that this computer is a new design so it has not been implemented on other product lines. I don't know what kind of information I can get from asking around other engineers.

Jason: Well, it never hurts to try. Let me know if you're running into more issues. Now we can update the schedule to accommodate the time lost during troubleshooting.

BACKGROUND

The IEM Company is a high-tech company producing customized Ion and Electron Microscopes. The applications of their products can be used in a variety of fields, from academia to high-tech industries. Their customers are given the options of customizing the product to meet specific process needs. The company's financial profile shows that their sales revenue for last year exceeds $400 million. The company is currently upgrading tools in the field for improvement in the imaging and wafer transfer system. This is required to grow the market size and to meet customers' satisfaction.

RISK PLAN

Julia: Jason, can you give a brief update on how we are on the schedule?

Jason: Sure. From the Gantt chart that I sent you yesterday, we are currently three days behind schedule. Rob Carter, the process engineer at RedGate, told me that the tool handover to production cannot be delayed due to production backlog. We may need to add a second shift for the upgrade to mitigate the scheduling issue.

Julia: What is the original upgrade timeline, Jason?

Jason: The tool is promised to be ready for production within two weeks.

Julia: Before we move forward on deciding what to do next, can we review your project scope and risk planning matrix? I've only seen your scheduling chart but did not have a chance to review the whole package when I approved this project.

Jason: Yes, I was aware that Marketing and Sales already promised the date to the customer before I finished creating this Gantt chart. I did not know the change on the timeline and date until I asked Markus if there were any changes on the project. If only our communication could be improved . . .

Julia: Well, that was in the past, now we have to create a mitigation plan for it. Have we figured out what the root cause of the problem is?

Jason: No, our original Systems Engineer for this project, Steve Huggins, left the company two weeks ago. He only gave a two-week notice and then was out of the office the last week before he left to use up his vacation days. Calvin James replaced him for this project; however, he is quite new to this product line so he is still learning on the go. He has not figured out the root cause of the problem yet.

Julia: Interesting. Can you describe what you included in the risk plan?

Jason: Yes, the core team brainstormed what should be considered risks for this project. We were focusing more on the design and supply chain, however. The implementation plan was assumed to be handled by the Technical Support Group (TSG). We grouped the risk plan per main category, for example, in Design, we split up the risk plans to Hardware and Software groups. We did it this way so that we can manage it much easier since the activity list from the WBS is quite big. We used discrete estimation for probability and risk impact. The qualitative data was per our best estimate learning from past projects. Steve Huggins had been involved in two major tool improvement projects in the past so he was very knowledgeable in this area. Here's the risk plan for the Software activity list (see Table 11.1).

Julia: So you excluded most activities in the WBS from the risk plan?

Jason: That's right. We thought we should only list activities with medium and high risk. What's the point of recording low-risk activities if they're not going to affect the project by much? Anyway, you can see that we did include computer testing as part of our WBS.

Julia: Yes, however, did you guys test for long-term reliability? I know we don't know why we had blue screen problem in the field but you might have caught it, if you tested the part for a long period of time.

Jason: No, we didn't do that. We only tested the first article for making sure most features worked because we didn't have time for a long-term reliability test.

Julia: Hmm . . . that's a little odd. Did you find out the plan developed by TSG?

Jason: No. When I submitted the project charter, our scope was only on the design, development, and testing.

Donna: The TSG are responsible for performing the task of upgrading the tool, however, since you are the project manager, you are the one responsible for connecting the whole areas, including field implementation. I was not aware that you assumed we were going to do the whole field implementation plan, while you were working on the upgrade schedule.

Julia: I can see there's a disconnect among core team members . . .

Table 11.1 Software Risk Plan

					Actions			Owner
Task	Risk Description	Prob-ability*	Risk Impact**	Risk Score P +(2×I)	Preventive	Trigger Points	Contingent	
Review marketing require-ment document	Marketing requirement does not include all customer's requests	2	4	10	Review marketing requirement document with the customer's process engineer before design phase kicks off	No customer's process engineer available for review meeting by June 30th	Escalate to the manager for their POC avail-ability	Project manager
Identify required changes on computer config-uration	Not enough knowledge on new process to be implemented	3	4	11	Review other tools' application to see if we can use their concept	Not enough data to determine what application can be used as an example by June 30th	Contact customer's expert to see if there's any current applications similar to the new process	Software engineer
Determine time and cost budget for new computer	Supplier can't meet our short timeline request to implement changes	4	5	14	Send specifi-cations prior to design reviews	Supplier's timeline is 2 weeks longer than expected	Send specifi-cations to a few suppliers and go with the fastest delivery timeline.	Sourcing rep

Task	Risk Description	Prob- ability*	Risk Impact**	Risk Score P +(2×I)	Actions			Owner
					Preventive	Trigger Points	Contingent	
Create success measures for computer tests	Success measures do not reflect all actual use cases	4	5	14	Review success measures with other software engineers for fresh-eye review	No feedback from other software engineers by July 10th	Escalate to project manager for help in gathering resources	Software engineer

*Probability (Discrete Estimation): 1 = Very Unlikely, 2 = Low Likelihood, 3 = Likely, 4 = Highly Likely, 5 = Near Certain

**Risk Impact (Discrete Estimation): 1 = Very Low Impact, 2 = Low Impact, 3 = Medium Impact, 4 = High Impact, 5 = Very High Impact

Discussion items

1. To some, these risk plans are wrong. What do you think might be wrong with the risk plan? Should Julia agree with the risk plan?
2. How can the project risk planning be improved in each of the following areas: project organization, implementation, strategy, leadership?
3. What would be the next step for the team to recover?

Monte Carlo in Italy

Meghana Rao

ABC is a high-tech company based in the United States. Recently the company had a major shuffle in its management and the new management is planning on expanding into global markets. The CEO realizes that the best way to achieve this goal is by acquiring and merging with smaller high-tech companies in key locations across the globe that provide products and services to the high-tech market. The CEO knows that process integration is an important issue of a successful merger and acquisition. This case discusses such an activity, especially the project risk management process.

GO STUDY THEM!

As a first step toward this mission, the top management calls for a meeting of all the Business Unit heads and the Program Managers. The agenda of the meeting is to identify feasible locations and companies for merger. At the end of the meeting, the management identified three countries in Europe and three in Asia as prospective locations for expansion, based on business opportunity, government policies, availability of skilled personnel, etc. The companies that were selected were all smaller companies with a good presence in their respective countries, offering high-tech products and services with a product portfolio matching ABC's range of products.

Additionally, ABC, being a CMM-certified company, has a strong emphasis on the processes which are followed toward being a better project-oriented organization. The management of the company understands that any merger or acquisition would result in aligning the company's policies and processes with the acquired company's. Therefore, it is considered important that the acquired company has an established set of processes for each phase of the product/project life cycle.

SEND OUR STARS

Since this is a critical step, the management decides that a selected group of business managers, who have a lot of project management experience and have a good understanding of the technology domain, would personally visit the companies that are short-listed and scrutinize their products and processes before finalizing any deal.

Peter Davis is one of the business managers working for ABC. He has been with the company for a long time and has moved up the management ladder, most recently from the project manager level. He understands the technology management aspects at ABC very well and has also worked for some of the top high-tech companies before joining ABC. He understands the dynamics of high-tech companies and also project life cycle management and is a renowned PMP-certified professional in the field. The management of ABC therefore chooses Peter as one of the managers to work with the acquisitions. Because of his prior work experience in Italy, ABC decides that Peter should go to Italy to verify the processes at PQR Inc.

Soon after his arrival in Italy, Peter schedules meetings with the business managers and the project managers at PQR Inc. He is really surprised at the amount of detail that has been given to every aspect of the project life cycle. He scans through the company's project selection process, project portfolio mapping process, project planning and control tools for the customer roadmap, scope (WBS, Change Coordination Matrix, Project Change Request/Log), schedule (Gantt Charts, Critical Path Method, Critical Chain Schedule, Milestone Prediction Chart, Slip Chart), cost (Analogous Estimate, Parametric Estimate, Earned Value Analysis), and quality planning (Affinity Diagrams, Quality Improvement Maps, Cause and Effect Diagrams, and control charts).

Of all the tools that Peter saw at PQR Inc., one tool caught his attention. This was the Monte Carlo Analysis (MCA) tool for risk planning. PQR Inc., being a small start-up company in a high-tech market, obviously faced a high degree of risk in terms of the new technology they were dealing with, competitors, markets, etc. So utilization of any risk planning tool would provide the company with a strategy to ward off any undesired events during the execution of projects. Given below is an implementation of how Monte Carlo Analysis was implemented at PQR Inc.

RISK MANAGEMENT AND THE PLANNING PROCESS

The risk management process at PQR Inc. has been iteratively defined by the following five steps:

- Risk Identification: Use past project experiences and data bases to uncover any risks that might occur during the execution of the current project. These are termed risk events.
- Risk Analysis: Identify drivers that might lead to the occurrence of risks identified above.
- Risk Priority/Impact Analysis: Each risk is given a severity score by assigning a probability of occurrence and impact of the risk. The risk severity is then calculated using a P-I matrix and a specific number of risks that score high on the P-I matrix are considered. For a detailed procedure of assigning probabilities to the risks, the Monte Carlo Analysis technique is used.

- Risk Resolution: Develop a Risk Response Plan to prevent identified risks. Contingency plans should be made in case of risks that cannot be prevented.
- Risk Monitoring: A constant monitoring of risks is done at regular intervals to prevent/mitigate them.

Using the above model, PQR Inc. has then implemented Monte Carlo Analysis for assigning probability to uncertain events such as schedule and cost probabilities. Below is an example of how MCA has been effectively applied to avoid the risk of incorrect scheduling on the project leading to late time-to-market, customer dissatisfaction, and loss in profitability.

A new product development (NPD) project at PQR Inc. has a network diagram showing the dependent tasks/activities in the project.

But seldom is the process of new product development so linear. Since time-to-market is critical for PQR Inc., most of the tasks are handled in parallel between Engineering and the Marketing departments (concurrent engineering), which induces a lot of interdependencies.

The uncertainty in assigning a final deadline to the project is the nature of the project itself. Since the company deals with high-technology products which are very new in the market, not many projects of a similar nature have been undertaken to estimate the schedule of activities with certainty. But accurate assessment of timelines for the individual tasks under uncertainty is both essential as well as critical. So then, how will the company provide a final estimate, so that the risks identified are prevented?

The next step followed in the process is assigning a range of possibilities for each activity. For example, say activity A can be completed in five days at a minimum, but can extend to 15 days at a maximum. An MCA simulation can be run on this range of possibilities to evaluate the mean likelihood of A's completion time. Single point estimates where available can be made use of without using MCA.

The Monte Carlo process thus gives project managers a more precise completion time of the project. In the case of companies that focus on new product development like PQR Inc., this gives a more accurate estimate for the time-to-market.

Peter, who had a PMP certification and had used a lot of project management tools in his long tenure as a project manager, was really impressed with such a great method for reducing risks due to schedule slippages. His audition of the company had given him a great impression of how well PQR Inc. had been managing its projects and their processes. In his audit results, he therefore gladly recommended that ABC acquire PQR Inc.

Discussion items

1. Discuss the advantages and disadvantages of Monte Carlo Analysis in project risk management.
2. Should Monte Carlo Analysis be used in every project? Why or why not?

Probability and Impact

Jovana Riddle

Salvatore Adamo was the head of Vintel Corp's Risk Management Department. His responsibility was to manage risks for many of the company's products. However, one particular product had major risk implications and mattered to Salvatore the most. It was Vintel's flagship product and had a very important risk event at stake. The risk at stake had a very low probability of occurring but had a huge potential impact on the company and its future. This was the first time in the company's history that a product with such a low probability risk event could lead to such enormous potential impact. This case presents how Salvatore adjusted Vintel's risk analysis procedure to address this type of risk.

RISK ANALYSIS STEPS

In order to effectively manage risk for Vintel's product, Salvatore typically followed the standard procedure. He started off with the identification of all potential risks. He then identified the probability and impact of each risk, using a 1-to-5–point scale. For each risk, the risk score was then calculated based on a set formula (Probability + 2 × Impact). He then used a Risk Rating table with predetermined thresholds to categorize risks into high (red), medium (yellow), and low (green) severity risks. Typically, the threshold for high risk is 12 and above. Medium risks have a score ranging from 8 to 11. Risks that score below 7 are considered low risks. Usually high risks will get more management attention and Vintel will perform further investigation and come up with appropriate response plans. Medium and low risks will get less management oversight. The use of this procedure would ensure that risks are analyzed in a systematic manner based on their probability and impact.

ADJUSTING THE THRESHOLDS

Salvatore knew from past experience that in general a very low probability risk event does not have as high a potential impact or a severe outcome. In such a typical situation, using the Risk Rating table in Figure 11.1 should work fine. But for this flagship product, a lot of very low probability risks have a very high level of impact, and using his normal Risk Rating table may not be appropriate. "Am I right?" Salvatore thought.

245

Figure 11.1 Example of Rating a Risk Impact on Schedule on a Five-Level Scale

Scale	1	2	3	4	5
	Very Low	**Low**	**Medium**	**High**	**Very High**
Risk Impact on Schedule	Slight schedule delay	Overall project delay <5%	Overall project delay 5–14%	Overall project delay 15–25%	Overall project delay > 25%

Probability	Risk Score = P + 2 * I					Key:	
NC = 5	7	9	11	13	15	▇	High Severity
HL = 4	6	8	10	12	14	▨	Medium Severity
L = 3	5	7	9	11	13	▢	Low Severity
LL = 2	4	6	8	10	12		
VU = 1	3	5	7	9	11		
	VL = 1	L = 2	M = 3	H = 4	VH = 5		
	IMPACT						

Probability Key
VU = Very Unlikely
LL = Low Likelihood
L = Likely
HL = Highly Likely
NC = Near Certain

Impact Key
VL = Very Low Impact
L = Low Impact
M = Medium Impact
H = High Impact
VH = Very High Impact

"I think I am," Salvatore talked to himself. He knew that he had to adjust his risk thresholds, such that those risks are in the category of high risk. Salvatore had a flashback to his experience with Ford Australia.

Several years ago when Salvatore visited Ford Australia for technical exchange, he participated in risk analysis training. One of the risk events they discussed was the SUV tire explosion. Typically, tire explosion is *a very low probability risk event*. But when it happens, it can cause the vehicles to rollover, which can lead to the severe injury or death of the passengers in the SUV. Besides the safety of the passengers at stake, tire explosion should impact the sales of the SUV by as much as $5 billion. This was definitely a high risk.

This flashback assured Salvatore that the risk thresholds needed to be adjusted. If not for this flagship product, a risk event like "tire explosion" would always end up being a medium-level risk and would never get as much management attention as it should. No management attention means no further analysis and no support. Then, if the risk happens, he could not imagine how huge the impact would be. "Probably not $5 billion but it would definitely be ugly and no more future for me at Vintel," thought Salvatore.

Discussion items

1. Would adjusting the threshold be the right move for Salvatore? If it is, what should the new threshold be to account for the very low probability and very high impact risks?
2. Would adjusting the thresholds be the only solution? What else should Salvatore introduce to Vintel such that risks get proper management attention?

Chapter 12

PROJECT PROCUREMENT MANAGEMENT

Project procurement management involves the activities related to the purchase or acquisition of products, services, or results needed for the project from outside vendors. Such activities could cause several problems that can impact the performance of a project. This chapter contains two critical incident cases relating to project procurement management, Chapter 12 of the *PMBOK® Guide*.

1. The $30,000 Frigidaire

   The $30,000 Frigidaire is a critical incident case. It illustrates an issue related to contract specification of a turn-key contract. The turn-key contract is not common in the United States, but it is used in other countries, especially for technology transfer.
2. Mountain of Iron, Mountain of Dollars

   Mountain of Iron, Mountain of Dollars is a critical incident case. It discusses the issue related to contract specification. This case portrays an incident that the owner and the contractors work off different sets of specification. Lack of communication also leads to further misunderstanding.

CHAPTER SUMMARY

Name of Case	Area Supported by Case	Case Type	Author of Case
The $30,000 Frigidaire	Types of Contract	Critical Incident	Dragan Z. Milosevic
Mountain of Iron, Mountain of Dollars	Contract Specification	Critical Incident	Dragan Z. Milosevic

The $30,000 Frigidaire

Dragan Z. Milosevic

Neither the contractor nor the owner had any idea that such a small issue could snowball into an incident of such proportions. At first, it did not look as though such a small issue could grow into a situation of uncontrolled words and behaviors, and almost bring the whole project to a halt.

And, such a small issue was called in the contract specification, "special cooling device." Its cost of $30,000 in a half-a-billion-dollar international project was almost nothing, but as the experienced know, the poison comes in small bottles. According to the contract procedure, every piece of the hard special cooling device ware that the contractor delivered for a project had to be physically accepted and taken in property by the owner's committee of experts. This process is called the physical inspection and acceptance.

This is a critical incident designed with the specific purposes of introducing a type of contract and the contract situation that seems to have both sides to react in a way as their contractual rights are breached. More specifically, the contract type is the turn-key contract that is not much used in the United States, but is appropriate for U.S. companies transferring technology to less developed countries.

THE TURN-KEY CONTRACT

The contractor is a $1 billion, European company that makes a living building industrial plants, like the one in this critical incident, mostly using turn-key contacts. The owner is a state-owned manufacturer from the Middle East.

To oblige with the contract, the contractor's engineers submitted what they called a "special cooling device." When the owner's engineers saw it, it is fair to say that they looked astonished. Also, there was rather unusual behavior. It more resembled some secret ritual than the work of an engineering committee. This was what the owner's engineers did: They spent about 10 minutes going around it, looking at it from different angles with intense curiosity, opening the door of it, putting their heads inside, making different facial grimaces, whispering to each other, all the time not asking any questions. Finally, they said they would be back in 20 minutes, without mentioning what the problem was, for it was obvious that

249

the whole situation smelled of a problem. After the 20 minutes they came back with a verdict in writing—the owner's committee was presented with an offer of an ordinary Frigidaire, asserting that it was a special cooling device, and the contract sum was to be reduced by $29,600, or to be determined on the level of the price of an average Frigidaire.

The heated debate began. Everyone on both sides formed an opinion, and shared it publicly and unselfishly. It appeared as though no one had control of the situation; at least neither management side did. In uncontrolled discourse, words like "the contractor is dishonest," and "the contractor is ripping us off" on the owner's side and "the owner does not understand the contract" on the contractor's side began to circulate. The conflict bubble was close to bursting.

This was a "turnkey" contract that was never really understood by anyone on the site. Basically, the essence of the contract is "function of entirety." That means that the contractor guarantees that the plant which is the subject of the contract shall function. It also means if the contractor "forgot" to include in the contract specification an item which is necessary for the plant's function of entirety, the contractor shall deliver the missing item free of charge. Moreover, the contractor is to "use and choose" subcontractors in a coordinated way in order to realize the function. For both of the essentials—entirety of function and coordination of subs—the contractor is entitled to a fee between 20 and 30 percent of the contract sum.

Regarding the dispute about the special cooling device, the owner's engineers thought that the contractor's delivery of a Frigidaire for $30,000 was a huge rip-off, and so reducing the contract specification item to the Frigidaire price would be the only fair outcome. They did not recognize that the contract was the turnkey. The contactor's opinion, of course, is based on their understanding of the meaning of the contract type. They based it on a recent example of delivering, free of charge, an overhead crane, which was not part of the contract specification. Namely, the contractor established the need for the crane in the process, and delivered it to secure the function of entirety. The contractor presumed that the item figure of $30,000 for the special cooling device was proper. The device was not a typical Frigidaire one could buy anywhere, it was a special industrial Frigidaire bought in Germany from a manufacturer of specialty cooling equipment. For that reason, the contractor was against the reduction of the item contract price. As there was no change in the total contract sum when the contractor delivered any item free of charge to secure the entirety function, there should be no change of the total contract sum in the case of typos or similar instances.

As the contractor held his ground, the owner contacted his headquarters, specifically, the corporate contract department, with the question of how to proceed in this case. To the owner's surprise he was told to leave the issue to the contract department, which is responsible for cases like this.

Discussion items

1. What are major traits of turn-key type contracts?
2. What are major traits of unit type contracts?
3. Contrast turn-key type contracts with those of unit type contracts.
4. Is the owner right in demanding the price reduction of the $30,000 Frigidaire?
5. Is the contractor right in refusing the price reduction of the $30,000 Frigidaire?

Mountain of Iron, Mountain of Dollars

Dragan Z. Milosevic

This critical incident case focuses on the special area in procurement-commercial discounts. Contract specifications contain one set of equipment data, while in actuality, a different set of data hold true of the delivered equipment. The equipment is mostly CNC tool machines. In the contract specifications, one number of grooves and their size are mentioned while those delivered are different. That is often a bone of contention between the owner and the contractor.

IBRAHIM

Mr. Ibrahim, a tooling engineer, had a dream job. He worked for the state, meaning he had connections. He worked out of the resident engineer's office, meaning he had money. He worked with foreigners, meaning he had connections and money. On the job site, he was with the maintenance department. As it was his job to inspect and maintain the equipment, he noticed that CNC tool machines, per contract specification, were delivered with a reduced number of grooves than they were contracted for. The grooves were shallow, and the total machinery weighed less. Ibrahim reported this to his boss, Abdullah-Kadir, who then ordered him to go ahead with the delivery as planned.

Ibrahim and Abdullah-Kadir represent the owners. They are also the new people in leading positions within the company. There weren't any former employees who knew about the history and people working under the contractor, which left Ibrahim and Abdullah-Kadir unfamiliar with past occurrences. So, a very difficult game had to be played by the ambitious players, even though they didn't know the history.

When Ibrahim returned with the data, which included the type of tool machine, the dimensions of the groove, and the number of grooves, he learned that all the machines delivered are different from contract specification. In fact, Mr. Ibrahim realized that the contractor worked off a different machine specification. After informing his boss about these results, he was given orders to ask the contractor to change the machine specification, meaning the grooves' number and size for future delivery. This resulted in a project change. This was the logic of the owner and represented how the owners viewed the issue. Since all the iron used to produce these machines was already paid for by Mr. Ibrahim's predecessor, the owner did not expect to pay any extra for this change. In fact, the owner expected to get the money back because the new specification could result in the use of less iron. Unfortunately, the contractor saw this as an attempt to introduce a new order. The contractor was not aware of the consequence of these project changes. He did not link these project changes with the cost, even though this was a significant change.

With the new specification, the contractor needed to use a better iron than the one Mr. Ibrahim's predecessor had originally paid for. In fact, the new iron was the technologically better iron, and therefore, it cost more. The iron of the higher technological value was used to produce a higher quality of grooves. The people working for the contractor did not succeed in explaining this to the owner.

So, these were two different views concerning the same issue. By the end of the project, the cost of the change was tremendous. The owner refused to pay the final amount. The contractor's representatives were astonished.

Discussion item

1. What can a contractor do to prevent such a mistake in the future?

Part II

CASE STUDIES IN PROGRAM MANAGEMENT

WHAT IS PROGRAM MANAGEMENT?

Proven to be highly effective in the management of complex product development efforts, Program Management has received more attention lately. The concept of program management perhaps originated from the U.S. Navy as the program management formation was reflected in the development of an underwater ballistic missile launch system during the 1950s. Looking at the structure of the missile launch system, a series of interrelated efforts including launcher, missile, navigation, operations, etc., was collectively and coherently managed to attain military targets. This, in fact, mirrors the program management definition generally known at present as the coordinated management of interdependent projects over a finite period of time to achieve a set of business goals.[1] The Project Management Institute also defines program management in a similar way, as the centralized coordinated management of a program to achieve the program's strategic objectives and benefits.[2]

There are a couple of widely used terms that are similar to program management; and although related, they are distinctly different disciplines. Project

---

[1]Dragan Z. Milosevic, Russ J. Martinelli, and James M. Waddell, *Program Management for Improved Business Results*, John Wiley & Sons, 2007, p.6.
[2]Project Management Institute, *The Standard for Program Management*, 2008, p. 6.

management is one good example in which many people use it interchangeably with program management. The two terms, however, are quite different. The difference lies in the focus of each discipline. While project management is *tactical* in nature, program management is more *strategic* and focuses on the business success rather than the execution success. The two disciplines, therefore, require different management mindsets. One is meeting time, cost, and performance requirements; another is achieving sustainable business results.

Project portfolio management is another term that often causes confusion, as it is also frequently defined as the management of multiple projects. Nevertheless, portfolio management is often regarded as a *process* to evaluate, prioritize, select, and resource new ideas, which may or may not be related or interdependent, that will help achieve the corporate business strategy. The program management is rather a *function* that determines the business and execution feasibility of a selected idea, which later turns into a plan successfully executed and delivered to the customer. Another way to look at it, the portfolio management process is to determine the business value of a product, service, or opportunity. The opportunity is then assigned to the program management function to turn these ideas into a tangible product, service, or infrastructure capability that delivers the value.

The Standard for Program Management from the Project Management Institute[3] suggests five program management process groups. They are initiating, planning, executing, monitoring and controlling, and closing process groups. These process groups are used as skeletons for organizing the case studies presented in Part II.

---

[3]Project Management Institute, *The Standard for Program Management*, 2008, p. 40.

Chapter 13

THEMES OF PROGRAM MANAGEMENT

This chapter provides case studies about major themes in program management. These cases address the program management methodology and program leadership in general. There are two cases in this chapter—one critical incident and one issue-based case.

1. KUPI

    This critical incident case describes a problematic situation when the unexpected rapid growth forced the company to lose its competitive advantage. In attempting to prevent a crisis, program management was considered as one of the potential responses. But would that be the right cure for the cause?

2. The Bounding Box Boxes You

    The Bounding Box Boxes You is an issue-based case that focuses on the differences between project and program management foci. It also discusses the use of the bounding box concept and gives examples of typical elements that might be bounded in a program.

CHAPTER SUMMARY

Name of Case	Area Supported by Case	Case Type	Author of Case
KUPI	Methodology	Critical Incident	Sabin Srivannaboon, Dragan Z. Milosevic, and Peerasit Patanakul
The Bounding Box Boxes You	Project and Program Management Foci	Issue-based Case	Sabin Srivannaboon

KUPI

Sabin Srivannaboon, Dragan Z. Milosevic, and Peerasit Patanakul

KUPI BUSINESS

Founded in 2005, KUPI is a start-up company in the high-tech manufacturing industry that provides a variety of build-to-order products. Because of the industry nature, new products are constantly being updated, and the slow movers usually get left behind. While other companies in the same market are struggling with the economic downturns, KUPI Business surprisingly grows fast. This is thanks to the management team, whose vision it is to encourage and cultivate new ideas for business and process improvements. One important factor contributing to the company's rapid growth is the organizational strategy, which directs the company to seize market opportunities whenever possible. Fast delivery is one of the company promises. This strategy works fairly well. In fact, a little too well because the company has grown unexpectedly too fast.

Recently, demands for KUPI products have risen over 150 percent. Although this is good news, the company's capabilities can't keep up with the rising demands, especially at peak periods. In attempting to deal with the growing demands, the company has created several more requisitions for customer service to take care of all new orders and customer relations. However, challenges in integrating various works among different functions are still overwhelming. That is because KUPI's build-to-order products require intensive assembling processes, which involve a number of people from electrical engineering, mechanical engineering, test engineering, and manufacturing, just to name a few. Each function possesses different specialties needed to carry the products to the final assemblies. At the moment, the work for each customer account is pretty much in sequence, following one silo to the next until the product is ready (see Figure 13.1).

The need for interaction among departments is minimal because of the way the products are originally designed. However, as the demands increase, so do the product complexities. This requires more and more interaction across the organization, which is difficult for KUPI's current structure. Plus, the existing operations worked fine only when there weren't too many customer accounts.

Figure 13.1   The Company's Current Work Structure

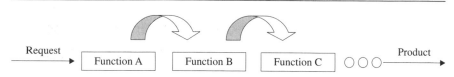

Now things are starting to fall apart. Just a week ago, multiple deliveries slipped, and the problem seems to be continuing.

John Rossman is the director of engineering at KUPI. He is concerned that the problem will soon become a company-wide crisis. He knows things will go south if there is no change in the company. But he isn't exactly sure what changes are needed. Changes in organizational structure to facilitate the communication and decision making? Changes in operations process to speed up the manufacturing? Or something else? The list keeps getting longer, and every solution seems workable.

One fine Sunday, John goes out for a walk and finds ads about program management training on a billboard. "Program management for improved business results: the coordinated management of interdependent endeavors to achieve a set of business goals." After reading the description of the course and short explanation about program management, he thinks program management may be what he's been looking for. With joyful excitement, John can't wait until Monday to discuss this idea with his peers.

Discussion items

1. Do you think program management is an appropriate approach?
2. What would be your recommendation? How would you approach the problem?

The Bounding Box Boxes You

Sabin Srivannaboon

Generally, a number of organizations perceive a program as an exceptionally large, long-range objective that is broken down into a set of projects. A project, in contrast, is a specific and finite task to be accomplished. Although these definitions have been widely used and adopted, some companies twist the concepts to suit their needs and culture. This case study takes a look at a leading manufacturing company, where project and program management are uniquely defined in a totally different way from what seems to be generally known; yet they are very successful. Perhaps the success comes from a clear distinction between the tactical role of project management and the strategic alignment role of program management.

THE STORY BEGINS . . .

It was 7:50 on a Monday morning. Bill Stevens was waiting in the lobby of Megatronic, a leading manufacturing company in the Southwest. Bill has been recently hired as a project manager with certain conditions. He was very excited and at the same time a little anxious.

Soon after he arrived, he was introduced to a number of people in the company, one of whom was Mike Fritz, the director of program/project management at Megatronic.

Mike: Welcome to Megatronic! I hope your first day is going well so far.

Bill: Definitely, it is my pleasure to be part of Megatronic.

Mike: The pleasure is actually mine. You know in your case we had six people interviewing you, and they all agreed that you were the best man for the job. You have a great deal of work experiences as the project manager for electronic platforms in the past, which I believe will benefit our company a lot. So we're all pleased to welcome you here.

Anyway, let me make it clear again at the beginning. You will start here with a title and salary of *project manager*. However, as we already informed you,

you will work in the role of *program manager*. And in six months, if you learn our program management and platforms, and successfully transform yourself to a business leader; we'll make you a program manager. If we see things differently in six months, we will let you go. In other words, you have six months of trial period. Is it clear?

Bill: Yes, it's clear. Thanks for this excellent opportunity.

These were the conditions that were causing Bill's anxiety.

BACKGROUND

Megatronic is a multi-million-dollar company, which sells all kinds of measurement equipment such as probes and oscilloscopes. The company employs over 12,000 personnel worldwide and has its corporate headquarters in the southwest area. The headquarters campus alone has more than 4,000 employees, and operates two major product lines of the company: Measurement Kits and Video Scopes.

The headquarters campus has three major entities, which include Manufacturing (Mfg), Sustaining Engineering (SE), and Product Lines (PL) (see Figure 13.2). Manufacturing and Sustaining Engineering are cost centers, where the majority of the work is running operations (Mfg) and/or sustaining the competitive products (SE). On the other hand, Product Lines are profit and loss centers. The program/project management department is a major office that upports all three entities with specific initiatives (projects and programs).

Figure 13.2   Organizational Structure

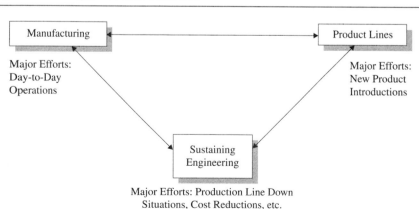

PROJECT VERSUS PROGRAM

Soon after he started, Bill was assigned to work on his first program. He was not a novice project manager—Bill had almost five years of project management experience. But it was his first time working as a program manager. Things were different here from where he used to work. He decided to call a meeting with Mike to clarify his roles and responsibilities.

Bill: Mike, thanks for your time. I know you are very busy.

Mike: Not a problem. How can I help you?

Bill: First of all, I am wondering if there are project and program management documents that I can look at. I want to learn more about them. It seems like project and program management here are different from what I know.

Mike: We do have project and program management policies. Yes, you can find them in our data base. But let me also walk you through these documents. As you know, Megatronic is a strong time-to-market company. So what we have created here is to facilitate the speed of our projects and programs.

At Megatronic, a program is an effort that produces new products to the market. We call it an "NPI" (New Product Introduction), whereas a project is anything else that is not an NPI. Simple—just like that. Therefore, programs are usually a new nomenclated product, which, for example, provides better performance than the existing one, and we are going to sell it to our customers by a different product name. Projects, on the other hands, include sustaining efforts, cost-reduction efforts, product conversions, or anything that does not produce new product nomenclatures. That's why we've created two different policies, program management policy and project management policy, to address these efforts differently. So it is pretty black and white when we talk about programs and projects here (see Figure 13.3). I would also add that NPI is our top endeavor because of our business nature. So, when they are fighting for resources or priorities, programs almost always win over projects.

Bill: That is quite an interesting definition. Program management here is different from what I know, which is the conventional definition, where a program is usually defined as a collection of interdependent projects with a common goal under integrated management (see Figure 13.4).

Figure 13.3 Megatronic's Program and Project Relationship

Figure 13.4 Conventional Relationship between Program and Projects

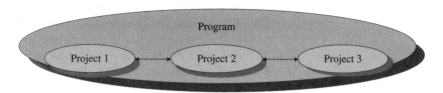

In my previous company, the program manager was responsible for ensuring that the overall program structure, and program management processes enabled the component teams to complete results. He or she usually interacted with each project manager to provide support and guidance on the individual projects, especially to convey the important relationship of each project to the bigger picture of the organizational objectives. The program manager, however, did not directly manage the individual projects.

Mike: That's also a very well-known and proven effective concept. But here things are unique. We tailor the concept to suit our culture and product types. However, the similar thing is that we expect our program managers to look at the big picture and focus on the strategic alignment. The bottom line is we need to deliver new products to make money. Completing programs in time thus is a must. We have a policy for program management, which is called "1103." This policy defines requirements and spells out authorities for the introduction of new products designed and manufactured and branded "Megatronic." Projects, however, are different. They are more of the tactical efforts, and may not need to be perfectly aligned with our business strategy.

THE BOUNDING BOX CONCEPT

Mike: Let me further explain it. Program management creates common languages used throughout the organization. A program manager here is a business leader, not a technical leader. You don't have to be a technical guru, but you must know how to communicate with people around you. Our goal is to accomplish desired business results, not just to get the job done within the time, cost, and performance requirements. Not simply just the "triple constraint." We have to think more of a business success rather than a firefighting type of work.

In order to achieve that goal, we have a concept, which is called a "bounding box." The box refers to boundaries, requirements, and limits established by management at program initiation and validated during the chartering phase by the program team headed by the program manager, and reviewed and approved at the charter review. The program team is free to plan and execute the product development program as long as the program meets the boundaries, requirements, and limits in the bounding box approved by management. If the box conditions are violated, the program will be considered "out of bounds." And if this is the case, the program manager will have to initiate an out-of-bounds review with the program stakeholders and sponsor. For many examples here, programs are primarily for time-to-market. So their schedules are the top priority in the bounding box. The schedule conditions and boundaries will be very tight with less room to make any mistake. On the other hand, cost and scope can be sacrificed in order to make the schedule happen.

Bill: That's an excellent concept. In my previous company, we didn't have it. We used only phased-gate reviews for our new product development process.

Mike: I see. We kind of combine them here. The bounding box concept also helps encourage communication among the key players, including stakeholders and program sponsors. It supports quick decision making and resolution of issues and problems at the appropriate management level and with the appropriate amount of documentation. Let me give you some examples of typical elements that might be bounded in a program.

- Market size
- Financials
- Schedule
- Alignment with strategy
- Customer requirements
- Required technologies
- Resources
- Speed of any technology development

The elements in the box should include the critical success factors, which should be scaled to scope and complexity of program. Equally important, the elements should be as objective and measurable as possible. For example, schedule and cost metrics can be direct quantitative metrics. You noticed that the box involves all metrics that are of a business nature and the person responsible for those metrics is the businessperson—the program manager.

The tightness of the box depends on a number of factors such as risk, size of the opportunity, potential strategic impact, skill of the team, and understanding of the market. The team and approvers from management, mostly major stakeholders and sponsor, determine the range of conditions that define the bounding box.

Bill: That's somehow related to the stakeholder management, the activities a business initiates to manage the relationships with its stakeholders.

Mike: Correct. The box concept depends heavily on the agreement among stakeholders.

Bill: What about the out-of-bound review? Can you explain it a little more?

Mike: The program team has the freedom to execute the program within the range assigned to each applicable element. Whenever the program is pushed and out of the specific limits, the out-of-bounds review is mandatory. Examples of the review may include the fact that a schedule slips more than four weeks, especially in time-to-market cases, or crucial components are no longer available from the suppliers, in some new product introduction cases.

Please note that when the bounding box status is presented at reviews, colored dots are used for each metric in the bounding box to denote status of that metric; green refers to a progressing well status, yellow refers to a heads-up, and red means that immediate actions are needed.

Bill: Well, what about the project guideline? Is it similar to the program guideline?

Mike: The guideline for projects, called "1104," is typically more flexible than the program guideline. In particular, the bounding box is not mandatory. Feel free to use it, but it is not required. Many project managers use their own spreadsheets to track their efforts. Some use the bounding box, or adapt the box concept. It's your choice.

Besides, project milestones are not as strict as program milestones. That is because projects focus more on execution success, like implementing something, rather than business success, which is being the first in the market and making money. Also, loose milestones mean less time to prepare documentation. Therefore, in running a project, flexibility is a luxury you have. But of course, when it comes to making a trade-off decision, programs always have higher priority and get more resources and attention.

Bill: What about the out-of-bounds review for projects? Is there anything similar to this?

Mike: Yes, we have a project review meeting. It's usually held at the beginning of each accounting period (AP). A project manager will provide project status to the sponsor and key stakeholders. Major items that need to be addressed in that meeting are:

- Brief project history (goal, start date, expected completion date, and resources needed)
- Project deliverables for each accounting period
- Project expenses update (labor and materials)
- Expected issues and problems

- Help needed, if any
- Activities last month
- Activities next month

Keep in mind that we also have the program review meetings. They are usually held in the second week of each AP, or one week after the project review meeting. The agenda is pretty much the same.

Bill: Now I have a better understanding of program and project management here. Thanks a lot. I will get back to work and do my best!

Discussion items

1. What did you learn from this case?
2. Compare the project and program definitions at Megatronic with those more generally known concepts. Which concept/definition is more effective in your point of view and why?
3. What are the keys to a successful bounding box approach?
4. How does the out-of-bounds review help avoid "big surprises"?
5. In what circumstances should boundaries be adjusted? What may happen when boundaries need adjustment?

Chapter 14

PROGRAM INITIATING PROCESS

This chapter presents two critical incidents related to the program initiating process, the first process groups in the PMI Standard for Program Management. These cases capture issues such as initiating the program, authorizing the program or a project within the program, and producing the program benefits for the program.

1. Business That Operated Without Knowing Where Its Profits Came From

    This critical incident concentrates on the issue of program development cost, more precisely on the labor cost. In particular, it explains that a program may get selected without knowing if it is really profitable, or to what extent it is profitable, if the labor cost is not taken into account.
2. Mega Security®

    This critical incident depicts a situation where initiating and selecting a program can be difficult. While multiple perspectives provide a proper path to optimize overall needs, managing them is undoubtedly challenging.

CHAPTER SUMMARY

Name of Case	Area Supported by Case	Case Type	Author of Case
Business That Operated Without Knowing Where Its Profits Came From	Program Financial Framework	Critical Incident	Dragan Z. Milosevic, Peerasit Patanakul, and Sabin Srivannaboon
Mega Security®	Program Selection	Critical Incident	Sabin Srivannaboon

Business That Operated Without Knowing Where Its Profits Came From

Dragan Z. Milosevic, Peerasit Patanakul, and Sabin Srivannaboon

CALLING IT QUITS?

Mathew Nordica became a vice-president of new product development (NPD) of Coastline Software (COS) three years ago when he was recruited for this position from a large electronics company. For all the three years, he has been trying to make what he calls "time-capture" be part of COS's NPD practice. In the long run, he wants all of his engineers to apply the same practice as a standard procedure used in the company. However, after the battling long and hard, he is ready to call it quits. However, he has suddenly received the help from someone whom he never expected—Pete Pitt, the CEO of COS.

BACKGROUND

Coastline Software is a market leader in construction project management software with annual sales of $62 million. COS was a program-driven business, so all of its income came from executing programs. Unlike in the early 1990, when COS's programs redefined the industry putting out an array of the trail-blazing software products, the beginning of 2000 brought only the software improvement programs. It looked like COS's engineers' imagination dried out, finding themselves without big ideas. It didn't go unnoticed in the industry. So, COS became a takeover target. At least, rumors had it so.

COS FOR SALE

Rumors were ceased when CEO Pete Pitt formally announced in the Board meeting that he was shopping COS around. To make COS more attractive for sale, Pete developed a list of improvements that managers of COS had to make. Mathew, like others, got his share of improvements from the list. Among the rest, Pete wanted Mathew to make the "time capture" part of his NPD department.

Currently, for every new program, NPD would have to prepare a so-called program brief, in other words a business plan. The plan goes to the Product Approval Committee (PAC) which selects the program proposed in the program brief or turns it down. They do that on the basis of the program brief elements that show how to:

- Integrate the operating functions within a company in the program
- Align the program with COS's business strategy
- Tame the fuzzy front end of the program
- Manage complexity
- Accelerate time-to-money
- Predict the program's software sales

THE TIME CAPTURE

While all this is fine, it is not enough. During and after the program implementation, there is no time capture whose job is to collect person-hours of all program participants. Multiplied by an hourly rate, the person-hours would yield the actual labor cost of the program. Without the prediction of this information in the program brief, PAC does not know the accurate, predicted profitability of the proposed program in the brief when they select a program. So, often they choose the wrong programs. What is worse, with the lack of the time capture, no one in the company could tell whether a program really made or lost money.

Mathew knew that he could tell his people outright that Pete ordered the time capture to be implemented. But he wanted to use that as a last resort, even though his people resisted for a million reasons the concept of time capture for years. Mathew wanted them to understand that COS can no longer operate as a business without knowing where its profits came from. He wanted them to truly appreciate the time-capture function. And more importantly, he wanted everyone's buy-in, rather than forcing them to do whatever they were told.

Although things seemed fine at the moment, the lack of time capture in fact could greatly jeopardize the company position in the long run. Mathew could foresee the potential disasters if his people continued to resist the time-capture idea. So he'd been thinking about a way or two to introduce the time capture. And he knew he did not have much time left.

Discussion items

1. What are the consequences for a business if it selects programs without knowing their estimated profitability?
2. What are reasons for people to resist time capture? Are any of them justified?
3. What is a good way to introduce time capture?

Mega Security®

Sabin Srivannaboon

JAMES DUFF AND HIS NEW JOB

James Duff has just accepted a position of Program Manager at Mega Security®, one of the leading safety equipment manufacturers in the southeast region, and a well-known importer and distributor in the country. After spending more than five years working as an engineer at another company on the West Coast, James finally decides to take this opportunity, and turns himself toward a managerial position. He is very excited about it.

Founded in 1980 as a family business, Mega Security® has grown in 25 years from a small vendor of seven employees to a well-recognized company of nearly 2,000. The company products include industrial safety equipment such as protective clothing, eye protection, face protection, head protection, respiratory protection, and the best-selling items of all: ear plugs and gloves. The company is currently expanding their businesses and is in the process of constructing a new building, which covers around 20,000 square meters of usable area.

Soon after he started, James was assigned to evaluate the fire extinguishing systems that will be used in the new building. As a result of the company's strategic plan to fulfill an initiative within a portfolio, the company now focuses on *law enforcement* and *employee safety issues*, and considers this effort a program as it is comprised of multiple related projects that must be managed in a coordinated fashion. James will have to make a recommendation and propose a system to his senior manager based on the available information. It is important that the right system is chosen. *That's the clear message from the company's CEO.*

THE MORE, THE MERRIER?

To ensure that he selects the right system, James decided to assemble a program team of selected key personnel including the Finance, Plant, Production, and Safety Managers, to help him with the decision-making process. James knows that these key personnel have different opinions and they prefer different systems based on their set of objectives and backgrounds. The Finance Manager once mentioned to him in the hallway that he would prefer a system with the lowest cost possible, while the Plant Manager requested a system that was easy for maintenance. He wanted to get the system up and running as soon as possible. Two weeks ago, the Production Manager expressed his interest in a system that was easy to use, while the Safety Manager was concerned about a variety of safety issues. As the program manager, James knows that he has to make sure the system chosen must meet the functionality required and comply with industry safety standards set by law.

THE PROBLEM

Last week James held a meeting with his team, but no progress was made because there were a lot of arguments about which system should be selected. All seemed to have reasonable needs. But satisfying everyone's needs at the same time? That would be a mission impossible. Besides, James knows he needs to avoid suboptimizations, and do whatever is best for the company as a whole. The serious question now is "how" he would optimize those individuals' needs for the company's greater good. And the "what-system-he-should-select" question is secondary. The team started fighting and ended up without a conclusion. Unfortunately, James has until the end of today to make a decision. Time is running out, and he still does not know which system should be chosen. The system has to be up and running within the next 60 days. Moreover, the system is expected to have around a five-year lifespan.

At the moment, four suppliers have provided quotes and additional information regarding their fire extinguishing systems (see Tables 14.1 through 14.4). These suppliers are Fire Extinguisher; Early Birds, Ltd.; Xtra Care, Ltd.; and Zebra Limited. They offer systems that have different functionalities, prices, maintenance costs, etc. The technologies used in these systems are also different (e.g., chemical-based, form-based, water-based), but they are all applicable to use in the new building.

Table 14.1 Fire Extinguisher

Model	ABC-1
Type/Technology	Chemical-based with the latest technology
Functionality/Law	Meets standard & complies with law
Installation	No hassle. There is no cost for installation
Unit Price	$85 per set
Spraying Area	Up to 100 square meters
Delivery Lead Time	15 days upon placement of order
Maintenance	$3 per set per year
Ease of Use	Easy
Relationship w/ Vendor	Never done business with before

Table 14.2 Early Birds, Ltd.

Model	DEF-1
Type/Technology	Chemical-based
Functionality/Law	Meets standard & complies with law
Installation	No hassle. There is no cost for installation.
Unit Price	$170 per set
Spraying Area	Up to 250 square meters
Delivery Lead Time	30 days upon placement of order
Maintenance	$550 after the first year, plus 10% increment for the following years
Ease of Use	Easy to some extent
Relationship w/ Vendor	Have prior business relationship, negotiations may be possible

Table 14.3 Xtra Care, Ltd.

Model	GHI-1
Type/Technology	Foam technology
Functionality/Law	Meets standard & complies with law
Installation	Cost of installation is around $4 per set
Unit Price	$115 per set
Spraying Area	Up to 200 square meters
Delivery Lead Time	50 days upon placement of order
Maintenance	$3 per unit per year
Ease of Use	Somewhat difficult
Relationship w/ Vendor	Never have business with before, but major stakeholder of Xtra Care, Ltd. is a close relative of Mega Security®

Table 14.4 Zebra Limited

Model	JKL-1
Type/Technology	Water-based
Functionality/Law	No information available, but most likely will not meet standard
Installation	No hassle. There is no cost of installation.
Unit Price	$2,430 per set
Spraying Area	Up to 5,000 square meters
Delivery Lead Time	55 days upon placement of order
Maintenance	No maintenance is required
Ease of Use	Easy
Relationship w/ Vendor	Never done business with before

When James looks at the quotes and provided information from the vendors, he has another problem. Some information is missing, and further information cannot be obtained at this time as the end of the day is approaching. "This is probably one of the worst days in my life," James said to himself.

Discussion items

1. How would you recommend James approach this problem?
2. What system would you recommend James choose? Support your recommendation, and state your assumptions.

Chapter 15

PROGRAM PLANNING PROCESS

Case studies of the program planning process are the central theme of this chapter. The planning process details structures for a program such as program plan, schedule development and risk management planning and analysis. There are four cases in this chapter—two critical incidents and two issue-based cases.

1. Quick Release

    Quick Release is a critical incident that discusses about a model called "Quick release" that is developed as a counterweight approach to mega-projects in order to avoid or minimize scope creep and/or major delays.

2. The Budica Program

    This is an issue-based case that focuses on program planning process and how the processes interact with metrics and tools, as well as other key business practices. This is the open-ended industry example where it doesn't disclose the outcome but leave the reader to pick what he or she believes is the likely outcome.

3. Best Practices Overview—Program Scheduling

    This critical incident explains the best practices of an electronics manufacturing company. In particular, the practices address a step-by-step process in scheduling a program.

4. Expect the Unexpected

    Expect the Unexpected is an issue-based case that describes an undesired event resulting from the company reorganization. In particular, it illustrates how a risk management plan was developed and improved over time to respond to the unexpected situation using the P-I matrix.

CHAPTER SUMMARY

Name of Case	Area Supported by Case	Case Type	Author of Case
Quick Release	Program Scope Management	Critical Incident	Dragan Z. Milosevic, Peerasit Patanakul, and Sabin Srivannaboon
The Budica Program	Planning Process Group	Issue-based Case	Diane M. Yates and Dragan Z. Milosevic
Best Practices Overview—Program Scheduling	Program Schedule Development	Critical Incident	Sabin Srivannaboon, Dragan Z. Milosevic, and Peerasit Patanakul
Expect the Unexpected	Program Risk Management	Issue-based Case	Sabin Srivannaboon

Quick Release

Dragan Z. Milosevic, Peerasit Patanakul, and Sabin Srivannaboon

Mega-projects are projects with complex scope or feature sets, and are usually characterized by large size and a substantial time frame. Such projects are inclined to suffer from mega-project syndrome, which is a condition that represents scope creep and/or major delays due to uncontrollable project changes.

The quick release model is developed as a counterweight approach to mega-projects such that scope is intentionally limited in terms of complexity, size, and duration, to avoid or minimize any need for change.

BACKGROUND

SSA is a 21-person unit specifically established to develop a proprietary software management tool for a 700-person software company, also known as the SBA. In particular, the SSA unit is in charge of internal software development tools used in testing microprocessors manufactured by the business units of this multi-billion-dollar company.

HISTORIC PROBLEM

In the past, SSA would usually spend up to six months collecting requirements from SBA; sometimes accepting requirements midway through the project, and trying to make everyone happy. In doing so, SSA would normally end up with a project that had a long list of features to accomplish. As a result, many projects took a very long time to finish. In many cases, promised features were dropped, but still usually with a huge scope creep.

There were so many complaints of SSA's handling of its projects' scope and timeline, literally from all sections of SBA, leaving SBA no option but to act. To help lessen the negative consequence, SBA put together a team made up of personnel from its major departments to analyze the situation. After a long study, the team proposed that SSA change its business model into a "quick release" model.

The quick release model was intended to be used as a way to improve quality in new product development for several years, but was just recently employed for the first time ever.

WHAT IS QUICK RELEASE FUNCTIONING?

Mandy Bock, manager of the SSA group, explains his team's approach: "Contrary to our existing model, the quick release approach favors frequent, smaller projects, each one implementing a small number of features only. We call these small projects quick releases. In particular, scope of each quick release is frozen at the beginning of the development phase of the program life cycle (PLC). This happens after the discovery phase, which is the first phase of the PLC. Each quick release scope typically has only two or three features, and must be approved by a program manager. Let's say we have a big program called program X. Instead of running it the way we did, we break it down into smaller projects (X_i), each with a few features. We begin to work on project X_1 (programming part); we plan its execution in detail, then transfer programming to the Russian part of the team. We then begin to collect requirements for project X_2. So, all of us from SSA here in Hillsboro, and all the other guys in Russia work on only small but frequent projects. Each quick release usually takes around three months, and is officially part of the program upgrade. In other words, any change aside from those related to the specified two or three features is difficult. And those two or three features are usually clearly defined because we don't have to focus on too many things at the same time. So without major changes we usually cruise through the program, almost never seeing scope creep or change in scope. Less is more!"

Discussion items

1. Generally, what do you think about the quick release approach *vis-à-vis* the conventional approach? Put together the list of strengths and weaknesses for both.
2. Is the quick release more suitable for large or small software development organizations? Justify your choice.
3. Explain how the quick release approach prevents large delays typical of mega-projects.

The Budica Program

Diane M. Yates and Dragan Z. Milosevic

This case illustrates how Digital Solutions uses standard program management practices to define, plan, and execute a program. The program team applies a traditional phase/gate development process, with key decision checkpoint approvals provided by senior management. Additionally, monthly program reviews are conducted to review program status with senior management.

The effective use of program processes is important for the successful planning and execution of a program. However, as this example also shows, they are not always instrumental in guaranteeing business success when the external market or environmental factors change significantly. The program manager must also look beyond the operational aspects of a program, and focus on the "big picture" that surrounds every program.

DIGITAL SOLUTIONS

Digital Solutions is a global leader in the electronic test and measurement industry, and is known for its customer-centric business strategy. Most of the functionalities built into the instruments are actually recommended by their customers. The Budica program is intended to deliver a new derivative product as requested by a small but strategically important customer. The program is geographically dispersed with the core team divided between the United States and the United Kingdom, with the customer located in Japan. The program finds itself in deep trouble, failing to meet the intended program strategy.

A BUSINESS OPPORTUNITY MISSED

George Wellington, program manager for the DPWO260 Millennium digital oscilloscope division at Digital Solutions Inc., remembers the day he realized that the program he had been managing became untenable and would not meet its goals. Not only was it late and over budget, but nearly the entire market had shifted to a competing product.

"When I took the program over from Jeremy Williams (the previous program manager)," George said, "I had my hands full just coordinating the program team.

Every time I talked to one of the project managers in the United States, they complained about how difficult it was to coordinate the work with the UK team. When I first talked to Ian McClellan, one of my project managers in the United Kingdom, he said that the marketing team hadn't provided his team with a firm set of specifications or features to work with. Worse, he said his team was not motivated and were functioning more as individuals who simply had a job to do instead of a team who owned the project and were invested in its outcome.

"One of the first changes I made as program manager was to give the UK team the authority to make decisions locally. This increased their motivation and made the difference with regard to project ownership." However, according to George, the problems didn't end there. "We couldn't get our customers to agree on the features and functions required, and there were problems understanding the standards that the Japanese customers were requiring."

Some programs are doomed for failure before they even reach the planning stage of the program life cycle. The key for program managers and senior managers is to recognize this fact as early as possible and take corrective action or terminate the program in order to minimize the losses. This program is an example of a poorly defined and executed program that was allowed to expend precious company resources beyond what was feasible from a business perspective.

PROGRAM BACKGROUND AND BUSINESS GOALS

Digital Solutions is a leader within the test and measurements industry, and is known for its ability to obtain customer input and feedback, and then implement it on its products. The program George Wellington was referring to, code-named "Budica" (the queen of Icene, who bravely fought the Roman invasion of the British Isles), was one of several programs in the digital oscilloscope business unit.

The business unit's primary objective was to create competitive advantage in the digital oscilloscope industry. It's strategies to obtain this objective were to establish a presence in a new market segment, maintain market leadership in the market segments it currently serves, and introduce new technologies within its products. It also emphasized a commitment to customer satisfaction, both throughout the product development life cycle and lifetime support of the final product. The customer vision was incorporated into the unit's business strategy and enabled the company to obtain a competitive advantage in the industry, since most of the functionalities built into the instruments were actually recommended by their customers.

Budica was initially seen as a program that would meet its business unit's objectives. The product itself was classified as "derivative," meaning that incremental changes were added to an existing product, the DPO260 Millennium oscilloscope. Customers were requesting additional features and interfaces that were consistent with market and technology trends. Although it used some technologies that were new to the company, for example wireless technology, they were not new to the industry.

Says George, "Normally we don't compete in the wireless technology space, but the decision was made that we would do so in order to prevent our existing customers from looking to our competitors for solutions to their needs." Bill Walsh, the VP of Strategic Business Planning, felt that the successful implementation of this program would create a new market in the United States and prevent Digital Solutions' competitors from making inroads in the Asian market. The program strategy was primarily defensive in nature—provide a compelling product that would prevent current customers from moving to a competitor's product.

"I preached strategy to my project managers," George said. "I made sure they knew the importance of this program and how it aligned with our business goals. I was told time-to-market was the first priority, and if features had to be removed or postponed in order to meet the schedule, then that was what should be done. I left it up to my project managers to coordinate schedules and make sure their respective teams bought into their projects, and that key deliverables were completed on time."

DEFINING THE PROGRAM: FUNCTIONAL AND BUSINESS ALIGNMENT

Trina Tektondi, marketing manager for the existing DPO260 Millennium oscilloscope, said that one of the features clients wanted was a wireless interface to their standard digital oscilloscope. A wireless interface would enable users to port waveforms and records from one oscilloscope to another without having to save the data on intermediate media, even if the other oscilloscope existed in another part of a building or at another location entirely. Additionally, the new product provided additional waveform monitoring capability more economically than the competitors' oscilloscopes.

Trina had clients in both the U.S. and Asian markets. One goal of the program was to meet expectations of both markets, but implement only one set of standards. However, from the beginning there were discrepancies on how to implement the wireless interface, and what standards were going to be used. Trina, located in the United States, had salespeople and technical marketing engineers (TMEs) working for her in Asia, and both salespeople and TMEs complained that meeting the needs of the Asian market was proving to be very challenging.

"The problem, as I see it, started right from the beginning," Trina said in a conversation between meetings. "Product definition was a problem because wireless interface standards were proving difficult to define. Lengthy communication with our customers put us behind schedule from the very beginning. But because of the short time-to-market target this program had, we were forced to align the product definition with some customers and not with others. Not having the product details nailed down in the beginning really cost us," she lamented. "I'll give you an analogy: We thought our customers wanted a chocolate chip cookie, but in the end they really wanted a blueberry muffin."

BUSINESS STRATEGY AND THE ROLE
OF THE PROGRAM MANAGER

"I feel my role as program manager was compromised from the start," George mused. "I was brought in during the planning phase of this program, so much of the program definition work—summarizing market data, assessing the customer's needs, analyzing the program feasibility, defining the macro plan, performing a financial analysis—had already been completed. Not only that, but the teams had already been assembled by the previous program manager. He handpicked the program core team and had the functional project managers pick the most suitable people to form the project teams." Because of the short cycle time of the program caused by the time-to-market goals, George had very little time to review all of the work and analysis that had been done previously.

"Still, I feel responsible for the outcome of this program. It's my job as program manager to make sure that our company's business strategies and the objectives of our program are aligned. I should have caught on sooner that certain things in the product weren't well defined. This very important shortcoming most likely cost us the market and should have been identified in the risk analysis during the defining phase. Incredibly, there were no specific risk areas identified for this program, because it was expected to be a relatively short program with established technology."

Digital Solutions has an excellent program management office within the company. "All of our program managers are trained in the latest program management techniques, and have the latest processes and tools at their disposal," Sherri Woodward, Director of Program Management said. "Furthermore, whenever one of our program managers is having trouble with a program, support is available to help him or her until the predicament is resolved and the problem taken care of. No one is left to struggle on their own."

According to Sherri, what Digital Solutions expects in a program manager is a skill set that includes good interpersonal skills, a working knowledge of the functional aspects of the program he or she will manage, good business acumen, market and customer knowledge, and strong leadership qualities.

Digital Solutions program managers are not expected to understand every aspect of the individual functional departments in detail, but they should have a broad working knowledge of the technical aspects of each. Having a technical background in a discipline such as engineering is not necessary, but it is helpful. Since programs are designed to be cross-project and interdependent in nature, managing programs successfully is more than managing multiple projects. Each project deliverable is dependent upon the successful delivery of all project deliverables, and is an element of the total solution, or whole product.

"The role of the program manager is to provide a focal point for ownership and accountability for business results," said Sherri. "He or she is responsible for championing product development, and is responsible for achieving specific business objectives. This individual works closely with the product line manager to make sure that business objectives for the program align with strategic objectives of the business unit and company."

George Wellington had been with the company for a long time, and had successfully managed several other key programs before being assigned to the Budica program. "We should not have originally assigned an inexperienced program manager to a program like this, due to the unique challenges it presented," stated Sherri. "I think the communication problem between the U.S. and UK project teams, plus the short duration of this program made it extremely challenging for anyone to manage. Unfortunately, there was not much George could have done once he was brought on board."

PROGRAM IMPLEMENTATION

Like many companies today, Digital Solutions uses established company processes and procedures for managing its programs. It is the program manager's responsibility to establish the program vision based upon the objectives established by senior management and create the appropriate links to the firm's strategic plan. The program vision is one of the key tools used by the program manager to provide focus and motivation for the cross-discipline program team.

Program Budica followed a traditional phase/gate development process, with key decision check points at the completion of each phase. In addition to the formal decision check points, monthly program reviews were conducted to review program status with senior management.

Work schedules were developed in the planning phase by the program manager and the core team, and were integrated via a bottom-up process. Trade-offs between project deliverables and delivery milestones were made during the integration process.

The program manager and his cross-functional project managers estimated the cost throughout the program's phases, including the defining, planning, designing, and ramping-up phases. A final financial performance was analyzed at program closure.

Risk analysis for Digital Solutions' programs begins in the defining phase and continues through all phases of the development cycle. Enough cannot be said about the importance of performing an adequate risk analysis process. The fact that the Budica program did not follow the established risk management process was a key factor in its execution challenges. Having a cogent analysis would have uncovered the cross-project confusion and weak product definition near the beginning of the program, where it could have been dealt with more successfully.

"We were asked to do a risk assessment, but honestly, our team didn't have much to go on because the technology was new to us, and the duration of the program was so short that we basically had to try to do an assessment on the fly,"

George said. This situation in fact is a strong indication that aggressive risk management is needed on a program. When a program has a short cycle time, and new technologies are being introduced, the program team should immediately establish a high risk level for the program. Then, as more is learned about the program unknowns and mitigation plans are put in place, the risk level of the program can be lowered if appropriate. "It was a corner that shouldn't have been cut. Like I said, everything was coming together really fast—product definition, risk assessment, cost—you name it," George stated.

The Budica program core team also discovered that their work breakdown structure (WBS) was incomplete, and as a result, product testing ended up being problematic near the end of the development phase. Testing wasn't started early enough in the engineering cycle to allow time to fix problems that were uncovered. Also, the lack of a robust change management process contributed to quality and functionality problems that emerged late in the program. Product definition continued well into the design phase and the lack of a requirements change management resulted in the lack of test cases being written for some key functions of the product.

Tools are used in program management to facilitate processes and measure progress as a program moves through its phases. Tools are a mixture of procedures, software, and techniques. Some of the tools used on the Budica program were the WBS, Gantt and wiggle charts, and spreadsheets to track costs and risks. A program strike zone was also established to help the program manager and his team monitor progress toward achievement of the program goals and business objectives.

Team culture is also important for effective management of programs. The program team members may be from several geographies and site locations. In the case of the Budica program, the teams were located in the United States and in the United Kingdom and at four different sites. The program manager was responsible for ensuring there was close alignment with the entire core team, no matter how many sites and geographies were involved. This was accomplished through a well-structured vision for the program and a decisive set of objectives. Additionally, well-managed and focused meetings, follow-up of action items, and followthrough by team members for completion of the action items were critical aspects of effective program execution at Digital Solutions.

FACING A DIFFICULT DECISION

For programs like Budica that have a short program life, it is important to accelerate time-to-market. "Adopting the program management model helps our company manage complexities such as this," George Wellington said. "The model encourages concepts like project ownership and concurrent development. It has worked well for this company in the past, and I'm sure it will continue to work well in the future."

However, one of the negatives with concurrent development is that problems are often not discovered until very late in the program. This is where misalignment between the work of project teams tends to show up, as was the case with Budica. This is also known as the "big bang" event. Since concurrent development involves functional project teams working simultaneously, the lack of integrated planning, clear project ownership, poor cross-project communication, and late product definition was problematic.

Digital Solutions was faced with the decision to finish the program or terminate it prematurely. Program metrics presented at the latest program review via use of the program strike zone tool revealed the following status:

- The program was four months behind the planned schedule
- Current program expenditure was currently $500,000, compared to $470,000 total budgeted
- Final cost was estimated to be $1 million
- A number of customers had moved to a competitor's product, and had indicated that the Budica product features are not what they wanted

In evaluating the program status, it was clear to Digital Solutions' senior management that the Budica program had gone dramatically off-track. The finance manager for Budica strongly advocated for the cancellation of the program due to the high cost overrun and estimated cost to complete. The marketing manager, however, advocated continuation of the program due to the fact that a strategic customer had the potential to become a multi-million-dollar contract in the future, and pulling out at this time would alienate them and cost the company future revenue.

The senior management team was faced with a tough decision, as both the finance and marketing managers made strong arguments in support of their respective positions. In the end, the senior executives took the correct approach—they evaluated the business objectives driving the program to make their decision.

To reiterate, the program strategy was to provide a compelling product that would prevent existing customers from moving to a competitor's line of oscilloscopes. This was in support of the business strategy to maintain existing share in Digital Solutions' current market segments.

Clearly, the Budica program was failing to meet the program strategy intended—existing customers had already begun to cancel their orders for the Budica oscilloscope and were instead ordering a competing product. Customers were indicating that the product features and functionalities were not in alignment with what they had requested. Additionally, even though the product still had the opportunity to capture a market segment that was new to Digital Solutions, the market was not large enough to warrant additional expenditure of company resources to complete the product development.

PROGRAM COMPLETION OR CANCELLATION?

George Wellington understood the challenging state that his program was in, as well as the positions of both the Budica finance and marketing managers. George also believed in his program and what the resulting product could offer his customers, and did not want to leave the program unfinished. After all, he was the leader of the program team who had worked diligently and tirelessly on the program despite significant challenges.

George was now due to give his recommendation for the future of the Budica program to his senior managers—should the program be cancelled, or completed?

Discussion items

1. What did you learn from this case study?
2. What program management processes were addressed in this case?
3. How did the Budica program team use program management processes to manage the details of the program?
4. How are trade-off decisions between project deliverables and program milestones made and why?

Best Practices Overview—Program Scheduling

Sabin Srivannaboon, Dragan Z. Milosevic, and Peerasit Patanakul

TRAINING

"Welcome to the training session," remarked John Lee, Engineering Program Manager and a former scheduling consultant. "Today, we will talk about one of the most important elements in program management. *Scheduling!* The course will cover our company's best practices in program scheduling."

"What is the difference between program and project scheduling? Simple. The answer can be found in the program and project definitions. While project scheduling focuses on the project's deliverables to a preplanned schedule, program scheduling focuses on coordinating all of the project schedules within the program and integrating them to ensure the program itself completes on schedule. Therefore, in order to have detailed schedules for the individual projects, we must create the initial program schedule first," said John.

"These steps were initially developed to help our program team create a fully integrated, cross-functional schedule for a program, which could be used as a maintainable tool. The key is to repeatedly question the team over and over again: '*Do you believe in the schedule, does it capture the work we are really doing?*' Importantly, the schedule is a tool for the team, not just for the program manager. We need all of your input," said John. He continued, "Here is the best practices overview:

Objective: All team members fully support the program objective/mission statement specified in the program charter, which provides the mandate to execute the program within a certain timeline.

Program Work Breakdown Structure (PWBS): Using a 'Post-it® Note Exercise' identify individual projects and dependencies needed to achieve the program objective. Then, chronologically list these projects under the appropriate milestone to complete the PWBS. Make sure you pay attention to the constraints posed by various factors such as funding constraints, resource availability, technical constraints, and hard deadlines.

Project Manager Assignment: Every individual project must be assigned a project manager who will take responsibility for ensuring the project is managed, resourced, and linked properly in the program schedule. The project manager will also take responsibility for estimating project duration and completing the project on time.

Connect: Link all projects together based on the dependencies identified from your 'Post-it Note Exercise'

Cleanup: The initial program schedule, also called Program Master Schedule, is now complete. It is time to fix the errors and optimize the schedule. For example, you may identify activities in the schedule that could be eliminated such as those activities without successors.

Weekly Schedule Updates and Pull-in Meetings: The program schedule should be updated on a weekly basis with the teams' input. In addition, weekly pull-in meetings will be held with the key project managers that control the program critical paths. It is recommended that you use the pull-in ideas for optimizing your schedule. Here are questions you need to repeatedly ask your team:

- Do we have to 'make' it? Can we 'buy' it instead?
- Could we establish a development/alliance partner?
- Is it possible to change the product definition/functionality?
- How about reuse and/or use of common parts?
- Can we challenge the base technology assumptions?
- Is there/should we set a common reference architecture?
- Can we eliminate (nonvalue) activities?
- Could we make more activities take place concurrently?
- Could we find more resources internally?
- Could we find contract resources outside?
- Could we realistically reduce the duration estimates?

Periodic Program Scrubs: Conduct periodic program scrubs with the program teams to revisit the PWBS, look for missing tasks (or projects) and blocks of work that should be done differently.

"Now, let's practice!" said John.

Discussion items

1. From your point of view, what are major challenges in scheduling a program under this best practices overview?
2. Using a Post-it Note approach to identify individual projects and their dependencies, what are the major benefits and potential disadvantages?

Expect the Unexpected

Sabin Srivannaboon

AutoX is a medium-sized company in the electronics design and manufacturing industry. Founded in 1995, the company has grown in 10 years from a small company of seven employees to a well-recognized company of nearly 200 employees. AutoX products include a variety of electronics platforms concentrating on Professional Instrumentation and Industrial Technologies. The company's annual turnover is more than $50 million with a balance sheet total of approximately $25 million.

Program management at AutoX is an entity with highly increasing demands. The department was established in 2007 to support business requests from multiple product lines. There are standard program management methodologies and essential tools available for any program manager to use. However, more than half of them need refinements due to the immaturity of the program management knowledge that the department currently has. Their level of knowledge doesn't keep up with modern thinking of program management just yet.

Jim Solo is a program manager at AutoX. He has been managing a number of programs for the company, one of which was a new product launch effort that would introduce several new features to the existing product without changing its form, fit, or function for the company's internal uses. The program code name was "XT+1."

Sue Lancer is an engineer at AutoX. She has been assigned to help Jim on many programs, including the XT+1 program. Sue has long been interested in making a career switch from technical engineering to program management. However, she has no prior educational background in program management. So, she often communicates with Jim not only to understand her program status, but also to learn program management from Jim's elite experience.

DECEMBER 2, 2008

Jim Solo looked pale and anxious. He was on the phone with his engineering manager, who was telling him some painful news that greatly impacted his program schedule. After the phone conversation, Jim called an urgent meeting with Sue to discuss potential responses.

Jim: Thanks for coming and sorry for such short notice. We have a situation here. I've just received a call from the engineering manager, explaining a possibility of slippage of the XT+1 schedule. It seems that there's a reorganization of the division and so priorities are shifting. And our program sponsor decided to leave the company. We now have a new program sponsor, who was recently transferred from the other campus. I don't know exactly why, but he's requesting our program be completed a whole month earlier than our previously agreed timeline. It's almost impossible. If we don't, or can't, our program will be put on hold until further notice.

They don't tell me much of the reason behind this, but I believe it's something to do with their funding and timing, or maybe politics, I don't know. But that's how it is. So, let's brainstorm on how to solve the problem.

While listening to Jim, Sue's mind drifted back to one of the very first meetings when the program team was developing the risk management plan.

THREE MONTHS EARLIER

Jim: The agenda today is to focus on a risk management plan for our XT+1. The good news is that the company has a template for a risk register with a few examples that we can modify for use in our program. So we don't have to start from scratch. We just need to come up with a list of all of the risks and some initial ideas about how we'd respond to them. And the list will need to be updated regularly since there might be unidentified risks that are not included from the beginning or the priority and/or urgency of each risk changes as the program progresses.

Sue: Well, Jim, I was never involved in this kind of process before. I just want to make sure I clearly understand it. Could you please explain this template to me?

Jim: Sure, Sue. Identified risks are risks that our team believes will impact our program somewhere during the program life cycle. Potential responses are how and by what means we will react if we run into these risks. Some responses may not be available right away, since they do not have an obvious response. So don't worry if you can't identify all responses at this stage. Root causes are the reasons for the occurring risks. Each of the risks will be assigned to a team member who will own the response plan. Category refers to the classification of your risks based on groups such as schedule risk and technical risk (see Table 15.1).

In addition, *priority* means how important a risk is, while *urgency* means when you need to take actions. Some risks are high priority but low urgency, and vice versa. For example, a production line down because of part shortages is often

Table 15.1 AutoX Risk Management Plan

	Identified Risks	Potential Response	Root Cause	Risk Owner	Category	Priority	Urgency
1	Program shows sign of not being completed on time	Request more resources and funds	Limited time to execute	Program manager	Schedule/ Overall	High	Medium
2	Critical components from approved vendor list become unavailable	Search parts from gray market with customer approval	Parts reach end of life stage	Procurement	Schedule/ Component	High	High
3	Quality testing delay because other higher-priority programs get testing resources	Do product safety and transportation testing first	Multiple programs are in queue for quality testing	Testing engineer lead	Schedule/ Overall	Medium	Low
4	Testing results are negative	N/A	New design to the company	Design engineer lead	Technical	High	Low

high priority in nature, especially in the manufacturing industry. Let's say there is a potential of having two component shortages, A and B. Both of them will be on a high-priority list since their absences impact company revenue. However, component A may have a five-day lead time versus a 30-day lead time of component B. Component B therefore will be on a high-urgency list since it takes longer to get the part, so we have to place an order as quickly as possible.

Sue: I see. We have to understand *where* and *when* to focus. So it is all about how to deal with potential problems. They may or may not become our problems, but if they do, we better have some plans about how to mitigate them. That's interesting. I also heard the term "watchlist." Is it the same thing?

Jim: Watchlist is a list of low-priority risks that you want to be monitoring as the program progresses. This list may not need to be in the risk register, but you have to make sure that you are watching them to see if conditions change, especially in a case that causes these conditions to make the low-priority risks higher priorities. If that is the case, you will need to bring them to the risk register.

Sue: Okay, now I understand.

Jim: Good, then, help me complete the plan. I need your technical expertise.

BACK TO PRESENT DAY, DECEMBER 5, 2008

After three hours of brainstorming, Jim and Sue reached an agreement.

Jim: There is no way to crash the XT+1 schedule without cutting its scope. We looked at other alternatives, but we don't really have one. There are no additional resources we can use or spend. The company is in a financial crisis mode. Plus, no activities can be overlapped to speed up the program. And at this point in time, the remaining activities are on the critical path. So it seems like the only thing we can do is to take out some of the XT+1 promised features.

Sue: Yes. From a technical perspective, we should consider removing features C and D. They are just fancy features, nice to have, but not required liked features A and B. I know we promised these features at the beginning, but I think this is the only way we can keep our program alive. So I agree. Let's propose this idea to our sponsor.

TIME TO REVISE—DECEMBER 6, 2008

Jim: Good news. Our program sponsor signed off on our proposal. He agrees with our plan, but on one condition. We will have to add those features back with the second product revision, which will be due next year.

Sue: That's good. I think this is the right decision.

Jim: Yes, I think so, too. Anyway, from looking at our existing risk management template, I think we can improve it. What we missed is the fact that we did not quantify the risk, nor look at the probability of its occurrence. So we did not know which ones we should pay attention to. Everything seems to be so important, and we don't have enough resources to monitor all of them at the same time.

So I've done some research about risk management. And there is an interesting concept that is called the P-I matrix where risks are to be assessed by quantifying them based on their Probability (P) and Impact (I) (see Table 15.2). One example is to use a five-point scale to rate a program. For example, to assess the probability, we can use 1 = very unlikely, 2 = low likelihood, 3 = likely, 4 = highly likely, and 5 = nearly certain. To rate the impact, we use 1 = very low, 2 = low, 3 = medium, 4 = high, and 5 = very high. We assess all risks in this manner and calculate the risk scores. Linear formulas can be used here, such as Risk Score = Probability + 2 × Impact.

Sue: That's interesting. I think we can incorporate this concept into our existing risk management planning process. Let's do it.

Table 15.2 The P-I Matrix

Probability (P)	Risk Score = Probability + 2 x Impact				
Nearly Certain = 5	7	9	11	13	15
Highly Likely = 4	6	8	10	12	14
Likely = 3	5	7	9	11	13
Low Likelihood = 2	4	6	8	10	12
Very Unlikely = 1	3	5	7	9	11
	Very Low = 1	*Low* = 2	*Medium* = 3	*High* = 4	*Very High* = 5
			Impact (I)		

12–15 High Severity   9–11 Medium Severity   3–8 Low Severity

Discussion items

1. What did you learn from this case?
2. What else can be revised/improved in their risk management plan?
3. What are your recommendations in developing the P-I matrix?
4. What are benefits, advantages, and disadvantages of the P-I matrix?

Chapter 16

PROGRAM EXECUTING PROCESS

The focus of this chapter is on the use of program management tools during program execution. In particular, these tools are enabling devices for program managers to reach an objective and produce a program deliverable. There are three case studies in this chapter—two critical incidents and one issue-based case.

1. Program Strike Zone

    This critical incident shows a tool used to track a program's progress toward achievement of the key business results by identifying the critical success factors of a program. The tool is often referred to as the program strike zone.

2. Program Map

    This critical incident describes a tool that provides an illustrative overview of the program, which includes cross-project interdependencies on a horizontal timescale. The tool is known as the program map.

3. Using Tools on a Mercedes

    Using Tools on a Mercedes is an issue-based case that focuses on various tools used for monitoring and controlling programs. It also describes tools often used during the selection and planning phases, and examples of metrics, which can be further discussed.

CHAPTER SUMMARY

Name of Case	Area Supported by Case	Case Type	Author of Case
Program Strike Zone	Distribute Program Information, Engage Program Stakeholders	Critical Incident	Sabin Srivannaboon, Dragan Z. Milosevic, and Peerasit Patanakul
Program Map	Manage Program Architecture, Manage Component Interfaces	Critical Incident	Sabin Srivannaboon, Dragan Z. Milosevic, and Peerasit Patanakul
Using Tools on a Mercedes	Direct and Manage Program Execution	Issue-based Case	Sabin Srivannaboon and Dragan Z. Milosevic

The Program Strike Zone

Sabin Srivannaboon, Dragan Z. Milosevic, and Peerasit Patanakul

AJ ELECTRONICS

AJ Electronics is an electronics manufacturing company, specifically founded to produce a measurement device for the healthcare industry. Known as a small but reliable company, AJ Electronics has customers in different regions including Europe and Asia. These days, their product brand is getting more popular as a result of relentless efforts of the management team to continuously improve their business in almost every aspect. Last year, the company revenue soared by more than $5 million. Considering the company size, this is a big growth. The management team is extremely happy about it, and wants to see their business continue to thrive.

At the beginning of this fiscal year, the management team discussed additional improvements they could make to their business. Because AJ Electronics is a program-driven company, one major challenge is definitely in the program management area, where its Achilles' heel is in the monitoring arena. So the company formed a team to study a number of monitoring concepts, and decided to change the ways their programs were monitored and controlled to improve its efficiency. To be more specific, they wanted to introduce the "program strike zone" concept to every major program in the company.

THE PROGRAM STRIKE ZONE

The program strike zone, also known as the bounding box, is an effective tool used to track a program's progress toward achievement of the key business results by identifying the critical success factors of a program. The key is to build the measures of a program that are important to the company, and then allow the team to function freely within those boundaries. In other words, the program team is empowered to plan and execute the program as long as the program meets the requirements or threshold approved by management. This program status is usually

299

represented by green (G). However, if a critical target or limit is (or is about to be) compromised, the situation must be brought to the attention of the governance council/management. These statuses are usually represented by yellow (Y) or red (R), depending on the severity.

The program strike zone is intended to foster excellent communication on program objectives, expectations, and status throughout the program life cycle from program initiation through program closure. Equally important, the tool is used to focus team and management attention on the critical program issues. Therefore, the specified conditions should be clearly stated, objectively measured, and understood and agreed to by both the program team and management at program initiation and each major review until the program reaches closure. The zone should be maintained and updated as necessary to reflect the current objectives, expectations, and critical program issues. If appropriately executed, the zone will provide useful and timely management guidance for better program monitoring and controlling purposes.

AN EXAMPLE OF THE PROGRAM STRIKE ZONE

In implementing the program strike zone concept, AJ Electronics requested that every program that forecasts more than a certain margin percentage incorporate the zone with its plan and address the zone in every major review meeting.

Figure 16.1 shows the program strike zone of one of these programs at the design and verification phase.

Figure 16.1   An Example of the Program Strike Zone

Value Proposition

- Program review if target market changes significantly   G
- Fast and accurate measurements   G

Business Strategy Alignment

- Support solutions for Network Element Manufacturing Test   G

Product Features and Functionality

- External tunable laser input   G
- Three receive data electrical outputs   R

Customer Driven Milestones and Schedule	Target	Threshold	
• Customer review of initial specification	Nov 3, 08	Nov 10, 08	G
• Customer demo	Feb 3, 09	Feb 10, 09	Y
• Customers review of final specification	April 9, 09	April 16, 09	Y
• Prototype available	April 13, 09	April 23, 09	G
• Prototype delivered to customer(s)	May 26, 09	June 9, 09	G
• Final system delivered to customer	Sep 1, 09	Sep 15, 09	G

Financial Assumptions/Forecast	Target	Threshold	
• Projected Program Spending:	FY01 ($1, 580K)	10% or $160K	G
	FY02 ($1, 664K)	10% or $170K	G
	Total ($3, 244K)	10% or $324K	G
• Forecasted Orders:	FY02 $8.6M	+/− 15%	G

G	Progressing well	Y	Heads up	R	Help needed

Discussion items

1. What are the major benefits of the program strike zone—major advantages, and disadvantages?
2. What are important criteria to identify critical success factors and their thresholds?
3. What are major challenges in implementing the program strike zone concept?

The Program Map

Sabin Srivannaboon, Dragan Z. Milosevic, and Peerasit Patanakul

TRAINING

Ken Sanford is a senior consultant at Titan consulting firm. He is recently hired by a small company that sells sensors and electronics platforms to conduct program management training to its senior staff and managers. The training is in a workshop format, which particularly emphasizes a concept of program management for improved operations competence.

The training is a five-day session. Ken has already spent two days talking about the program management framework, and the strategic aspects of program management and its tools. Today, he is planning to address the operational aspects of program management tools. One of the tools on his list is called "the program map," which is an enabling device for managers to reach an objective or a program deliverable. He plans to give the audience hands-on experiences in developing one.

WHAT IS THE PROGRAM MAP?

The program map is a tool that provides an illustrative overview of the program, which includes cross-project dependencies on a horizontal timescale. The map includes the critical deliverables of each involved project team throughout the program life cycle, where it uses arrows to represent the dependencies of cross-projects.

The goal of the map is to enable the program team to understand the deliverables and dependencies among the project teams on the program. The mapping of deliverables from one team to another helps the program team to determine and fully understand the cross-project dependencies that exist on the program.

The key steps in building the map are as follows:

Step 1: Prepare information inputs (e.g., key elements of the program strategy and requirements, project managers' input, and knowledge of the program technology).

Step 2: Identify primary project deliverables from the program WBS (PWBS) and the detailed requirements. For each deliverable, document the critical information needed for the mapping process—deliverable name, time to develop the deliverable, and dependencies required to complete the deliverable.

Step 3: Create vertical partitions that represent the lowest level of schedule tracking that will place on the program (days, weeks, months, or quarters). Create the horizontal axis that represents the different project teams involved in the program. Then, enter deliverables for each project team into the horizontal lanes, matching the expected timescale on the vertical partitions. Continue the mapping exercise until all program deliverables are entered on the program map.

Step 4: Build cross-project dependencies by using the arrows to connect independent milestones with dependent milestones. Cross-project dependencies will determine the sequence of development, interfaces, and responsibilities, as well as the initial timeline.

THIRD DAY OF TRAINING

Ken: It's almost 9:00 AM. I think we should start the session. What we will cover this morning is a tool that is called the program map, which I also briefly explained yesterday. The instruction of how to create the map is in your folder. I hope you had a chance to study it last night because we are going to do an exercise about the map today.

Table 16.1 Cross-project Dependencies

Project Teams	Major Deliverables	Cross-project Dependencies	Expected Completion Date
Software development	(1.1) Power control software	N/A	Workweek 1
	(1.2) BIOS	N/A	Workweek 3
	(1.3) Software verify report	(4.1), (4.2)	Workweek 8
Hardware development	(2.1) Hardware design files	(1.2)	Workweek 4
	(2.2) Hardware verify report	(4.2)	Workweek 8
Enclosure development	(3.1) Enclosure design files	(1.1)	Workweek 2
Manufacturing	(4.1) Manufactured enclosures	(3.1)	Workweek 4
	(4.2) Manufactured circuit boards	(2.1)	Workweek 6
Product test	(5.1) Test case	(2.1)	Workweek 5

Now let's assume that you guys are managing a program initiated to introduce a new type of electronics device. Because of its complexity, the program consists of multiple project teams; namely software development, hardware development, enclosure development, manufacturing, and product test. Each project team is expected to produce major deliverables at different times. And these deliverables may have cross-project dependencies. Here is the summary.

Discussion items

1. Construct a program map of the given information in the case.
2. What are the major benefits, advantages, and disadvantages of the program map?

Using Tools on a Mercedes

Sabin Srivannaboon and Dragan Z. Milosevic

BACKGROUND

RollingSys is a privately held corporation that provides computer solutions to help customers design, build, deploy, and manage next-generation numerically controlled tool machines. These products require a lot of customization, resulting in each order being organized as a program. For this reason, the products are well known and have earned many outstanding awards.

RollingSys has seen an increase in orders from Asia and Europe, which puts a lot of pressure on their six program managers. After a long discussion, a decision was made to hire Keith Richardson, an experienced program manager. Mick Beggar, the director of the program management office (PMO) for RollingSys, prepared a plan for Keith to transition to the new job, including familiarizing him with RollingSys's program management system. Following is part of the familiarization relating to the strategy and program management tools and metrics. Taking part in the discussion are Keith; Mick; and Charles Waters, a longtime program manager.

ALL ROADS LEAD TO THE BUSINESS STRATEGY

Mick: I want to make something clear from the very start. RollingSys's choice and application of program management tools, like all managerial processes and actions, is driven by the business strategy. Therefore, we should first talk a bit about RollingSys's real strategy—not the company's public relations' word on strategy—but such things as the company's strategic uniqueness and what makes it successful.

As director of the PMO, I often interpret the strategy of RollingSys to my program managers. So, at this time, I feel obliged to put on the hat of the director of PMO and explain the business strategy. First of all, RollingSys is unique in terms of breaking down the components of the business and understanding what makes it successful and what doesn't make it successful. Our products are often recognized as the best products on the market. If you think of an airplane seat analogy, they are in the first class section. Also, the ability of RollingSys to get customer input makes a huge difference. What our customers want to use the product for often becomes clear through the ability of the company to get engineers in front of customers. Most of the functionalities of RollingSys's products are actually used by customers. Our business is moving from the old traditional "here is the computer we build" to "what are the features you need in the computer we are going to build?" Moreover, we have repeatability across different product lines. The repeatability allows the company to shift the program manager from one product line to another with an adequate understanding of performance criteria and his or her responsibilities. This makes us unique and is a factor in our success.

What, then, is RollingSys's business strategy, and what is the role of program managers in the context of such strategy? RollingSys is often a first mover, so it is technology innovation and to be first in time-to-market. In that context, the program managers are responsible for more than just getting from program start to program finish. They are required to deliver the program on time, meet all the objectives of the program, and are responsible for business results. Therefore, you are expected, as any other program manager, to be a very visible and seasoned business manager aligning your programs with the strategy.

Keith: That having been said, can we now talk about tools? I would like you to take an example of a program and tell me how using specific tools in that program helped make it successful. Give me the background of the program.

MERCEDES

Charlie: I suggest using a program called Mercedes as an example. The customer had all sorts of requirements, one being the program name. They said, "In our country, a Mercedes car is the ideal of high quality. We want this program product to be of high quality, like a Mercedes car. In order for you to keep our high-quality expectations in mind at all times, let's call this program Mercedes." Basically, they wanted to have a capability added to RollingSys's existing product. RollingSys got an opportunity to win a large competitive sale in Spain if the program was delivered on the particular date. So, RollingSys formed a team to execute this program. I was the program manager for it.

The program went from conception to completion in eight months, which is not normal. It is probably the best example of how our process works at optimum. On top of that, it brought the company several million dollars in revenue to date. In RollingSys, managers believe that the program was successful partly because it used all the tools of the program management knowledge base that are available at the company. So there is a belief that the program management system as a whole works very well here.

SELECTING AND APPROVING A PROGRAM

Keith: Can you explain what strategic tools were used in Mercedes to make sure that it was well aligned with RollingSys's business goals and strategy?

Mick: RollingSys uses both strategic and operational tools, divided per major program activities. But there is a word of caution here. Typically, program managers would not be involved in the development of strategic tools. Generally executives do them. However, each program manager needs to know them well because he or she will use them in communicating with executives about program status.

The whole alignment process begins with the strategic plan and continues with portfolio maps and the business case. RollingSys's practice differs from some companies in, for example, the business case. The strategic plan drives the alignment. In other words, it is a tool or mechanism to ensure the quality of the alignment in RollingSys.

RollingSys's strategic plan usually includes the product roadmap, technology roadmap, customer technology roadmap, and the business model for the next three years. It addresses things like mission, objectives, long-term strategy, market size, segmentation, competition assumptions, and market share. In particular, the product roadmap proposes products within the three-year time frame and addresses those currently in development. For each product, it includes start and completion dates, milestone dates, total nonrecurring expenditure, and the three-year sales forecast.

RollingSys's strategic plan drives its formal portfolio management process, where programs are prioritized in terms of the program portfolio needed for customers and for the business. Mercedes is no exception. It was a program selected from the product roadmap and prioritized in the portfolio process, and its implementation was sped up by the customer requirement.

Generally, executives focus on the strategic plan to analyze the growth plan and determine what the right markets are for the company, where the company is the most successful, and where the customer gets the greatest value for the products. Then, the programs are planned in alignment with the strategic plan over the next three years with their expected sales and profit dollars are identified. By looking at them, we are able to see the growth from different directions, such as its existing business (extend or upgrade), new products in new markets, and pure technology transition. As a result, the importance of certain products becomes more obvious than others. Then, depending on product complexity, market pressure, and other significant factors, programs are initiated and selected into the program portfolio. We don't really use any specific tools for the selection of a program into the portfolio, but there is a lot of discussion. Once a program is selected into our portfolio of programs, we use several types of a tool called a portfolio map to display all our selections. Each one has different parameters on the x and y axis, for example, net present value (x axis). The portfolio map helps us to balance the selected programs by visualizing all of them, comparing them, and seeing where we have to intervene to achieve product balance. This is very important for our organizational success.

Now, we begin to develop the business case for programs close to their implementation time. Some companies, as I said, use the business case differently than we do. They use it to select programs into the program portfolio. We use it to approve a program for actual implementation in the concept phase. If we do a good job using this tool, i.e., the business case, we will give the go-ahead to run good programs, and kill poor programs. So, the business case tool is of make-or-break importance to our program success. Choose a poor program, and you are kind of doomed. Choose the right one, and you are given an opportunity to succeed. Our program managers tend to be assigned to a program shortly after a concept is approved, and are responsible for the successful completion of the program. They will make trade-off decisions on features to make sure that the plan is actually aligned with its objective.

PROGRAM PLANNING

Keith: Once you select a program into the portfolio and a program manager is assigned to execute the program, what tools did you use to ensure the quality of the alignment during the program planning, and how did they contribute to the program's success?

Charlie: A lot of tools were used in Mercedes. First, a tool called the program strike zone was used. The strike zone is simply a set of agreed upon program critical success factors and business results established to help executives and program managers monitor the programs by specifying quantitatively the boundary conditions under which the program can operate. Metrics such as time-to-market, target market, net present value, and key milestone dates were included in the strike zone. In Mercedes, the priority success factors were schedule, features, profitability. You see here how the business strategy (i.e., time-to-market) shapes the program strike zone, making schedule its first priority. Simply speaking, it was our primary alignment tool and it helped me to develop a program plan and, at the same time, make sure that the program met the business needs. By doing all this, the program strike zone contributed to a successful program.

During a one-day workshop called Map Day, the core team developed a program map, which is a tool showing critical cross-project dependencies related to the program schedule and the critical deliverables of each project team throughout the program life cycle. The program map was used for two things—to do a preliminary program work breakdown structure (PWBS) and a preliminary master program schedule. Actually, the critical deliverables went into the PWBS, which then served as a guideline to project teams to develop their project WBSs, which I took back and, after a thorough review with project managers, merged into a detailed PWBS.

Based on the critical cross-project dependencies from the program map, the initial program schedule was developed. Once the master program schedule was completed, it was decomposed into the seven project schedules. In this way, a hierarchical schedule with two levels was obtained—the first being the master program schedule, the second being the project schedules.

In terms of scheduling, standard project management tools were used for the master program schedule and the project schedules. The business strategy of being the first in time-to-market had a key role in determining which tool was chosen.

PROGRAM MONITORING AND CONTROL

Keith: What about tools used for monitoring programs?

Charlie: One of the tools we use is called the program dashboard. The dashboard is a management tool visualizing the status of programs, by using red, yellow, and green indicators. Red means management intervention is needed, yellow means warning, and green means that the program is progressing well. Tools like the dashboard and the program strike zone are commonly used everywhere in the organization to help executives communicate with program managers. The executives want to see if the programs are still aligned with the business strategy, and determine if any corrective actions are needed to recover them from misalignment. In many instances, when there's an issue that pushes the program out of the success criteria limits, one of the program indicators will turn red. Then, executives and program managers will have to develop corrective actions and/or adjust the success criteria limits.

In Mercedes, we also used the program review, which is a tool used to communicate program progress or to involve senior management when they need to step in and make some tough decisions. We used the periodic program reviews in addition to the phase gate milestone reviews.

Lastly, a mandatory tool we use is the wiggle chart. The chart anticipates the expected rate of future program progress, focusing on predictions of major program events, like milestones and program completion. Let me show you an example (see Figure 16.2).

Figure 16.2   The Wiggle Chart

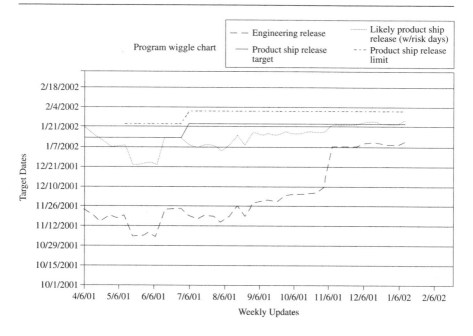

The vertical axis shows the team's predicted completion date for a specific program milestone, for example the engineering release or product ship release, while on the horizontal axis we can see the actual date the prediction was made. The beginning point on the horizontal axis is the time when the schedule baseline is prepared and the high-level program milestone dates are marked on the vertical axis. After the beginning point, the program work is kicked off, and the horizontal axis represents the actual program timeline. The team reviews progress regularly and makes milestone predictions. By connecting all predictions for a particular program milestone into a line, we can obtain the milestone trend line. Should the line go upward, the program manager would know that there is a milestone schedule slip. Delivering the program milestone on time would produce a horizontal line. If we estimate an early program milestone completion, the trend line would go downward. Although it is effective in predicting milestone progress, the chart is even more effective if used to develop actions required to eliminate any potential deviation from the baseline program milestone schedule. In general, the wiggle chart acted like a compass, helping in Mercedes by warning us early of potential schedule slips so we could take corrective action and navigate toward success. Just imagine navigating the troubled program sea without a compass!

METRICS

Keith: I read somewhere that man is a tool-using animal. In Mercedes, you seem to confirm that. Now, what about metrics? Could you please elaborate more on the topic with regard to balancing the program and company needs, and how they helped the program succeed?

Charlie: Program performance is mostly measured by the ability of the program to meet major milestones, such as the launch target date. Since RollingSys wants to be a first mover/technology innovator, other metrics are important, but not as important as schedule metrics. But there are some differences among programs and the metrics they focus on.

For example, in Mercedes, the priorities of metrics were schedule, features, gross margin, PI (profitability index), development expenses, manufacturing costs, market share, and staffing levels. The schedule was so important that progress was measured by the ability to meet the milestones and delivery date, together with the feature sets requested from customers. In parallel with satisfying customer needs, the company's bottom line is of primary concern. So executives use the program strike zone to specify the boundary conditions that match the company's business needs, like return on investment (ROI), in the form of the PI, or profitability index. The other metrics such as performance to development cost, manufacturing cost, and market share were of second level of importance compared to time-to-market, feature set, gross margin, and PI.

Overall, what I am trying to say for Mercedes is that customer needs, time-to-market, and feature set are balanced with our business needs—which is to make money, to be brutally honest. We combined those things together and created the business results. That also has another loud message: our metrics, like tools, were driven by the business strategy. We did this properly, and metrics contributed to the program's success by letting us know if we were heading to success and should stay the course, or if we needed to take corrective action.

Discussion items

1. What strategic and operational tools were addressed here? How were they used differently?
2. Explain the alignment process at RollingSys.
3. List the major advantages and disadvantages of the fact that program managers tend to be assigned to a program shortly after a concept is approved, not at the beginning.

Chapter 17

PROGRAM MONITORING AND CONTROLLING PROCESS

The central interest of this chapter is the program monitoring and controlling process. To monitor is to measure a program, whereas to control is to fix whatever problems we have once we measure the program. We offer cases in two different industries: the information technology field and the mobile service industry. There are three cases in this chapter—two critical incidents and one issue-based case.

1. I Have Only Three Minutes a Month!

   I Have Only Three Minutes a Month! is a critical incident that talks about the importance of a progress report for program management and its brevity.

2. OSSOP!

   OSSOP! is an issue-based case that focuses on monitoring and controlling programs using a dashboard approach. The case also describes a guideline for program classification, which can be further discussed.

3. That Which Is Not Earned Is Never Valued

   This critical incident case discusses the Earned Value concept for monitoring and controlling program status. Particularly, it shows an example of the unsuccessful program where the earned value concept was applied without a deep understanding of the way it worked.

CHAPTER SUMMARY

Name of Case	Area Supported by Case	Case Type	Author of Case
I Have Only Three Minutes a Month!	Progress Report	Critical Incident	Dragan Z. Milosevic, Russ J. Martinelli, and James M. Waddell
OSSOP!	Monitor and Control Program Risk	Issue-based Case	Sabin Srivannaboon
That Which Is Not Earned Is Never Valued	Monitor and Control Program Performance, Program Schedule, and Program Financials	Critical Incident	Sabin Srivannaboon

I Have Only Three Minutes a Month!

Dragan Z. Milosevic, Russ J. Martinelli, and James M. Waddell

Al Petroff, CEO of DirectConnect, a world premiere producer of interface cable, was very expressive, almost rude: "Look guys, my time is very expensive, and I am sick of wasting my time reading poor reports. I need you to design a report showing the monthly status of my 40-plus programs going on at any time. It must be a one-pager showing the most important things about my programs; it need not contain words, only numbers and graphical symbols, and I must be able to read it in three minutes because I have only three minutes for that purpose a month. Is it clear?" A consultant who Al was talking to nodded and said, "Yes, it is clear."

Seven days later the consultant was back with a one-page report. He began by saying, "We included five metrics, covering program management as it relates to measurement from strategic management, portfolio management, program management, and project management:

LEVEL: STRATEGIC MANAGEMENT AND PROGRAM PORTFOLIO MANAGEMENT

(1) *Alignment of programs to business-unit strategic goals*: Percentage of total program portfolio that is compatible with documented business-unit strategic goals. It appears difficult to find a program that does not support specific business-unit goals. But if so, an explanation should be provided.

LEVEL: STRATEGIC MANAGEMENT AND PROGRAM MANAGEMENT

(2) *Projected future income from program road map*: Fraction of future net income by year projected from programs on the program roadmap over multiple years; the probability times the net income for accomplishing each program goal will also be provided.

LEVEL: PROGRAM PORTFOLIO MANAGEMENT

(3) *Program portfolio distribution*: A way to express fractions of the total program portfolio among various dimensions that are important to program stakeholders. The metrics help determine how to modify the program portfolio if the programs are not in balance.

LEVEL: PROGRAM PORTFOLIO MANAGEMENT AND PROGRAM MANAGEMENT

(4) *External customer satisfaction survey*: Average value of ratings given by key external customers, on a Likert scale of 1 to 5, with 5 being the highest value and 1 being the lowest value. It measures various dimensions such as timeliness of the program completion and customer value of the program output.

LEVEL: PROGRAM PORTFOLIO MANAGEMENT, PROGRAM MANAGEMENT, AND PROJECT MANAGEMENT

(5) *Percent of the program milestones accomplished*: The percent of all program milestones in the portfolio of programs achieved within appropriate time. It reflects the in-process timelines of the program portfolio, individual programs, and projects within a program as the metric acts as an early warning signal for a company's time management system."

Then, he went on to say, "Each metric shows one number for the month and one for the cumulative value, where applicable. Quality of the monthly status, cumulative value, and overall trend are shown by colors. A green status signifies progress as planned, a yellow status indicates a heads-up to management of a potential problem, and red requires management intervention. We have tested the time needed to read and interpret the report with our executives, and they need an average of three minutes."

Al took a long look at the one-page report, paused, and said, "Okay, let me test it."

Discussion item

1. Do you agree with the metrics suggested on the one-page report? Why or why not?

OSSOP!

Sabin Srivannaboon

WHO'S WHO?

Mr. John Jackson was Executive Vice-President of Information Technology Group at one of the leading life insurance companies in the region. He was a renowned programmer who had more than 25 years of work experience in various technical fields, and had been with the company since his career began. Two years ago he was promoted to a management position after the sudden and tragic death of the previous vice-president, who was also his mentor.

As the new executive, John oversaw the big picture of software projects, and led more than 70 programmers, computer engineers, and technicians in his department. Historically, more than 30 software projects were executed each year mainly to support the company's business units. However, more than half of them failed either because they were completed late, over budget, missed functions, terminated early, or combinations of any of these. John urged changes as he saw a lot of room for improvement, one of which was definitely in the project management area. He decided to hire an experienced consultant to give him recommendations on how to start.

Mr. Sammy Lee was a young independent consultant, and expert in project/program management. He was hired to help John develop a project management methodology and a simple tool to track all project status in the IT department. Sammy was born in Singapore, and was not familiar with the corporate culture at all.

THE COMPANY BACKGROUND AND CULTURE

In 1969, the company was founded as one of the very first locally owned businesses with the absolute goal to provide security and protection for families in the region. The company underwent major changes in the organizational structure and management system during the 1980s, leading to a new foundation for modern and efficient operation in many respects. The company has seven divisions including claims, human resources, finance, investment, and accounting departments, a medical center, and an information technology group. In 2008, the company assets were around $300 million, maintaining their ranking as the leading life insurance business in the region.

Under John's supervision, the Information Technology Group (ITG) focused on solving the challenges faced by growing insurance businesses. The ITG organization was structured into two main subgroups: software and hardware teams, with a total of 80 employees. The uniqueness of the ITG laid on a strict policy of using open source software, which was the original source code available to the general public for use and/or modification "free of charge." In particular, no commercial software, especially those with license fees, was allowed in the company at all. John called it the "OSSOP!" policy (**O**pen **S**ource **S**oftware **O**nly, **P**eriod!). And because of the OSSOP! policy, the company saved millions of dollars last year. And the CEO was very pleased to see the OSSOP! continue.

FIRST MEETING WITH THE EXECUTIVE

John: I'm very pleased that you decided to take this job. Our group is in need of fresh ideas, especially with regard to the implementation of a system for project management. I am sure you can help us.

Sammy: Thank you. It's my pleasure. First of all, if I understand the contract correctly, you'd like me to develop project management methodology and tools in your group for improved business results. You'd like to standardize the way your project managers manage their projects, and be able to regularly track the status of each individual project. Is that right?

John: Yes, currently we do not have a systematic tool or standard for managing projects. Each manager manages his or her projects from personal experience. And if you notice, there is no person designated as "project manager" in my group. Senior programmers who have more than 10 years of work experience are usually assigned to be responsible for success of the projects. That's the way we have been doing it. But now I want to change it. Here are the formal documents that we have for project management. Not a lot, as you can see. Hopefully, they will give you some ideas of our business and how we run it.

Sammy: Okay, so let me go through these documents tonight. Also, I'd like to talk to your people sometime this week. Would you please arrange that for me?

John: Sure, I can do that.

Three days later, Sammy met seven people who had assumed project manager roles. He discovered several interesting things. First, each manager had his or her own distinct way of managing projects. Second, all of them claimed that they had important projects, and the resources were inadequate. Third, there was an inconsistency among the tool utilizations. Some managers used open source spreadsheet programs[1] to schedule their projects. Many did not do scheduling at all. They simply forecasted the schedule based on their experiences. Why didn't they use an open source program to do that? Mainly because they believed open source software was not sophisticated enough to do scheduling. Their complaint was "It's darn slow!"

Sammy learned from the interviews that John held the monthly meeting to get project status reports from all managers, and that was the only meeting associated with project management. If there was an urgent problem that couldn't wait, most managers went directly to John's office and requested support informally. Most of these were sorted out, but some weren't because as an executive vice-president, John was very busy. What happened was that some issues were neglected, and so no follow-up was initiated in many cases, especially those with low priority projects. Of course, this was not John's fault. Simply, it was the system and approach that prevented John from providing support to everyone every time it was needed. Fortunately, John was more than ready to make a move. In fact, it was the best time to make changes because people perceived this time as crisis, due to the high number of projects that failed last year. And with crisis often came opportunities.

SECOND MEETING WITH THE EXECUTIVE

Sammy: Now I understand both your concerns and your people's concerns. Overall, I think your managers are frustrated with the speed of open source software for project management simply because it's relatively slow. They'd actually like you to consider purchasing commercial software licenses because they believe the commercial ones are much better. But they didn't want to say it out loud because they know it is against your OSSOP! policy. They wanted me to talk to you.

But you know what? I don't think commercial software is so much different from open source in terms of the critical features. Although not sophisticated, many open source software can very well handle constructing a network diagram, identifying a critical path, and so on. So I believe your people think open source software is slow because they don't know how to do scheduling properly. They've never been trained. This is fine. I can set up a training session for them.

But first of all, I strongly suggest you to consider establishing some criteria for project classifications. Your managers don't have consistent ways of managing their projects. Big or small projects were managed differently depending

---

[1]In this case, the term "program" refers to written programs, procedures, or rules pertaining to the operation of a computer system, and wasn't used to indicate a set of interrelated projects.

on each person's experience. To develop and implement project management methodology, we need to have a *documented* approach for performing project activities in an *accountable, consistent,* and *repeatable* manner. We need everyone's buy-in on this.

According to the information I learned from the interviews, I would suggest dividing your projects into two groups. One group is called "high priority" project bucket and the other "low-to-medium priority" project bucket. And we can use these criteria to filter our projects:

- Business Alignment
- Regulation-related Effort
- Security Impact to IT System
- Business Impact on a Large Scale in Terms of Company Revenue

These criteria were identified during the interviews days ago, waiting for your approval. Any projects that fit into one or more of these criteria will be called the "high priority" projects, and these are what we need to pay attention to the most. Those that do not fall into any of these criteria, which are the majority, we'll call them "low-to-medium priority" projects. I drafted the initial guidelines for managing these projects. Please take a look.

Guidelines for a Low-to-Medium Priority Project (Subject to Change)

1. A project manager is required to identify an expected completion date.
2. A project manager is required to report the status at the end of the project life cycle.
3. A Gantt chart is optional.

Guidelines for a High-Priority Project (Subject to Change)

1. A project manager is required to identify major milestones (and dates) *and* expected completion date for a project.
2. A project manager is required to report the status at each major milestone.
3. A Gantt chart and critical path determination are mandatory except if a project has a very short timeline (less than two months for software development projects and three weeks for hardware projects)

John: I think we will have to refine it a little. Let me call a meeting with my people next week. But this is a good start.

Sammy: Sure. I'd like everyone to get involved and agree on the approach. Second, I'd like to recommend a concept of project dashboard as the monitoring and controlling tool for all projects. The dashboard concept has been widely used recently as an indicator to show the status of each project using colors.

Basically, the commonly used colors are green, yellow, and red, like a traffic light. A green means projects are doing fine. A yellow indicates a heads-up. And a red means management help is immediately needed. We can also use as many colors as you wish. There are many commercial software packages on the market that do the work very well. But I know this is against your policy. So I suggest we develop one of our own on the spreadsheet (see Table 17.1).

Of course, you can add more information that you think appropriate to the dashboard like the project customers or the size of the project team. This is just a draft. This dashboard should be stored in the network location that everyone can access. Each week, let's say every Thursday, you require all of your project managers to provide their input to the dashboard. Then, Friday morning you'll be able to view the status of each project online.

John: How are we going to do this? If we store this dashboard in the directory that everyone can access, it means someone or everyone can mess with it.

Sammy: That's true. So I recommend you assign someone to compile all the inputs from the project managers. Let's say Project Manager 5 (PM 5) is assigned to be responsible to collect the dashboard data from Project Managers 1 to 4 (PM 1 to PM 4). PM 5 will put the collected information on the dashboard every week. Once the file has been updated, it will be made as

Table 17.1 Dashboard Example

Project	Brief Scope	Status If Yellow or Red, Explanation & Corrective Actions Are Needed	Priority	Responsible Person (Project Manager)	Expected Finish	Budget
Barcode Initiative	Create a barcode system to store employee information and use with the employee badges	Green	High priority	Mr. John Doe	December 2009	$76,000
Online Statement	Create an online system for customers to view the statement and request for help	Green	Low-to-medium priority	Ms. Jane Doe	September 2009	$62,000

Figure 17.1   Reporting Concept

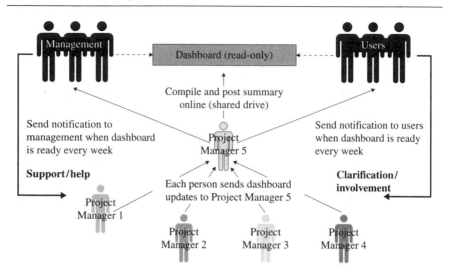

read-only before being saved to the network to prevent further corrections and/ or accidental modifications. Then, PM 5 will send a notification to you that the dashboard is ready to be viewed. This is also a way of limiting the requirement changes during the development phase without your approval. Here is the idea (see Figure 17.1).

I also strongly encourage opening this dashboard to your customers. They are internal customers, right? So they should not have any problem in accessing this file, if you grant them access. You know, there's a lot of research identifying customer involvement as one of the top success factors in software development projects. This dashboard will be one way to increase the degree in which the customers will get involved. They can come to the dashboard any time, and if they have questions or suggestions for your project teams, they can contact your teams directly.

John: Okay, that sounds doable. Let me call a meeting, and see what my people think about it.

Discussion items

1. In managing software projects, to what extent do you agree (or disagree) with the OSSOP! policy? Why?
2. What do you think would be the next step? What would be the team's reaction to the new approach to project management tools?

3. Do you think Sammy's recommendations would work? How would you amend these recommendations? What sort of additional channels could be used to make the dashboard more visible in addition to storing it on the network drive?
4. What would be the major challenges in implementing the project classification method and the dashboard concept for monitoring and controlling purposes? How would you overcome such barriers?

That Which Is Not Earned Is Never Valued

Sabin Srivannaboon

DXT

Famous for its reliable service and low price campaigns, DXT was one of the well-known mobile operators in the area. The company customer base comprises prepaid customers, and it was doing well in this territory. To take the company into another level, the management team started looking at expanding the company business into an uncharted area—the postpaid customer segment.

DXT was a program-driven company. Last year, DXT initiated more than 40 programs, addressing various issues from the organizational structure to the competition perspectives. This year, many programs were dedicated to the postpaid customer segment as efforts in diversifying the company business. However, the postpaid customer segment was something that DXT wasn't familiar with. And the company was still unclear as to what competitive advantage it would and should provide to its postpaid customers. As a result, several programs were moving in different directions, and the company ended up nowhere. The deadlines for many programs were clearly defined from the beginning. Nevertheless, DXT failed to provide a clear idea of what kind of strategies and actions they really wanted to see. Eventually, many programs failed. The lack of the alignment seemed to be a major issue at DXT these days.

XTRA AND THE EARNED VALUE CONCEPT

Xtra was a program that was carried out at the end of the year. Because of its strategic importance, the program delay wasn't acceptable. To make sure the program would be completed on time, the program team decided to use the Earned Value concept. Although it was a good intention, the program team faced one big challenge: the earned value concept had never been implemented in the company before! And as one could expect, the team ran into multiple problems and difficulties not only in managing the program itself, but also understanding and using the earned value concept.

Figure 17.2 is the Earned Value Chart of the recently closed-out Xtra program. The program objective was to develop a set of marketing projects for DXT's postpaid customer segment over a period of three months.

WHAT WENT WRONG?

Trying to understand what went wrong and the earned value concept better for future programs, the team looked back at the history of the Earned Value Chart.

Although the Xtra program status appeared to be ahead of the schedule and budget plans at the beginning, the turning point was around December 21, the long vacation period, because of a lack of the company's clear direction. Since then, things got worse as the program progressed. As an effort to recover the program status, the Xtra program team tried to cut down several program scopes with hopes of improving the earned value. But problems never ceased coming. In the end, the program was not fully able to recover, and less work was accomplished than planned. In other words, the program was not able to deliver its full results. Eventually, the program was called off. At the termination time (February 22), the program was behind both schedule and budget. The Xtra program ended up spending $10,000, but only accomplished $8,600 worth of work, while the forecasted budget at completion was $13,000 (March 8).

One major lesson learned was that in implementing the earned value concept both the Schedule Performance Index (SPI) and Cost Performance Index (CPI)

Figure 17.2   The Earned Value Chart of the Xtra Program

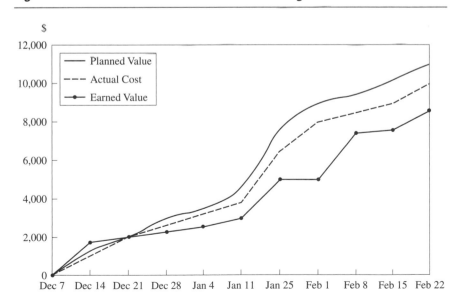

were important indicators to watch, but the CPI was clearly the more sensitive factor because a poor CPI was likely to be nonrecoverable. The program team should have monitored closely the trend of the CPI throughout the program life cycle. The SPI, on the other hand, was more important during the early phases, but became less significant as the program neared completion.

Discussion items

1. Analyze the program history and calculate the program schedule performance index (SPI) and the program cost performance index (CPI) at different major points in time of the program life cycle.
2. What could have been done when the cost and time overruns were detected?
3. What would be the major challenges in implementing the earned value concept? How can such challenges be overcome?
4. Why does the SPI become less significant as the program neared completion?

Chapter 18

PROGRAM CLOSING PROCESS AND PROGRAMS IN ACTION

This chapter addresses the program closing process and gives several examples of program management in inclusive settings. The chapter consists of one critical incident and four comprehensive cases, each one with different areas of focus.

1. A Checklist

    This is a critical incident case of the program closure and the expectations during the closure phase.

2. General Public Hospital

    This comprehensive case describes a situation when the program team wants to make what it considered to be a big change. However, the program hits a stalemate at the baseline gate review, in which senior management is considering the cancellation of the program. Creative thinking on the part of the program manager breaks the stalemate and brings the possibility of approval to progress to the next phase of the program life cycle.

3. American Shogun

    This comprehensive case shows how program management is made simple in a high-tech company when the fundamental principles of the discipline are followed. The teams' dedication to coordinated collaboration between projects, focus on business goals and the bottom line, understanding of cross-project dependencies, and effectively using horizontal and vertical management techniques are the keys to the program's success.

4. Planet Orbits

This comprehensive case offers a story of the possibility of extraterrestrial life. However, conflict between scientist and organizational management is highlighted. Senior management ignores the cosmic glory of "to boldly go where no one has gone before," and focuses on earthly issues of schedule delay and cross-site organization. After waiting for more than a decade for funding of the program, the scientists are not willing to see it stopped and are ready for a fight.

5. ConSoul Software

This comprehensive case demonstrates good practices of the program management discipline, as well as a couple of opportunities for improvement. This example shows how business strategy drives the program management practices, structure, methods, metrics, and tools. In particular, the case demonstrates how business strategy influences trade-off decisions on a program. This example also shows the impact of not using consistent scheduling and budgeting processes to manage both projects and programs.

CHAPTER SUMMARY

Name of Case	Area Supported by Case	Case Type	Author of Case
A Checklist	Close Program	Critical Incident	Sabin Srivannaboon, Dragan Z. Milosevic, and Peerasit Patanakul
General Public Hospital	The Standard for Program Management, The Program Management Knowledge Areas	Comprehensive Case	Peerasit Patanakul and Dragan Z. Milosevic
American Shogun	The Program Management Framework	Comprehensive Case	Bjoern Bierl and Andrea Hayes-Martinelli
Planet Orbits	The Standard for Program Management, The Program Management Knowledge Areas	Comprehensive Case	Peerasit Patanakul and Dragan Z. Milosevic
ConSoul Software	Business Strategy and Program Management	Comprehensive Case	Andrea Hayes-Martinelli, Dragan Z. Milosevic

A Checklist

Sabin Srivannaboon, Dragan Z. Milosevic, and Peerasit Patanakul

Similar to a project, a program has its own life cycle from start to completion. It is born; it lives; and eventually it dies. And when it dies, the joy of discovery and the excitement of team compositions are about to be history. Nevertheless, the closure process is never easy, as administrative dislocation is often an issue. A program faces termination either because its charter has been fulfilled or conditions arise that bring the program to an early close. In the former, the closure begins after a phase-gate review of the delivery of program benefits, where the product is delivered, accepted by the customer, and/or transited into an operation. In the latter, the program is stopped because it may be unsuccessful or has been superseded.

VACATION TIME, ALMOST

James Powell is more than ready to take a long vacation in Hawaii with his family. He just needs to finish his work, go back home, and catch the flight! This sounds simple, but the work still keeps James busy at his office even now, at 6:30 PM on Friday. James is a program manager, who has been managing a new product development program for six months, and now it is about time to cease it. Even though the program is almost completed, James still needs to prepare a checklist of what needs to be done during his program closure. He wishes it could be just a list of things to see in Hawaii.

James knows that projects under the program expect to be closed before the program is terminated. And the program closure should capture important information such as lessons learned and customer's sign-off. He also knows that the formal acceptance of the program should be achieved by reviewing the program scope and the closure documents of the program, and by reviewing the results of any verification of deliverables against the program requirements. All of this will help James learn about things that lead to success and/or failure for future programs in the company.

As he is going through the documents, James starts jotting down some notes:

- Assure all deliverables have been completed and the program completion criteria have been met.
- Obtain customer sign-off or an agreement that the program has finished and that no more work will be carried out.
- Review significant feedback from customers.
- Release the program resources to other programs.
- Analyze the program results including lessons learned, which address the following:
  - Did the delivered product/solution meet the business requirements and objectives? What did we miss? What did we learn from this program, strategically and operationally?
  - Was the customer satisfied? What did they like? What didn't they like?
  - Was the program schedule met? Could schedule pull-in opportunities be identified for future programs?
  - Was the program completed within its budget forecast? Could cost reduction opportunities be identified for future programs?
  - Were the risks identified and mitigated? Could it be used for future programs?
  - What could have been done differently?
- Assure the lessons learned results have been shared in appropriate venues.
- Assure all required documents have been archived.
- Celebrate the program completion!

James thinks his list is comprehensive enough. Now it's time to go home.

Discussion items

1. How would you change the list?
2. Who should be involved in the closure review?
3. To what degree should the postmortem be comprehensive? Why?

General Public Hospital

Peerasit Patanakul and Dragan Z. Milosevic

INTRODUCTION

The time is 6:30 AM, midweek, in the month of February.

Julia Skown is the program manager for the TKS Program. She is in a meeting room looking at the agenda for a program review meeting to assess the status on the TKS Program scheduled to begin at 8:00. In the room with Julia are Stacey Cook, a payroll specialist, and Tom Black, an information technology (IT) project manager. They are members of the program core team, and all employees of General Public Hospital (GPH). TKS is a program that was initiated to replace the current Time Keeping System, and it is obvious that all three are very nervous regarding the potential outcome for their program.

There is good reason for them to be nervous. The purpose of the meeting today is a program baseline review (Go/Kill/Hold) with senior management. This represents a big decision: *Go* means the program is ready to proceed; *Kill* means the program will be terminated; *Hold* means the program needs to be reworked and brought back to the review. The team is concerned that senior management may decide that their program must be put on hold. This decision will mean that the TKS program will be sent back for rework and lose its priority for available resources. In other words, TKS may be put on the back burner.

PROGRAM BACKGROUND

The IT organization for GPH took the lead in championing the TKS program. Operating under the Chief Information Officer (CIO), the group provides various types of services and solutions for both internal (doctors, nurses, students) and external (patients) customers. Some of those services and solutions include the registration system, employees' clock in/out system, payment system for patients, and computing power for medical research, which is among the best in the United States. The vision of the hospital is to be a national and international leader in healthcare, education, research, and technology development. To help accomplish the vision, strategic goals and objectives for the IT group are provided as the basis for the development of specific strategies and programs. Generally the

Figure 18.1   Organizational Chart and TKS Program Team Structure

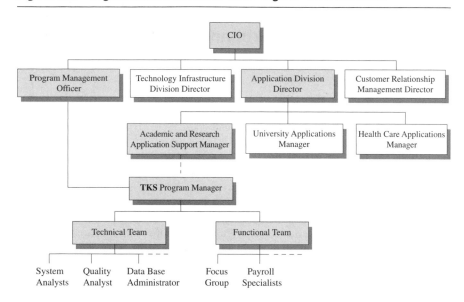

uniqueness of the IT organization includes the fact that it is operating in one of a few academic medical centers in the Northwest United States that employs a "best of breed" strategy. This means that the group achieves high-quality solutions by using best-in-breed commercial products, rather than developing their own and integrating the solutions in a patchwork fashion. Says John Menegy, CIO, "We work with individual vendors, trying to find the best solutions, and then we successfully integrate the solutions with our existing products."

There are more than 300 employees in three major groups under the Applications Division Director of the IT organization. These three groups include Academic and Research Applications Support, the University Applications, and Health Care Applications. The TKS program resides in the Academic and Research Applications Support group as depicted in Figure 18.1.

TKS is a formal program which has a program manager and a team of skilled and knowledgeable members. All the members are motivated and committed to the program.

POTENTIAL PROGRAM SCOPE CHANGE

It had been emphasized since the beginning of the program that the objective of TKS was to implement a new time keeping system for GPH. The system's replacement represented an operating necessity. It was a $1 million program and was anticipated to take 18 months to complete. The motivation for the program came from the fact that the vendor which provided services for the current time keeping system had sold its system to another company. As a result, GPH was forced to make a change because the support for the current time keeping system would be discontinued by the new vendor. Management was faced with two options: (1) purchase a new system from the new vendor, or (2) seek other vendors for time keeping systems. The first option appeared to offer fewer risks (familiarity with the previous generation of the system) and enabled GPH to keep the front end system for its employees (e.g., tax system, clock in, clock out).

Julia, Stacey, and Tom came to the office early to do last-minute preparations before the management review. Still concerned about the outcome of the review meeting, they discussed the major issues.

> **Julia:** Stacey, let me ask you this. Frankly, what do you think about our program so far?
>
> **Stacey:** I think we are doing great. There have been problems here and there, but I think we took care of them.
>
> **Julia:** Okay. But Stacey, you are the team member representing our customers. You are going to be the one who uses this new system to manage the payroll activities. If you anticipate any problems that we can prevent from happening, please let us know. I know I've said this maybe a hundred times already, but we really want the new system to function properly.
>
> **Stacey:** Don't worry about this, Julia. I will tell you if I see anything wrong. I know that the quality of the new system is our top priority. Plus, I don't want a system that cannot pay people accurately. That's why I am here. Thank you, though, for involving me in this program and valuing my opinions so far.

The discussion turned to adding a new transaction inquiry feature of the TKS system, which will make the system much more user friendly, a feature that currently is not in the program plan. However, the new feature imposes more technical difficulty than what the team had expected and will cause an increase in the program cost and schedule.

> **Julia:** So, the question here is whether we will move on to the next phase without this feature, or do we want to spend more time adding it to the new system? How long will it take us to add this feature, Tom?
>
> **Tom:** On the technical side, it will take us at least two months to do, meaning that total program delay will be two months.
>
> **Julia:** And you said that we have to buy new hardware since the current system is not compatible, correct?

Tom: Correct. But we already budgeted for most of the hardware; just a couple of pieces that we didn't know we needed at the beginning of the program.

Julia: How much will this cost us?

Tom: Our vendor told me that this additional hardware will cost us about $100,000.

Julia: $100,000 is a lot of money. What is their lead time?

Tom: Well, it will take about three weeks. I already included this when I told you that it will take us at least two months to add this feature.

Julia: All right. Do our existing testing and implementation plans account for this new feature?

Tom: Mostly, yes. But you never know what to expect with this new hardware. It is a program risk, but probably only of medium severity. Our vendor is familiar with this new hardware and should be able to help us in a timely manner if we encounter any problems.

Julia: Okay. So, in a nutshell, if we add this new feature we will have about a two-month delay and be $100,000 over the planned budget. This means that instead of us implementing the new system at the beginning of July, we will have to wait until the beginning of September. Plus, we will end up well over budget.

Stacey: But if we do not add this feature now, we will be on schedule, right?

Julia: Yes, if everything goes according to our plan. Our options are to add this feature later when we do a future system upgrade, or we can spend more time and cost now.

Tom: Julia, what do you think we should recommend to the Review Committee?

Julia: First, I don't think they will let me go into a lot of discussion on the topic. They pretty much know what is going on in TKS because as you know I talked to them a few times in the last month. Second, there are not many options to consider. Third, I will tell them to consult the focus group because they represent our customers and other stakeholders. We have arranged to have the focus group stand by so they can come in five minutes into the review, assuming the Committee agrees that they should participate.

Stacey and Tom headed out to get coffee. With just a few minutes remaining before the review meeting, Julia used the time to review the various elements of the TKS program.

TKS PROGRAM TEAM

The structure of the TKS program was a dedicated, autonomous team, including representatives from several functional groups (for example, hardware engineers, internal and field technical services personnel). The technical team consisted of three system analysts, a quality analyst, a project manager, and a database administrator. Customer representatives were also part of the team. Payroll specialists were considered as primary customers. Also representing the customers was a focus group that was chosen by the functional sponsors. Most team members were selected based on their skills and knowledge. The team members were motivated and worked well as a group. The TKS program structure was typical for programs in the organization. In addition, the TKS program also had strong support from the program management office (PMO). The PMO was responsible for developing, implementing, and continuously improving the program management processes and tools to better program performance, and to assist the IT organization achieve its business objectives.

PROGRAM STRATEGY

The healthcare industry places a large emphasis on the quality of the products and services it develops. It is translated into the business strategy of GPH, where it primarily focuses on maintaining a stable, limited line of products or services; it offers best quality with low cost; and it delivers what its customers want. Influenced by the business strategy, program management elements are directed to satisfy customer needs through the delivery of quality products and services. The first priority of the TKS program is to install the system and make it work correctly with all other existing systems. The major constraints are cost and schedule, since the program has limited resources and a strict go-live date, previously announced to all employees.

Objectives: The objective of the TKS program is to implement a new time keeping system to the organization. The target customers are payroll specialists and the users of the new system.

Product definition: The product is a time keeping system that stores all employee-related transactions. More precisely, TKS has to integrate employee data, leave accrual balances, labor schedules, and modified pay rules into a new system. This includes developing new interfaces, training 500 time keepers, and deploying the application to all personal computing devices. Even though it is an off-the-shelf product, the team had to work on the details with the vendor, especially the product interfaces with the existing hardware which required some modifications by the vendor. The inputs from payroll specialists are vital for getting initial set-up and configuration of the system. Resources are required from the Payroll office, Field Technical Services, Network Applications, Unix Administration and Database Administration, and a focus group consisting of time keepers and managers from a cross-section of the hospital.

Value proposition: The advantage of the new solution include the fact that other products available are not nearly as mature or advanced as the one the company is employing. TKS intended to give GPH the best solution available today. In addition, the new system has the advantage over the previous product generation in terms of reporting capability, scheduling capability, robustness, and various types of features. From a cost perspective, the new system helps reduce the number of part-time employees due to the efficiencies gained from the new generation of product.

Success criteria: The primary success measure of the program was the accuracy of the system. Other success criteria include the following: user-friendly interface, July go-live, and $1 million total cost.

These success measures were recognized upfront and are well articulated in the program. To meet these success factors, the team understands that the program must have clear goals and scope, a high level of communication, effective system testing, and a high level of stakeholder support and buy-in. In addition, the effective use of program management practices are understood as factors contributing to program success.

PROGRAM PROCESS AND TOOLS

Closely following the standard process the PMO created, TKS has four major phases to complete—Integration and Request, Program Planning, Program Execution, and Implementation and Support—and three major decision checkpoints: functional review, baseline review, and customer approval (see Figure 18.2).

The TKS program is currently at the end of stage two—the program planning stage. This process was influenced by the organization's strategy on the basis of having a stable, standardized, and tested process with quality and customer focus. This standard process is referred to as the Program Management Solution Development Life Cycle (PMSDLC). To ensure the use of the process,

Figure 18.2 Program Phases

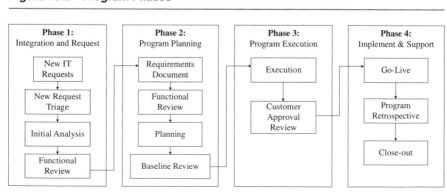

the management demanded a two-day PMSDLC process training class where all senior management, program managers, and major team members were required to attend. Customers also play a major role in the program process to drive the process flow. Program and project tools focus on quality assurance, cost, schedule, and performance criteria.

Not all programs within the IT organization are required to go through all the standard phases and milestones. The program life cycle is "scaleable" upon the type of the program with keen emphasis on achieving quality with minimum cost. Maintenance programs, which are those that require minimum efforts (a couple of weeks to a couple of months in duration) and are initiated to fix some glitches in existing products or services, go through an accelerated life cycle. Minor programs are those that require in-depth planning since they involve some degree of risk and impact on the organization. A minor program normally lasts from two months to a year in duration and will pass through all phases of the PMSDLC, but require fewer Phase 1 steps. A major program like TKS is required to go through the entire program life cycle. A major program normally takes from six months to three years in duration.

The scope of the TKS program was defined based on business requirements. The program schedule was developed on the basis of standard hour estimates of previous programs. A Gantt chart was created to support the development of the program schedule. The program manager worked with the technical sponsor, the vendor, and the program core team to determine the budget. This included the cost of hardware, software, consulting fees, licensing fees, resource hours, and training. Even though TKS is considered a low-risk program since it involves few new technologies, the team follows the standard risk management process. The program team documents the probability of multiple risks that may occur in the program in terms of dates and/or dollars. The program team created the risk list and developed contingency plans to address and resolve each risk as needed.

At TDK, the program manager is responsible for achieving the program goals. The program manager holds regularly scheduled meetings and documents the minutes from these meetings with action items for ongoing follow-up by the core team. In terms of quality, the program has a quality analyst whose responsibility it is to ensure that the program will satisfy the needs for which it was undertaken. Activities such as measuring, examining, and testing were performed to determine whether the results conformed to the requirements. Actions are taken to bring defective or nonconforming items into compliance to the specification requirements. Checklists are used to track quality.

In the TKS program, the means of communication include program meetings, program review presentations, emails, and phone. The program manager meets with the program steering committee in order to achieve program approval. After approval, the program manager reports the program progress back to the committee periodically through formal meetings that are held once a quarter. The program manager meets with the program team daily while involving the customers on a regular basis. In addition, the program has an integration process to make sure that the various elements of the program are properly coordinated. It is done through an iterative process of multiple meetings and consistent communications on a regular basis.

PROGRAM METRICS

The significant program metrics tracked on the TKS program include milestone completion, risk mitigation, budget tracking, and a number of key program scope changes. Program schedule is tracked using milestones and tasks performed to the schedule baseline, forecast schedule, and actual schedule. Program cost is tracked by expenditure, account balance, hardware and software cost, consulting fee, and training and staff cost. This cost metric is tracked separately by the program manager in the program reviews and is not shared with all team members. The program risks are tracked by the mitigated risks versus exposure.

PROGRAM CULTURE

In general, the organizational culture is diverse, complex, formal, and relatively inflexible to change. Quality and customer focus are valued highly by all employees.

GPH has a wide array of employee types—physicians, nurses, administrators, engineers, and technicians—which make it very diverse in personalities. To control this diversity, GPH opted for a more rigid and hierarchal form of organizational structure. In addition, most employees in the organization are reluctant to change, especially as it pertains to new technology.

Because the time keeping system is an emotional subject for everyone in the organization, the TKS program team has to be sensitive to the needs of the customers and convince them to accept the new change. Quality and accuracy is the key in the minds of the customers. The program has consistent, regular, and open communication which contributes to team members and customer representatives understanding their roles and responsibilities.

THE PROGRAM BASELINE REVIEW MEETING

The door to the management conference room opened and Julia and her team were invited to join the program review committee. The committee consisted of John Menegy (GPH CIO), John Bacon (customer relationship director), and Art Counter (payroll director).

> **John M.:** This is the baseline review meeting in which we will decide whether to approve the TKS program to move into Phase 4 of development. Since we have discussed the status of your program with you several times over the past

few weeks, I think we understand what's going on. What is the agenda for today's meeting?

Julia: Here is the agenda:

- Summary of program status (program manager)
- Schedule update
- Financial update
- Risks mitigation activities
- Resource status
- Key customer events and interactions
- Presentation of the goal accomplishment

John M.: I don't think we need to cover all agenda items. As I stated, we pretty much know your status. But we want to hear if there is anything new to add.

Julia: Well, adding the new transaction inquiry feature we were asked to assess will add two months to the schedule and $100,000 in cost. It is a lot, we know. But we need to ask our customers and other stakeholders if they believe the new feature is necessary.

John B.: You mean to ask our focus group?

Julia: Yes, I think they are the right people to ask about this new feature. They are the ones who clock in and clock out every day. Plus they are the representatives from their functional departments.

Tom: Quality is our focus. All in all, we want the new system to work accurately and be easy to use and maintain. And we all know that the payroll specialists and the focus group are our customers. I think we need feedback from the focus group.

John M.: How long will it take for them to get here?

Julia: Five minutes—they are waiting for our call. All five of them.

John M.: Wow, you've done your preparation. John and Art, do you agree that we bring the focus group in here?

John B. and Art: Yes, agreed.

The focus group joined the meeting. For several minutes, the committee talked to each member of the focus group. They all believed that it was in the best interest of GPH to add the new feature now and proceed with the program execution. However, the committee members were well aware of the significant adverse ramifications to a slip in the schedule for completion of the program. The benefits of adding the new feature versus the costs of delaying the introduction were weighed carefully. John M. consulted with the other members of the committee and announced to the team that they will need to deliberate with the committee and will communicate their decision in half an hour.

With that announcement, Julia excused herself to go check her email. When she got to her office, she stood for a few minutes before opening her email account and thought to herself, *Half an hour—that feels like an eternity! I think the information from the focus group will save the program. At worst they will come back with a 'hold' decision, but no kill decision.* After responding to a few emails from her program team, Julia returned to the meeting to hear the executive team's decision.

Discussion items

1. What are the key takeaways from this case?
2. What do you think would/should happen?
3. What are the advantages and disadvantages of using the focus group?
4. Of all the various elements of the TKS program, what element do you think is the most challenging to manage? Why?

American Shogun

Bjoern Bierl and Andrea Hayes-Martinelli

GETTING STARTED

It was in late May 2002 when Jan Vesely, sales manager for Southeast Asia and the Pacific region at International Instruments, Inc., received a call from RisingSun, one of its key accounts in Japan. "They told us that they were interested in our 1001 series monitors if we were able to provide audio capability—a feature that our competitor already had implemented in their product," Jan said. "Additionally, RisingSun wanted us to deliver the product in 11 months, which was an aggressive time-to-market goal. Since RisingSun was one of our most important customers, we jumped into action." International Instruments, Inc. was a global market leader in the field of monitoring systems, and the 1001 series was their main product line of monitors addressing the biggest segment of the overall market.

The audio capability for the 1001 series monitors was previously discussed because, as mentioned, a major competitor had already brought a monitor with audio capability to the market. But Manuel Scriba, the segment manager for the 1001 product line, found the market too small to justify adding the audio feature, but the telephone call from RisingSun changed everything. As recalled by Manuel, "Suddenly the program, named Shogun, would help us to meet our financial, market share, customer relationship, and competitive business goals," he said. "First and foremost," he commented, "a new program had to fulfill our business goals. That's what it is all about—the business goals."

As for the financial goal, the order was large enough to cover the development and research cost associated with the program and make the desired contribution to the company bottom line. But it was more than that. International Instruments, Inc. was focused on market share and customer relationship as key strategic goals. It was clear that Shogun would support the achievement of these strategic goals. "We had excellent customer relationships before this program, and if an important customer wanted to have the new feature—and the program was financially viable—what else could we do but satisfy them?" asked Manuel. "On top of that," Manuel added, "Shogun would also provide gain in market share for this monitor product line. Not only could we increase our market share, but we could attain our competitive goal, which was to preempt our competitor from gaining more market share in Japan."

Manuel proposed the new program to Robin Weiland, vice-president of International Instruments, Inc. "We had the chance to increase our market share," recalled Robin. "It was tough, but feasible. So the question was no longer, *Does it make sense?*, but rather, *Can we get this done in only 11 months?*" The next steps were to assign a program manager and set up the program as soon as possible.

THE FOCAL POINT FOR BUSINESS RESULTS

Melanie Lehr came from a strategic marketing position and was new to program management. As she remembered, "I was new to the company and new to program management, but I knew the company pretty well from my former jobs. This program was about to become a great challenge for me and the company. But I was glad to have the support from the program management office."

The program management office, a knowledge base for the program management activities throughout the company, provided not only standardized but flexible processes and tools for each of the programs. "We don't expect naturally born program managers," said Bob Mitchell, head of the program management office. "Program managers are made; they are trained on the job and in classrooms. Through the years, we developed a skill set map for program managers. A program manager's task is strategic—the focal point for business success. They must have strategic skills. They lead all kinds of people—some easy to lead, some difficult. Therefore, they need leadership skills. They face tough times, requiring tough character. Hence, we expect them to have a special set of intrapersonal skills. Programs cost a lot of money and require many resources, so program managers need to have financial skills. Since we develop highly technical products, program managers should have a working level of technical skills. Finally, the customer must be understood inside and out, requiring program managers to have customer skills. In summary, we require six sets of skills from our program managers."

Like every program manager, Melanie was held accountable for business results. Therefore, her job was one of the most critical in the company. "She did not have all of the skills we require," said Bob. She was, however, an experienced engineer with an MBA, eight years of experience in design engineering and strategic marketing, and was involved with many programs as a member of extended teams and as a project manager. She had all of the required skills except for the program management set. "We planned to promote Melanie to program manager in the long term," Bob added. "But when Shogun was added to the product roadmap, we did not have a seasoned person available, so we assigned the program to her. New program managers need as much help as possible. Mentoring from others who have had the same challenges is the best way to teach them. We solicited the help of an excellent mentor, which provided the program management skill set Melanie was missing. Standardized processes and tools also helped her master the challenge. This example demonstrates that you do not need to reinvent the wheel. And Melanie did a great job."

"What helped me most," Melanie recalled, "was the mentoring part. Marcel Greenhill was a program manager for 20-plus years and did all kinds of programs. His mentoring and support taught me all of the small tricks and tips that helped during the start-up phase, which was the toughest for me. Setting up a program required tremendous effort because it involved multiple projects and disciplines in the company, such as engineering, marketing, finance, production, and sales. Cross-discipline development was really important for achieving the business goals."

INTERDEPENDENT PROJECTS

Preparations were undertaken to assemble the program team. A program requires a lot of different people with different backgrounds, making it difficult to coordinate them within one team. The usual program management approach at International Instruments, Inc. was to define two layers—a program core team and several extended teams. The program core team for Shogun consisted of 12 members. They represented the functions that had major involvement, such as marketing, mechanical engineering, manufacturing, purchasing, and finance. Additional required functions for the program were also represented—like systems, engineering hardware and software, quality control, promotion for the new product, and introduction support. Each member of the program core team led his or her extended team.

Melanie commented, "It wasn't like running 12 different projects and assembling them at various stages throughout the process. We had to break down the whole program into multiple interdependent projects. Each project manager managed a project within his or her own function or discipline. I, on the other hand, managed across projects, coordinating them to make sure that functional objectives were in tune with the business objectives. Essentially, my job was cross-project and cross-disciplinary."

The projects on the Shogun program were all interrelated and dependent on each other. For instance, if marketing could not finalize the required specifications with the customer, engineering could not start to develop required features, quality could not be controlled, and manufacturing would be delayed.

Managing the interdependencies between the projects was a real challenge for Melanie. "I knew that I would not be able to manage all of the projects in detail and take a close look at everything," she said. "So I needed to make sure that each project could function individually. At the same time, everybody needed to be aware of the consequences of their actions for the other project teams. Projects have delays, no matter how well you plan ahead. The only question is whether or not the team has enough of a chance to 'extinguish' the fire before it affects the other projects and finally the whole program. Transparency is important."

BUSINESS GOALS

Shogun was started with a specific set of goals—profitability index, market share, development and product cost, product performance, and time. Development cost was estimated at $5.1 million, while the performance was set by the customer's requirements. Product cost required more detailed input for the exact configuration of features. The timeframe was a tough constraint because the product needed to be delivered in 11 months

COORDINATED PLANNING

Jumping into action, Melanie worked out a rough plan with the program core team. Each project was addressed with initial budget and timeline identified. Cross-project interdependencies were identified in a very rough manner.

These outlines were then given to the program core team members, who, because they represented their functional projects, had the best insights. Each of the projects was then broken down into detail by the responsible project managers. They added their insights on scheduling and resources, and their experience of how to complete a program of this scope.

"I was given the outline for the broader marketing function," remembered Christian Foyer, who was the marketing project manager. "I had to address steps like customer validation and continuous screening for market requirements. The way to achieve these goals was up to me and my extended team. Melanie only needed to know our results, deliverables, and risks. I sat down with my team and we developed our project work breakdown structure (WBS) to deliver our part of the program. My extended team could thus focus on the marketing jobs, while I supervised the work and ensured alignment with the other projects."

After the project managers had created the detailed WBS for their individual projects, the program core team, under the guidance of Melanie, assembled a complete WBS for the program. The WBS outlined the specific tasks for each of the functions in respect to other projects, or more precisely, cross-project dependencies. Only those work packages that had an impact on other projects were considered important on the program level. More detailed outlines and scheduling were the responsibility of the program core team members.

"This was not always easy," Christian commented. "I had created the WBS for my team and, although it was a tough timeframe, we managed to get all of the parts completed in time—at least on the outline. But when we sat down in the program core team meeting and worked through the WBS, Melanie required some changes. These were mostly earlier deadlines because there were other projects that needed my input sooner. So we accelerated some tasks and left others with more time. But it made sense—not for the marketing function, but for the overall program. This understanding became crucial. We constantly reassessed the program progress in the program core team meetings and adjusted for problems."

The program core team then agreed on important milestones that were tracked. Tools like the program strike zone and the program dashboard enabled Melanie to effectively track the progress of Shogun and all of its projects. The dashboard provided an overview of the current status, and the program strike zone showed whenever a program-level milestone was significantly delayed. "Communication is the key for everything," Melanie stated. "But you always have a mentality of 'bad news never travels up.' Tools like the program strike zone and program dashboard helped us to have an objective measurement and information system that everybody could agree on."

COORDINATED MANAGEMENT

The program involved nearly every function within the company, and tight interdependency between all of the projects was a given. Communications became a must-do on a regular basis. "We had regularly scheduled meetings and communication patterns, as well as more informal and driven-by-demand communication," recalled Melanie. "If there was a problem, there needed to be communication. Nothing else mattered."

The program core team met once a week to discuss program progress and issues. If changes were needed, the whole team had to agree on them. Alignment of the project teams was important, especially with the short timeframe. "The whole process is not always straightforward, but it helps a lot if the members know each other very well and communicate across the functions to solve smaller problems on their own," commented Melanie.

In addition to the program core team meetings, some of the functions needed to work together more closely and more often. As an example, the engineering function required input from marketing for prioritizing the features. Prioritization of features drove prioritization of work for the design team.

The program core team members met with their project teams as required. The engineering project manager, Gregory Wolfe, recalled his approach. "I had the largest project team comprised of eight engineers," he said. "To keep them updated, we met every week directly after the program core team meeting. In doing that, I ensured that they had the most recent information and that they felt more involved with the program. Of course, I did not need to deliver all of the information. Other than the business view of the program, my project was focused solely on the technical work—to get the product engineered in time and in the desired cost range. But there was also a motivational aspect to keep everybody focused on the program vision and strategy."

Communication in the program did not only flow from the top down. In International Instruments, Inc., programs are critical to the company's business strategy and the achievement of strategic business goals. This importance was reflected in the program reviews for senior management which were held on a monthly basis. Several vice-presidents attended, headed by Robin Weiland, and spent the whole day listening to program managers' reports, dissecting issues, and removing obstacles—occasionally killing programs that were not accomplishing their strategic goals. The program managers' reports were based on the information contained in the program strike zone and dashboard.

Melanie commented, "What senior management did for the program was what I did for each of the projects. In the end, management is responsible for the program. So they need to know what's going on—not every single step, but the major milestones. So I kept them updated—just like the program core team and the functional representatives kept me updated. It was just on a broader scale. The closer the program came to its desired launch date, the more often there were management reviews. Management wanted to increase control. That was tough on one hand, because it sometimes took time that I could have spent on other things. On the other hand, it ensured their support."

ALIGNING EXECUTION WITH STRATEGIC OBJECTIVES

Since the program was initiated by RisingSun's demand for an additional feature, customer focus was very strong. Christian stated, "Besides the strict validation processes in the beginning of the program, we ensured that each major step and each feature we planned to implement was validated and approved by the customer."

But RisingSun was not the only customer in that sense. "The 1001 series monitors had additional capabilities in event logging and the like," Christian said. "This was partially demanded by RisingSun, but also determined within my department to be a major value-add to customers in general. So we looked at both RisingSun's specifications and the overall market demand. We wanted to have a competitive product for the whole market."

At the midpoint of the program, Gregory realized there was a problem for the hardware project team. "In order to implement all of the desired features, we weren't going to be able to meet our deadline," he said. "We pushed hard, but final feature requests from marketing were too late, and the additional quality testing required would take us too long. At that point we needed to make a decision."

Gregory deferred the question to Christian, who commented, "This was a tough problem for us. RisingSun wanted to have a portability feature in the 1001 series, but we needed to eliminate it in order to meet the schedule deadline. For us, it was either the portability feature or the deadline, and it was hard to determine."

Christian faced a decision. "To find a solution, we examined the marketplace and called some customers to ask if they would need the portability feature," he said. "In the end, it was determined that the market didn't really need the feature, so we talked to RisingSun who agreed it was desired but not essential."

FLEXIBILITY

Flexibility was really important for the program. "It was a complex program with an aggressive schedule," Melanie commented. "You can't just develop a plan and wait for the results. We tried to plan as well as we could, but there were a considerable number of changes that occurred. We sat together in the program core team meeting and asked, 'What can we do? What does it take to implement a change, and how much would it cost?' We were flexible because we needed to be. Having a team that worked so well together was really the key for everything."

Midway through the program it became obvious that the team would need more people and more resources to accomplish the program in the desired timeframe. Thus, management decided to support the program by having engineers from the company's Japanese engineering group work on Shogun. "There were problems in the beginning," Melanie remembered. "One of the engineers was very junior and didn't speak English. We had a politically sensitive issue, but senior management helped us by moving the engineer to a program with a less aggressive schedule."

FINISHING WITH FLYING COLORS

Bringing the product to launch was still a challenge due to considerable pressure from upper management. The team had an opportunity to prove whether or not the communication and, more importantly, the informal networks were working. "The end was brutal," recalled Melanie. "We needed to get everything out as quickly as possible, and although I did as much as I could, I couldn't have managed on my own. The project teams worked well together, and everybody headed in the same direction. We wanted to meet our goals. The program was finished within the 11-month target, with higher product performance, and within the given development budget. That was the easier, more tactical part of the goals. We had to wait an additional nine months after the product launch to see if we had achieved what we strategically planned in the business case: increase profits, achieve gains in market share, and preempt a competitor. We did."

Discussion items

1. What are the lessons learned in this case?
2. What are the things we've usually observed in well-managed programs?
3. What are important skills that every good program manager should possess?

Planet Orbits

Peerasit Patanakul and Dragan Z. Milosevic

INTRODUCTION

With $467 million in total budget and 144 months in duration, Planet Orbits is an ambitious program. Its objective is to build a spacecraft with a photometer for identifying the terrestrial planets in the universe. Scientists believe that this program will eventually help them understand the extent of life on the planets and across much of the universe. "It represents, fundamentally, a breakthrough in science that has the potential to change mankind's views about his position and place in the universe," according to Eric Anderson, the Planet Orbits program manager.

Next week the Preliminary Design Review (PDR) will be held. The review committee is expected to be tough, and Eric knows from experience that the program has to be in excellent shape in order to be granted approval to move to the next phase of development (from definition to the design phase). Eric believes the program is progressing well technically, but is aware of some interpersonal and human relations issues that may be a concern to the committee. Additionally, Eric is concerned that the latest schedule has some serious disconnects with senior management's delivery expectations. However, he feels that the program core team can push the team sufficiently hard enough to make up any schedule shortfall that they may encounter during the next few development phases. However, these issues could create barriers for a PDR approval decision.

Eric is excited to have the Planet Orbits program finally ready to advance to the next phase of development. He thought to himself, "It took us almost 10 years to achieve funding approval for this program. When we submitted the proposal to the program selection committee, it required several iterations to finally reach the necessary approval to move forward to actual development." It has been three years since the program was approved. Eric still feels that the objective of the program fits well with the mission of the space agency— exploring life in the universe.

Eric has been through the status of the program several times already in preparation for the PDR meeting with the committee. However, it may be useful to recap with some of his key team members one more time to ensure all issues and concerns have been properly addressed. Eric reviewed his status documentation in preparation of Charles Wright's arrival to his office. Charles is the project manager of the local site for the Planet Orbits program. When Charles arrived, they proceeded to review the program again in comprehensive detail.

PROGRAM BACKGROUND

The Planet Orbits program is a special space mission program sponsored by the space agency. The assigned program manager, Eric Anderson, is a senior scientist with the agency. The idea for the Planet Orbits program was initiated by Eric Anderson, and had been proposed to the space agency multiple times since 1992. *"Our idea is to use a transit photometer to build a device that will help increase the understanding of life in the universe and other planets like Earth,"* Eric described. On several instances, the program cost estimates and proposed scientific methodology were compared (by the program selection committee) with much larger and complicated missions of the past. As a result, with each selection cycle, the program was forced to respond to new committees' demands and requests. In particular, the program had to redo its costing several times to validate its accuracy and had to prove the viability of its chosen data collection technique via ground-based demonstrations. Additionally, they had to demonstrate that the Research Center had the capability for the end-to-end management of the program. It was not until December 2001, after persistent attempts to prove the idea for almost a decade, that the program was eventually selected.

PROGRAM ORGANIZATION AND MANAGEMENT

Because the breadth, depth, and complexity of the program are so large, the agency determined that all of the program components (program management, ground segment management, and flight segment management) were too much for the assigned program manager to handle individually. Therefore, in June 2002, the Planet Orbits program, whose main mission location is at the Research Center, selected the Spacecraft Laboratory as the mission management partner in an attempt to reduce risk. Additionally, the Technology Corporation was selected as the industrial partner of the program for development of the hardware. Under this arrangement, the Research Center was still responsible for roughly 70 percent of the direct cost via instrument development and delivery, and science management.

The structure of Planet Orbits is a program form of organizational structure involving multiple subteams in three organizations (see Figure 18.3).

The structure is atypical since the program is shared by two centers (Research Center and Spacecraft Laboratory). The program has a Program Management Office (PMO), which is an entity created for a specific period of time that is dedicated specifically to the Planet Orbits program. The PMO provides a variety

Figure 18.3   Program Organizational Structure

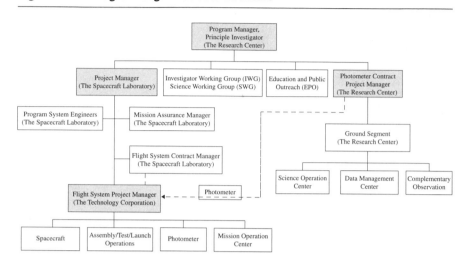

of managerial and administrative support pertinent to managing a major and complex program such as Planet Orbits.

Team members were selected by the program's core team, which represented the functional project managers reporting to Eric. The criteria for selection of the team members included experience, competency, dedication, and enthusiasm. External hiring for the program included employees from the Spacecraft Laboratory and the Technology Corporation, who were also picked by the Planet Orbits team.

The Planet Orbits team size varies across program phases as the demand for personnel fluctuates. At the end of the definition phase, the team at the Research Center was composed of 25 full-time employees; the team at Spacecraft Laboratory numbered 12 full-time employees, and 5 part-time members; the Technology Corporation's team included 80 full-time equivalent employees. Since the program is geographically dispersed across three different organizations, communication among the members is critical. The major means of communication include phone (frequent), email (frequent), face-to-face meetings (frequent), and teleconference (weekly). There are also the casual and informal communication channels, including periodic social gatherings, picnics, and barbeques.

A formal team-building exercise (referred to as Four Dimensional (4-D) training) was held somewhat late after the start of the program. The 4-D training, provided by an external organization, is a mechanism for the team to get to know one another, form a cohesive group, and create what is called "smoothness and harmony" across the organizations involved. This is a vehicle for cross-organizational, cross-site team building and integration of the Research

Center, Spacecraft Laboratory, and Technology Corporation Program personnel. This cross-organizational, cross-site team building was not as effective as planned. Many team members felt that 4-D, although a good initiative, came too late to significantly benefit the team.

As program manager, Eric Andersen is responsible for the execution of the entire mission and the scientific integrity of the investigation. The responsibility for day-to-day management of the program is delegated by the program manager to the project team at the Spacecraft Laboratory, where the project manager resides. The Planet Orbits Educational and Public Outreach (EPO) organization and the science teams, consisting of the Investigator Working Group (IWG) and the Science Working Group (SWG), report directly to the program manager. In general, the Research Center and its program manager are responsible for the development and test of the photometer and the overall ground segment. The Spacecraft Laboratory, and its project manager, Charles, are responsible for providing the flight system, which includes the spacecraft, flight segment assembly/test/launch operations (ATLO), and flight segment checkout. Technology Corporation is a subcontractor to the Spacecraft Laboratory, and its project team is responsible for developing, building, integrating, and testing the flight system and Mission Operations Center, in addition to launching the flight segment.

For the Planet Orbits program, top management of the agency is involved during the gate approvals and program reviews. Three formal review teams are established to evaluate the program. These include the program formal review (PFR) team—the standing review board picked from various key organizations; the independent assessment team (IAT)—selected from the agency; and the systems management office assessment (SMOA) team—selected from the spacecraft laboratory. The PFR has 21 members; the IAT has 6 to 9 members; and the SMOA has 2 to 4 members. Review teams are used as the mechanism to evaluate whether or not the program is in line with the agency mission. Generally, the review teams make recommendations (or so-called "findings") and evaluate the readiness of the program to proceed to the next phase. Reviews help the program manager and team manage the program and stay in line with senior management's expectations and the formal objectives for the program. The team must address any significant plan deviations and respond back to the review panels to ensure that they agree that the team has properly addressed and resolved their concerns. This mechanism is used during all development gate reviews.

PROGRAM STRATEGY

The program objectives must align directly with the mission and strategy documented by the space agency. The *objective* of the Planet Orbits program is to build a spacecraft with a photometer for space exploration. Its scientific goal is to conduct a census of extraterrestrial solar planets by using a photometer in a heliocentric orbit. The first priority or strategic focus of the program is science goals; followed by the clear mandate to stay within a number of constraints (schedule, cost, and technical integrity). The science focus is clearly associated with agency value and is reinforced by the program manager to the core team and other team members throughout the program.

Product definition: The major products of the program include a photometer onboard a spacecraft and its associated ground system. The photometer is an instrument for measuring the luminous flux of a light source to observe the periodic dimming in starlight caused by planetary transit. When a planet passes in front of its parent star, it blocks a small fraction of the light from that star. If the dimming is truly caused by a planet, then the transits must be repeatable. Measuring three transits all with a consistent period, duration, and change in brightness provides a rigorous method for discovering and confirming planets. Simply speaking, the program aims to build the instrument that will find the terrestrial planets in the habitable zone of other stars. This includes developing and testing the photometer and the overall ground segment, developing the flight system, and flight segment in-orbit checkout system.

Program value: The program is very well-aligned with the agency's strategic plan. The value of the program is to contribute to answering the fundamental questions of *Does life in any form, however simple or complex, carbon-based or other, exist elsewhere than on Earth?* and *Are there Earth-like planets beyond our solar system?*. The value also is to contribute to all the Agency's Office of Space Science Themes, addressing the questions that the formulation and evolution of planets pose and to provide exciting scientific results of great visceral interest to the general public about exploration.

Success criteria: Time, cost, and performance are the major criteria determining the success of the program. The agency specified that the program cannot drop below the minimum performance it promises and must launch within a certain time. With a target date of next year, the program will be initiated onboard an Expendable Launch Vehicle used to launch the spacecraft with the photometer, and the end of the baseline mission is five years after launch. In terms of cost, the program cannot exceed the development cost cap and cannot exceed the funding profile.

PROGRAM SCOPE

The Planet Orbits program is built around the following three major deliverables, allocated to three organizations, namely: Research Center, Spacecraft Laboratory, and Technology Corporation. The Research Center is responsible for developing and testing the photometer and the overall ground segment. The work done in the Laboratory includes the development of the flight system (spacecraft, flight segment assembly/test/launch operations) and flight segment on-orbit checkout system. The Technology Corporation is responsible for developing, building, integrating, and testing the flight system. The major customer who will benefit from the program is the space agency, the parent organization of the Research Center and the Spacecraft Laboratory.

PROGRAM PROCESSES

The Planet Orbits has six major program life cycle phases. The first phase is advance studies (Prephase A). The objective of this phase is to produce a broad spectrum of ideas and alternatives for missions from which new programs can be selected. Second in the life cycle is preliminary analysis (Phase A). In this phase, the team has to determine the feasibility and desirability of a suggested new major system and its compatibility with the agency's strategic plans. Third is definition phase (Phase B), whose objective is to define the program in sufficient detail to establish an initial baseline capable of meeting mission needs. Next is the design phase (Phase C). In this phase, the team completes the detailed design of the system. Then, the program goes to the development phase (Phase D) to build the subsystems and integrate them to create the system while developing confidence that it will be able to meet the system requirements; and to deploy the system and ensure that it is ready for operations. The last phase in the life cycle is operation (Phase E). In this phase, the team has to make sure that the system actually meets the initially identified need, and then dispose the system in a responsible manner. Figure 18.4 summarizes the program life cycle and the major milestones, including the timeline.

Major milestones include Mission Concept Review (MCR), Mission Definition Review (MDR), System Definition Review (SDR), Preliminary Design Review (PDR), Critical Design Review (CDR), System Acceptance Review (SAR), Flight Readiness Review (FRR), Operational Readiness Review (ORR), and Decommissioning Review (DR). These phases and milestones are consistent and standard to the agency. There is definition and understanding as to the decisions that need to be made at each milestone review (technical, schedule,

Figure 18.4 *Program Life Cycle and Major Milestones*

Prephase A: Advanced Studies	Phase A: Preliminary Analysis	Phase B: Definition		Phase C: Design	Phase D: Development			Phase E: Operations
▫ MCR	▫ MDR	▫ SDR	▫ PDR	▫ CDR	▫ SAR	▫ FRR	▫ ORR	▫ DR
			☆ Current status					

Early 90s	3/01	12/01	10/04	11/06	11/07	11/12

and cost). Planet Orbits is currently scheduled to pass the PDR stage by going through the three review teams—the PFR, the IAT, and the SMOA. The first review team is the PFR, which is a standing review board. Their job is to verify or recommend the program move forward on the basis of the completeness of its requirements and the understanding of the agency's requirements. The second review team is the IAT, whose job is to make the recommendation whether or not the program should be confirmed. Next, the SMOA team evaluates the aspects of the program based on the interest of the chief engineering office. After the program achieves PDR approval, it must meet a major constraint set by the agency that the program must then launch within 36 months, including one month of commissioning. This means that Phases C and D cannot exceed more than 36 months, collectively.

The tactical management of the Planet Orbits program includes a formal work breakdown structure (WBS), roughly seven levels, where specific team leaders are responsible for certain levels. Charles, the project manager in the Spacecraft Lab, primarily has control over level 1 (program level), 2 (segment level), 3 (system level), and 4 (subsystem level), whereas the project managers in the photometer and flight segments and functional teams have been responsible for the lower levels (assembly, subassembly, parts). Each level contains numerous activities. The spacecraft, as an example, has approximately 4,000 activities. The WBS was used as the baseline to develop the program schedule and cost.

The schedule was structured to conform to the boundary conditions which constrain (1) the funding profile for fiscal year 2003 through fiscal year 2005; (2) the start of phase C and phase D to be after the program preliminary design review and confirmation review; and (3) Phase C and phase D duration of 36 months or less. The cost estimation process involved triangulation of various methods such as a bottom-up, top-down process, independent cost modeling, and comparison with other similar programs (analogous estimating).

While a mitigation plan is developed for risks, the team has a mechanism with the mission assurance organization that tracks issues and problems separately. The agency has set a formal standard on risk mitigation (risk management policy document) in which the risks are rated based on their probability of occurrence and severity. A red, green, and blue chart is used to capture the varying degree of risks, which is discussed in the quarterly and monthly management reviews. Risks generically are owned by the program, and they are assigned to a responsible individual to resolve and report up the management chain. Planet Orbits is a fairly low-risk mission since it has a very high technology readiness level, and does not involve onboard human life.

The program is tightly monitored and controlled. All technical performance, cost, or schedule parameters that require approval by the agency administrator, the associate administrator, the lead center director, or program manager are identified in Table 18.1. These controlled parameters are well documented and shared with all team members in progress meetings and reports.

Table 18.1 Controls Requiring Program Manager and Higher-Level Approval

Controlled Parameter	Controlled by Enterprise Associate Administrator	Controlled by Lead Center Director	Controlled by Program Manager	Change Request Thresholds
Funding by year			Yes	Change in approved program budget
Level 1 Requirements (Technical Requirements)	Unless controlled by program manager		Yes	Anything beyond agreed upon de-scope options
Program Objectives	Unless controlled by program manager		Yes	Anything beyond agreed upon de-scope options
Program Plan		Yes	Yes	Changes in scope, schedule, or budget

PROGRAM METRICS AND TOOLS

Several project management applications are used in the program. A critical path schedule and Gantt chart are used for scheduling and are entirely based on the WBS. The number of activities in the schedule is in the thousands, disaggregated through the system of hierarchical scheduling based on WBS system levels. Standard enterprise management tools are used for budget and expense tracking. The team is *required* to do earned value analysis starting with Phase C for the Technology Corporation's contract. Off-the-shelf commercial products, such as Live-Link, Project, and Doors are used to collaborate among the three major organizations. However, all team members have not used them universally.

Program performance is measured and tracked very carefully. In particular, aggregate actual-to-plan metrics for the master schedule and total budget are required by the agency.

CONCLUSION

Eric and Charles completed their review of the program documentation and concluded that the program was proceeding well. There were a couple of issues that they discussed further as these had been the source of team frustration in the past.

The first concern was that cross-site coordination was not going as well as expected, creating several time delays and inefficiencies. Early cross-organizational training turned out to be relatively ineffective. Implementation of some selected software tools such as Live-Link have helped the situation considerably, but more effective management by the program manager and core team leaders will be essential.

The second key issue related to the schedule constraints imposed by the agency. For the most part, the completion within 36 months was, in essence, dictated by senior management. The program team accepted this date without the appropriate analysis and consolidation of schedule to support it. Additional schedule analysis since the last milestone review indicates that the target completion date may not be achievable and will need to be appropriately discussed at the upcoming PDR meeting.

As they concluded, Eric and Charles recapped where they thought they stood with the program overall. Eric summarized as follows:

- We have a strong team with high morale.
- We have competent personnel in all key positions who are very experienced and knowledgeable in their respective fields.
- We have clearly defined goals, focus, and strategy.
- The program aligns well to the strategic goals of the agency which helps in terms of management support.
- The program is quite stable in terms of its scope and science.
- The program baselines are set.
- The program schedule is a challenge, but still manageable.
- We have a strong program management team.
- We know what we have to focus on in the future to be successful.

Eric concluded, "I think we are very well prepared for the upcoming committee review. I'll review what we discussed today with the other core team members, and I think we are ready to proceed."

CLOSING

The Planet Orbits example demonstrates how the program management model can scale as a viable management approach for small, multidiscipline development efforts to very large and complex efforts involving multiple organizations. However, as program size and complexity grow, the capabilities and experience of the program manager become more important. Fundamentally, this program is sound; however, it is evident that cross-project and cross-organization collaboration and synergy is breaking down. This shows the importance of leadership and other "soft skills" on the part of program managers.

Discussion items

1. Summarize the key takeaways of the case.
2. Why do you think it took Planet Orbits 10 years to get selected?
3. How should the earthly issues and the cosmic glory of Planet Orbits be balanced in terms of the success measures?
4. What are the major challenges of the Planet Orbits structure?
5. Explain why soft skills are so important, especially in this case.

ConSoul Software

Andrea Hayes-Martinelli
and Dragan Z. Milosevic

"Wait a minute!" said Bali Balebi, the Silverbow program manager, while passionately waving his hands. "Do I understand you correctly that senior management is saying that my program must hit the release date, and if that requires dropping the two automation features, it is okay?"

"Yes," responded Christine Smiley, the PMO director, "you understood me. But please calm down. We need cool heads now."

Bali continued, "So, first we add the automation features despite the program team telling us that the planned program duration of 21 months would only allow for the 8 original features. Now, we are in the integrate phase three months before we get to deployment, and I'm being asked to drop the automation features because we're a month behind schedule? I'm sorry if I'm having a tough time keeping a cool head, but we can't do that."

"They are not *asking* you, they are *directing* you to remove the features in order to get back on schedule. The delivery date is crucial," replied Christine. "Again, please calm down and tell me why we can't drop the automation features. Give me a logical argument that I can take back to senior management. I can't just go back to them and say we can't remove the features because the program manager is passionately against it."

"Okay," said Bali. "Two reasons: First, I have already made an announcement to our lead customers that the automation features will be included in the next release. Second, the features have already been integrated with the other features in the release. It may take longer to back them out and redo the integration than to just continue with the integration as is."

"Oh, now I understand the problem," responded Christine. "Let me talk to Matt Short (vice-president of enterprise software), and you sit down with your team and review all possible options to make the original delivery date, both with and without the automation features. Also, call the lead customers who we know plan to purchase the new release. Tell them about the possible delay of the two automation features until the next release and see how they react."

COMPANY CONTEXT

Before we see what program actions were taken on the Silverbow program, we need to evaluate ConSoul Software. Program decisions need to be put into the perspective of the corporate context in which they exist, as the company environment heavily influences program decisions and strategy.

OVERVIEW

ConSoul Software was a late entry in the facilities and construction software industry, which was dominated by two primary competitors. Matt Short, vice-president of enterprise software at ConSoul, recognized that, although the competition had a foothold in the industry, their products were not user friendly. Matt described the competitive environment: "The competition had tremendous advantages," he said. "Their products were just packed with features. The problem was, their products were extremely cumbersome to use and required large amounts of user training." ConSoul's strategy to enter the industry and to continually gain market share was to provide customized solutions that focused on ease of use and required little or no user training. "It had to be that easy, or no one would buy our products," Matt said.

ConSoul is now an award-winning software product development enterprise, which produces integrated software solutions for accounting, payroll, fixed asset management, human resources, customer relations, and e-commerce applications. The company specializes in developing integrated and customized software solutions for more than 20,000 customers in the facilities development and construction industry.

This case describes how ConSoul struggled with the management of a program, Silverbow, in its attempt to release an upgrade and grab more market share in a dominated market and the challenges it faced.

ConSoul Software was founded in 1978 as a start-up company. Its mission is to empower its customers to succeed by providing them with extraordinary software and services. It does so by focusing on four primary business postulates:

- Providing best-in-class products: By developing customized software solutions packaged as office suites, ConSoul gains both repeat business and a continually increasing customer base.
- Implementing a rapid software release cycle: ConSoul delivers software products with new features and functions every six months for most product lines.
- Developing ease of use and integrated solutions: Compared to its competitors' products, ConSoul's software solutions focus on ease of use, which results in little or no user training and tight integration with customers' existing business systems.

- Providing competitive pricing: A unique bundling strategy and integrated feature set allows ConSoul to provide its customers with a total solution at a fair and competitive price.

BUSINESS STRATEGY

The strategy that ConSoul employed is best described by what Snow and Hambrick call a customer-intimacy strategy.[1] A customer-intimacy strategy has proven to be an effective way for companies to enter markets that are already dominated by a few competitors. Competitive advantage can be gained by delivering what a specific customer wants, not what the market as a whole may want. This type of organization demonstrates superior aptitude in advisory services and relationship management.[2] Its business structure delegates decision making to employees who are close to the customers and its management systems are geared toward creating results for carefully selected and nurtured clients. The corporate and program culture embraces specific, rather than general, solutions and thrives on deep and lasting client relationships.

ConSoul Software has three principle strategic objectives, listed in priority order:

1. Maximize customer satisfaction
2. Provide high product and service quality
3. Maintain competitlve tlme-to-market dellvery

These strategic objectives guide the work of the company and its program teams, as demonstrated in the following sections.

STRATEGIC PROGRAM PLANNING

As stated earlier, a program manager within a business unit is charged with delivering an entire software solution through multiple software releases to ConSoul's customers. Program release planning is accomplished and communicated through the use of a program roadmap tool. An example of a program roadmap for the enterprise solutions business unit is shown in Figure 18.5.

The roadmap represents the software releases planned for each program over a two-year timeframe within each of the four product lines. Over the two-year timeframe, each program consists of multiple release projects. It should be noted that even though the releases are shown as sequential events on the program roadmap, work is performed in a concurrent manner with overlapping activities and resources.

---

[1]Snow, C.C. and D.C. Hambrick, "Measuring organizational strategies: Some theoretical and methodological problems," *Academy of Management Review*, Vol.5, No.4, 1980.
[2]Ibid.

Figure 18.5   Program Roadmap

	Q1'07	Q2'07	Q3'07	Q4'07	Q1'08	Q2'08	Q3'08	Q4'08
Financial Releases		▲ 1.0		▲ 2.0		▲ 3.0		▲ 4.0
Operations Releases	▲ 2.0		▲ 3.0		▲ 4.0		▲ 5.0	
Estimating Releases		▲ 1.5			▲ 2.0		▲ 2.5	
Corp. Solutions Releases		▲ 1.0		▲ 2.0		▲ 3.0		▲ 4.0

ConSoul Software is composed of two business units, enterprise software solutions and office software solutions, each led by a corporate vice-president. The business units are supported by centralized finance, marketing, and manufacturing organizations, as illustrated in Figure 18.6.

Figure 18.6   ConSoul Software Organizational Structure

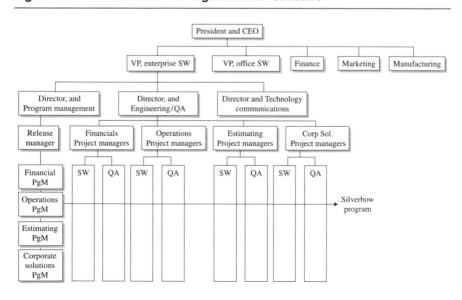

Each product development business unit is organized as a matrix structure and is led by three functional directors: director of program management, director of engineering, and director of technical communications. Each business unit is further segmented by product line. For example, as shown in Figure 18.6, the enterprise solution business unit is segmented into financial, operations, estimating, and corporate solutions product lines. A program manager is assigned to each product line and is charged with delivering the entire product solution through multiple software releases. Each release is organized as a separate project and is managed by a technical project manager.

The software development and software test/quality assurance (QA) resources report directly to the engineering functional organization and are loaned to the product line program managers and release project managers as needed. Representatives from other centralized support organizations are also assigned to the product line programs to represent their respective functions.

URGENT SILVERBOW, TEAM MEETING

After Bali's conversation with Christine, he called an urgent meeting with the Silverbow core team. The agenda was short-plan for options to finish the program. Bali briefly explained, "You know that the program plan shows we'll be a month late if we include all the 10 features. Senior management has sent us a message that we must hit the release date, and if that requires dropping the two automation features, it is okay." He continued, "Remember that senior management requested that we add the automation features, despite us telling them that our schedule wouldn't support it. Now, they have reversed their position. The trouble is we have already made an announcement to our lead customers that we have 10 features in the next release, including the two automation features. Dropping them may significantly lower our sales.

"Christine instructed us to put our heads together and review all possible options to meet the original delivery date, with and without the automation features. We will call the lead customers who we know plan to buy the new release as well, and see what kind of damage the removal of the automation features may cause.

"Now, let's review our program from top to bottom, which may help us consider all the possible options. If you have any questions about senior management's request, you can ask them as we go through the review."

THE SILVERBOW PROGRAM

Silverbow is a code name for a program that was funded and executed in the office solutions business unit of ConSoul Software. The Silverbow program supports ConSoul's operations segment and provides a comprehensive set of document control capabilities. The program consists of six projects, each structured as a software release package delivered to customers in consecutive

Table 18.2 Silverbow Program Characteristics

Industry	Facilities and Construction Software
Product Description	Software upgrade (project management software)
Program Description	Provide solutions for common document control processes (handling of meeting minutes, daily reports, submittal packages, notices, issues)
Product Novelty	Derivative
Technological Uncertainty	Low technology
System Scope	Full system
Pace	Fast and competitive
Business Unit	Operations
Customer	External
Strategic Goal	Extension of current product line for increased life cycle revenue
Program Size/Duration	20 people/22 months

release cycles. Total duration of the program was 22 months. It was estimated that company revenue would increase 53 percent in the first year and 46 percent in the second year due to software licensing agreements associated with the Silverbow program. Table 18.2 summarizes the characteristics of the Silverbow program.

The Silverbow program used a PCT structure, which is common for all ConSoul product development programs. The PCT structure was highly matrixed, with many interfaces between the engineering, test and quality assurance, technical communications, and marketing functional organizations. The PCT was led by the program manager and consists of the current release project manager, the subsequent release project manager, the engineering manager, and the software test/QA manager. Project managers report directly to the director of engineering and indirectly to the program manager for the duration of their involvement in the program. The program manager, in turn, reports directly to the director of program management.

Each project was segmented into multiple subteams that corresponded to primary product features such as drawing logs, meeting minutes and reports, as shown in Figure 18.7. Each feature had an engineering lead, a test and QA lead, and a technical communications lead, as well as a marketing product manager who managed the technical design and development work for his or her respective function on the project.

The PCT is responsible for the product and business results derived from the delivery of the family of releases that constitute the program. The project teams are responsible for designing, coding, and QA testing of the various features that comprise the software release.

Figure 18.7   Silverbow Program Structure

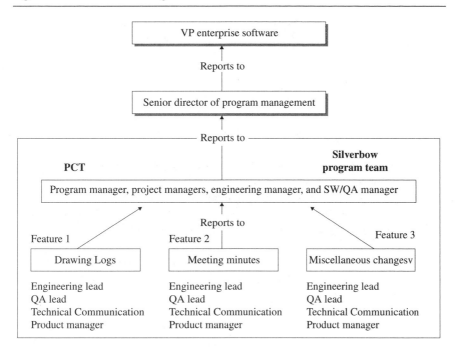

THE PROGRAM PROCESS

Silverbow was one of the first programs to use the new product development process adopted by ConSoul. The process is broken into the following elements: program phases, kitting, schedule management, cost management, risk management, resource management, and communications.

Program phases: As illustrated in Figure 18.8, the PLC consisted of seven phases: predefine, define, plan, design, integrate, deploy, and evaluate.

Generally, a program is started by creating what ConSoul calls the story, which describes the end product in terms of the usage model. The usage model is a description of the interaction between the user and the program product that identifies the product's benefit to the user.[3] From there, use cases (sequence of interactions between the user and the product), primary features for each use case, high-level resources and timeline estimates, and a release plan are developed. A feasibility analysis is then performed.

---

[3]Treacy, Michael and Fred Wieresema. *The Discipline of Market Leaders: Chose Your Customers, Narrow Your Focus, Dominate Your Market*. Reading, Ma: Addison Wesley, 1995.

Figure 18.8   The ConSoul PLC Phases

Once feasibility has been proven, each project, and the program as a whole, is scoped, functional specifications are developed, and development and resource allocation plans are created and presented to executive management for approval. If approved, the program moves into the design phase and execution of the project plans begins.

Throughout the design phase, customer change requests are evaluated and approved changes are added to the program and project plans. During the final integration phase, the software code is integrated, software test and QA is performed, release documentation is generated, and the current release product is evaluated for deployment. Once a product reaches deployment, focus shifts to manufacturing, order fulfillment, and customer support.

The predefine, define, and plan phases are executed only once for a given program. The design, integrate, and deploy phases are executed once for each project and, therefore, multiple times throughout the life cycle of the program.

Once the final project is deployed, the program enters the evaluation phase, in which a program retrospective is performed and customer feedback information is evaluated. Table 18.3 summarizes the primary activities associated with each phase of the ConSoul PLC.

Kitting: Project scope was managed by a process referred to as kitting. The idea was to segment software deliverables into five-day work packages. At the end of five days, or one cycle, each subteam within a project released a deliverable for each of the primary features. Each deliverable consisted of one element of the product that was functional and fully tested. This process of iterative software development is known as Agile development.[4]

The kitting process was used for the first time on the Silverbow program and proved to be a challenge for the project managers because it did not lend itself to the linear process of historical project management methods. Tracy Brooks, the release project manager, described the problem to Bali: "Kitting doesn't address anything about logical units of work of a software project," he said. "Software development is not manufacturing, as the kitting idea is. A software process doesn't chunk up the same way a manufacturing process does."

The source of the problem was having two planning systems in place, which were incompatible. Per the first one, the program and each of the projects were planned by major milestones corresponding to large deliverables and phase completion. Per the second one, small deliverables of five-day increments (kits) were scoped. As a result, kitting performance and milestone performance were

[4]Schwaber, Ken. *Agile Project Management with Scrum*. Redmond, WA: Microsoft Press, 2004.

Table 18.3 PLC Phases and Activities

Phases	Major Actions
Predefine	Perform feasibility analysis with marketing, product development, technology; perform financial analysis; get high-level estimates
Define	Identify product definition team; create research plan; perform customer task and needs analysis; risk management; develop product definition; create initial estimate
Plan	Develop the functional specifications; resource allocation; develop plans to support all the business needs; present plan to marketing and make sure that this is what they wanted before going into design phase; negotiate with marketing; go/no-go decision
Design	Execute projects; do change requests from customers for better design solutions; integrate code; system test
Integrate	Generate release documents; finalize products; assess project
Deploy	Duplicate software disks; update internal systems; new order fulfillment; release fulfillment
Evaluate	Program retrospective; customer evaluation; overall evaluation of phase assessments

not aligned, leaving the performance-to-schedule metric ambiguous. As changes to the program scope were approved, the feature changes were either incorporated into existing kitting plans or new software kits were developed. Bali was able to determine that the dichotomy and lack of knowledge of how to coordinate the two planning systems were a major reason for the program delay of one month.

Schedule management: The schedule was developed based on time estimates from the engineering and software test/QA functions, not as a program team. High-level time estimates were first developed in the predefine and define phases, then detailed estimates were developed during the kit planning process in the plan phase.

Milestones were based on major project deliverables and standard program decision checkpoints and monitored every two weeks by the PCT. Example milestones included software detailed design complete, user-interface complete, code complete, code freeze, and product ship release.

Cost management: Cost of the program was expressed in terms of the number of people on the team and the number of hours to complete their tasks, not on a budgeted dollar amount. ConSoul sets budgets based on business solutions, rather than individual programs. The program manager, therefore, does not know the cost of the program in dollar value. Cost is managed through the program resource management process.

Bali believes this is an area of potential improvement. "I think the current program cost management process can get us in trouble," he said. "Nobody realizes how much money it actually takes to develop a product."

Risk management: Risk management was a shared process on the Silverbow program, both by phase and by team. During the program predefine phase, the marketing function was responsible for identifying potential risk events, evaluating the risk level, and developing necessary risk management plans. During the other phases of the PLC, the PCT drove the risk management process.

Risks contained to a single project were managed by the project manager and his or her project team, while risks that had potential effects on other projects, or program success as a whole, were managed by the PCT.

Each risk event was quantified according to the following two variables: probability of occurrence and severity of impact. Priority-one risks were the highest risk events, while priorities two and three were relatively lower. All priority-one risk events had to be communicated to all internal stakeholders, and a mitigation or avoidance plan had to be created. Priority-two and -three risk events were monitored only on a recurring basis.

Resource management: Resource management was one of the primary responsibilities of Silverbow's program manager; the program manager also focused on ensuring that the project teams, as well as the PCT, were fully staffed with capable people to complete the work. Once a program plan is approved at ConSoul, resources are committed to the program by the various functional managers. The Silverbow program was fully resourced according to the program plan as it entered the design phase of the PLC.

However, resources were a primary constraint on the program. Because the program manager did not have a monetary budget to use for managing program constraints, the majority of scope, schedule, and quality issues were addressed by changes in program resources. For example, the program manager had to continually adjust resources to accommodate the additional features requested by customers or other stakeholders and address any quality issues associated with the current release or previous releases already operational in the field.

Communications: The PCT met once a week, generally on Monday mornings, to review program and project status, current issues, potential risk events, and planned activities for the coming week. Project status was focused on the kitting progress for each of the feature teams on the project. A review of each of the feature team's five-day kitting progress and deliverables was conducted. Program status focused primarily on progress toward major program milestones, resolution of current problems, and review of risk mitigation or avoidance plans for each of the priority-one risks.

Each project team conducted their own team meetings once per week at the beginning of the program, then two to three times per week during the later stages of the design and integration phases. The focus of the project team meetings was on the technical details associated with each feature team.

Communication of program status to internal stakeholders was accomplished by the program manager in the form of a formal status report once every two weeks. The report was sent directly to the senior management team and posted on the program website for all other interested parties.

PROGRAM METRICS AND TOOLS

The Silverbow program employed standard metrics and tools for all ConSoul programs and focused on measurement of the following processes: scope management, schedule management, risk management, and quality management. As mentioned in the previous section, program cost is not measured.

Scope management: Program and project scope was measured by the number of kits associated with each feature in a project. Performance was measured by the number of kits completed versus kits planned and required a mitigation plan for variance recovery. Wall charts were used to manage scope and performance by tracking the number of kits completed by marking the kits off as they were finished, then monitoring the kitting progress against the original plan.

Schedule management: Performance to schedule was focused on progress toward completion of the key program milestone dates. As mentioned earlier, there was no correlation between completion of program milestones and completion of the five-day kitting activities. A standard Gantt chart was used to manage the program schedule.

Risk management: Risk management metrics included the total number of risk events for each priority level (high, medium, or low) and progress toward mitigation or avoidance of each high-priority risk. Timely development of a risk mitigation or avoidance plan for each priority-one risk and the completion date were closely monitored. A risk management spreadsheet was used to track progress. Key fields in the spreadsheet included risk entry date, risk description, root cause, severity of impact, probability of occurrence, priority level, mitigation/ avoidance plan, targeted completion date, and status.

Quality management: Software quality was measured by the number of defects recorded, resolved, and outstanding for each release. Software quality is the primary indicator used for the ship release decision. ConSoul utilizes an in-house developed tracking data base to manage software defects on all of its development programs. The tool categorizes defects by severity level and resolution owner.

Table 18.4 summarizes the program metrics and tools used for the Silverbow program.

Table 18.4 Program Metrics and Tools

Program Processes	Metrics	Tools
Scope and Performance Management	Number of kits completed versus planned	Kitting and wall chart
Schedule Management	Performance toward meeting key program milestones	Microsoft project
Risk Management	Total number of risks per priority level (high, medium, low)	Risk management spreadsheet
Quality Management	Time to develop risk mitigation plan for high-priority risks	In-house tracking data base
	Number of defects recorded	
	Number of defects resolved	
	Number of defects outstanding	

ALIGNMENT BETWEEN BUSINESS STRATEGY AND PROGRAM EXECUTION

As stated earlier, ConSoul Software uses a customer-intimacy strategy to compete and gain competitive advantage in the facilities and construction software industry. Each business unit focuses on delivering particular features with specific, rather than general, solutions and does so with a lot of attention paid to maintaining the highest quality, customer involvement, and satisfaction possible.

All elements of the program management discipline explained previously are aligned with the three principle strategic objectives: maximizing customer satisfaction, providing high-quality products and services, and maintaining competitive time-to-market delivery schedules. Each element of program execution and its alignment to strategy is shown in summary form in Table 18.5 and is described in detail thereafter.

Program organization: The Silverbow program organization is aligned to the business strategy by the use of a matrix structure that enables a high degree of cross-discipline interaction and collaboration. Interdependencies between engineering, software test/QA, technical communications, and product managers are focused on maintaining customer satisfaction and delivering quality products. The team was made as small as possible to facilitate open communication.

Product managers were the representatives and voices of customers, and their responsibility was to make sure that the product matched the customers' needs. The software test/QA team was responsible for assuring the highest-quality software possible. Mike Billard, the engineering lead for the Minutes Meeting feature, told Bali, "We do set dates, but we are not hard set on achieving the dates because customer satisfaction and quality factors take precedence. We would drop everything if a customer had a problem with one of our products in order to fix it for them."

Table 18.5 Program Elements Aligned to Strategy

Program Organization	Matrix structure with many interfaces between engineers, software testers, technical communications, and product managers
Program Process	Milestones are deliverable-driven. Change requests often come from marketing and customers. Risk is based upon uncertainty toward meeting strategic goals
Program Metrics	Measure program details toward achievement of strategic goals. Focus on meeting major program milestones, quality of the product, management of customer requested changes and management of risk
Program Tools	Scope, performance tracking, and risk tools are important. Schedule tools are standard and flexible, whereas there is no specific tool for cost
Program Culture	Customer driven, collaborative, wide-open communication, high degree of respect, pride, and teamwork

Program process: The program process adopted by ConSoul is highly focused on satisfying their customers. Customers are encouraged to be involved in the process by defining their needs and wants in the predefine phase, evaluating preproduction releases during the design phase, participating in the integration process, evaluating the product under operational conditions, and providing feedback to the program teams.

Designing the software is accomplished through a process of iteration, allowing for continual improvement, maturity of the product over time, and an opportunity for customer change requests to be evaluated and incorporated.

Program metrics and tools: The program metrics and tools used were highly aligned with the three primary strategic elements by focusing on the measurement and tracking of software quality, management of program and project risks to success, feature content, and performance to committed schedule.

Program culture: The program culture was highly influenced by the business strategy of the company. Therefore, there is excellent alignment between business strategy and program execution. Strategic objectives are clear, concise, and measurable, and the program strategy, organization, processes, metrics, and tools all support attainment of the business objectives. The cross-discipline program structure provides clear lines of responsibility and authority, and the PLC process has well-defined phases and phase-exit criteria. Some of the culture's other elements include the following:

- Cross-discipline collaboration and open communication. There are no walls between engineering, software test/QA, program and project managers, or high-level executives.
- The organization is driven by the customers.
- The status of employees' work is tied to how well they know the product.
- The environment encourages a high degree of respect, pride, and teamwork.

WHAT ARE THE SILVERBOW PROGRAM OPTIONS?

Team Silverbow spent nine hours in two days planning for review of the options. They rechecked the status, reviewed their strategy, organization, process, metrics and tools, how they were applied, and how the relevant decisions were made. Multiple iterations of what-if analyses were performed, resource reallocation was evaluated, pro and con analyses of each option were carried out, and rough comparisons established. Eventually, three final options emerged. Bali will present the following options to senior managers:

Option 1: Stay on the current course.

Strategy: Keep the scope, all features intact; shoot for the same deadline, using the same plan; and pursue quality as goal number 1.

Pros: We know the game.

The team feels comfortable playing the way it knows.

The goals are reachable.

Cons: We will most likely be late by one month.

We will lose sales.

Team members may start finding new jobs as the program is nearing the end.

Option 2: Keep all features intact and hit the announced release date.

Strategy: Crash schedule, fast-track, and outsource. First, in some parts of the program, we will crash the schedule by adding more resources, retaining activity dependencies, and shortening critical path activities: Second, we will fast-track some parts of the program by breaking activity dependencies and creating new ones to overlap and speed up the schedule. Third, we will outsource the software testing to an external, top-notch company to accelerate the schedule.

Pros: It will be a feat.

The product will be released as announced.

Customers will be happy.

Cons: We may not be able to deliver on time with the quality desired.

Outsourcing was never tried before.

Fast-tracking is a new approach for us.

Option 3: Drop the two features and finish on time.

Strategy: Drop the two features. That enables the elimination of time to integrate the two new features, making it much easier to deliver on time and also to pay attention to the quality of the remaining features.

Pros: The product will be delivered on time with good quality.

The product will be released as announced.

It is still the best product in the market.

Cons: We may lose about 20 percent of the product sales (according to a phone survey of lead customers).

Because we failed on our product feature promises, we may lose some customers permanently.

We may have a tarnished reputation.

MEETING WITH EXECUTIVES

After getting word from Christine that the Silverbow program manager was ready to present a set of options, senior managers agreed to see Bali for a brief review. The senior managers were organized as a Product Approval Committee (PAC) who have the authority to approve all product programs, perform gate reviews, and make major strategic program decisions. The PAC included vice-presidents of marketing, enterprise software, office software, finance, manufacturing, and the director of program management. The PAC was headed by the vice-president of marketing, John Biffin. When Bali entered the conference room, the PAC members signaled that they were ready to start the meeting.

"Bali, Christine told us that you have some new options to offer," stated John Biffin.

"Yes, I have three options—some old, some new. Bear with the details, although I will be brief," replied Bali. He took seven minutes to present his three options, not forgetting to mention the source of his data, whenever he could. Finally, he wrapped up, with what he believed was his punch line: "After you decide which option to pursue, the program team will make it happen."

John Biffin's response took Bali by surprise. "Actually, Bali, we won't choose the option, you will. We saw you today for one reason only. We wanted to make sure that you have a reasonable plan, and you do, and tell you that we will give you whatever resources you need. Choose your option and go for it. But you should know—we will hold you accountable."

Stunned, Bali tried to think about how to respond. But it was too late to say anything because the PAC members had already adjourned the meeting. Bali hurried out and was ready to reconvene his team members to decide which option to pursue.

Discussion items

1. What did you learn from this case?
2. In your own words, how can project management be used to achieve the business strategy?
3. Which of the three options would you choose? Why?

Part III

CASE STUDIES IN ORGANIZATIONAL PROJECT MANAGEMENT

WHAT IS ORGANIZATIONAL PROJECT MANAGEMENT?

As several projects and programs are implemented, to maintain excellence in project and program management, an organization should establish a systematic approach for their management. Such an approach is referred to as Organizational Project Management (OPM), which supports the implementation of projects and programs and facilitates governance and alignment of them with the organization's strategic goals.[1] Enterprise Project Management is another term used to describe this approach.

OPM can consist of several elements that support project and program management, for example, the establishment of an appropriate organizational structure, organizational and project cultures, and information systems. The establishment of project portfolio management and a program management office and the development of standardized project and program methodologies and project and program management competencies are also part of OPM.

---

[1]Organizational Project Management Maturity Model (OPM3), Project Management Institute, 2008, p. 9.

The Project Management Institute (PMI) has published *Organizational Project Management Maturity Model* (OPM3) based on PMI standards for project, program, and portfolio management. In addition to the standards, OPM3 also includes organizational enablers, which are structural cultural, technological, and human-resources practices that can be leveraged to support and sustain the implementation of best practices in projects, programs, and portfolios.[2] To best practices, OPM3 provides methods for organizations to establish their organizational project management process and to assess the capabilities for future improvement. Each best practice has its identification, name, and description.

---

[2]Organizational Project Management Maturity Model (OPM3), Project Management Institute, 2008, p. 30.

Chapter 19

ALIGNMENT AND PORTFOLIO MANAGEMENT

This chapter contains three cases relating to strategic alignment and portfolio management. The issues discussed in these cases support the definition of OPM and several best practices in OPM3 such as Standardize Portfolio Categorize Component Process and Standardize Balance Portfolio Process. There are two critical incidents and one issue-based case.

1. LorryMer Information Technology

   LorryMer Information Technology is an issue-based case, discussing the alignment of IT programs to business strategy. The case also details a specific process including some tools that are used to facilitate such an alignment.

2. Who Owns the Portfolio?

   Who Owns the Portfolio? is a critical incident. It illustrates a battle between marketing and engineering regarding who should own and manage the portfolio. Should it be marketing or engineering?

3. Our Portfolio Stinks

   Our Portfolio Stinks is also a critical incident. It discusses a major portfolio problem of one organization—an unbalanced portfolio.

CHAPTER SUMMARY

Name of Case	Area Supported by Case	Case Type	Author of Case
LorryMer Information Technology	Strategic Alignment	Issue-based Case	Sabin Srivannaboon and Dragan Z. Milosevic
Who Owns the Portfolio?	Portfolio Ownership	Critical Incident	Dragan Z. Milosevic and Peerasit Patanakul
Our Portfolio Stinks	Portfolio Balancing	Critical Incident	Dragan Z. Milosevic, Peerasit Patanakul, and Sabin Srivannaboon

LorryMer Information Technology

Sabin Srivannaboon and Dragan Z. Milosevic

LorryMer Corporation was a leader in a specialized motor vehicle industry in North America. In the early 2000s its sales began to suffer substantially due to the recession in the U.S. economy. As a result, the company was forced to change its primary strategic objectives to focus on cost efficiency. In particular, it needed to cut operating costs to improve the company's competitive position. This case shows how LorryMer's IT programs were aligned with the company's strategic goals and business values. The alignment process was tailored for LorryMer's new business strategy and an effective alignment tool was used. The tool is simply a chart designed to present the alignment between business goals, business values, and the IT programs. The alignment between business strategy and program execution begins at the top, where strategy drives the desired business results. Effective program management practices then deliver the business results.

IN TROUBLED WATERS

LorryMer specializes in the design, development, and manufacture of a complete line of technologically advanced motor vehicles. The company has been in business for more than six decades, and now operates seven major vehicle manufacturing plants and one parts manufacturing plant in North America. With more than 14,000 employees, its mission is to provide the highest standard of technological innovation and premium quality to its customers. LorryMer remains committed to the highest degree of innovation and quality, and is dedicated to meeting its customers' needs.

Saul McBarney, LorryMer's information technology (IT) program management officer, has been with the company for 25 years. Saul is proud of the company's history, strategy, and background. "We are unique in terms of listening to the customers. We find out what customers' business needs are first, and then develop products for them that meet their needs."

However, 2001 was a painful year for LorryMer. As a result of economic hardship in the motor vehicle industry, the company's sales and profits dropped, and its losses exceeded $1 billion. In September 2002, the Refrigerated

Transporter reported that depressed used automobile prices caused many carriers to extend trade cycles because they owed more on their automobiles than they could sell them for.

As a response to the crisis, the company embarked on a major cost savings initiative in late 2001, which later produced a number of cost savings programs. Dingo Ostar, a LorryMer business value account manager and a member of the business advisory group, said, "It was about cost, cost, cost and nothing else."

Nevertheless, LorryMer continued to experience substantial business challenges in the following years. With the economy still suffering, LorryMer's automobile production in 2002 did not meet 2001 production levels, which were significantly lower than 2000. Funding for any new programs was minimal, especially during 2003. LorryMer needed to change in order to survive.

OVERCOMING STRATEGIC OBSTACLES

John Mennon, chief information officer for LorryMer said, "The business environment for all automobile manufacturers remained depressed and was not expected to regain its historically high levels until late 2004. We knew we had to make some changes to be ready to compete." LorryMer focused on reducing costs quickly. As a result of relentless cost-cutting programs, LorryMer's operating costs were dramatically reduced, and breakeven profit was achieved in 2003.

"However, we knew those changes were tactical and not sufficient," John said. "We had to make additional changes." Therefore, during 2005, LorryMer modified its strategy by focusing on five new core values: providing market leadership and brand coverage; pursuing technological innovation; partnering with operators for maximum productivity; focusing on the needs of its customers, employees, communities, and the environment; and being an advocate for their industry.

ALIGNING INFORMATION TECHNOLOGY PROGRAMS TO BUSINESS STRATEGY

During the past several years, the IT strategy at LorryMer has been to maintain legacy systems. Most of the new IT investments were dedicated to e-business websites for after-market sales, and marketing. Dingo said, "Eighty-five percent of what we sell is a complete commodity available from thousands of other people. So the only thing we really have to sell to increase profit is service." The remaining incremental IT spending supported enhancements to keep the legacy systems compliant with government regulations and provided some basic level of additional functions.

To improve the company's competitive position, radical changes needed to take place to create an IT environment that was better positioned to support all five of the new core values. As part of the improvement efforts, the IT department was reorganized and made much more agile in 2005. That was done by replacing the functional structure with a matrix organization, consisting of application (IT application developers and service providers) and program management (program managers and business system analysts) competency centers. Then, a business advisory group concept was introduced. It consisted of value account managers who provided IT focus in business units and acted as the IT voice of the customer, which was something LorryMer lacked previously. Also, a joint strategic planning session of business units and IT was initiated in the middle of 2005. Its purpose was to determine the business needs to be supported by the replacement of legacy IT systems. Saul commented, "This was previously unheard of. In the old system, IT goals were either left to IT or imposed on us by the company. Now, our strategic goals and direction were determined by the end users in the joint strategic planning."

In order to ensure IT strategy and business strategy alignment, IT activities and programs had to complement the needs of the business. The IT team used an alignment chart tool to accomplish this (see Figure 19.1). John explains, "The alignment chart provides a strategic mapping of the business goals, the business values (initiatives), and the IT programs. As part of the strategic changes, we were tasked by the company to design the alignment process, part of which was accomplished by the alignment chart. The chart was designed to help us visualize the alignment between business goals, business value, and the programs."

Dingo offered an example of how to read the chart. "A bubble on the intersection of Business Goal 4 and Business Value 3 means that they are aligned. Further, a bubble indicates that Business Value 3 intersects with Program 3, meaning they are aligned. In summary, Program 3 delivers Business Value 3, which achieves Business Goal 4. The alignment is all about IT programs producing business value by helping us achieve our business goals. That is the language we want everyone to speak."

Figure 19.1   Sample Alignment Chart Tool

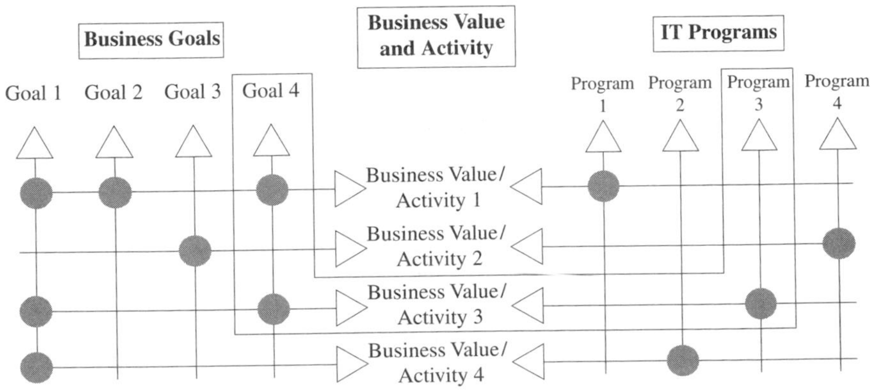

The alignment chart is a useful mechanism that not only helps visualize the alignment efforts, but also builds a standard language in the company where every IT investment must be aligned to an organizational need.

ALIGNMENT BEGINS AT THE TOP

On paper, the alignment chart is straightforward, but the execution of the steps in the alignment process is more complex. LorryMer has a mix of formal and informal processes for ensuring proper IT strategy and business strategy alignment. The processes are internally embedded within the business strategy formulation and throughout the program life cycle.

There are six steps in the alignment process (see Figure 19.2):

- *Strategic planning*
- *Informal portfolio management*
- *Envisioning*
- *Planning*
- *Development*
- *Deployment*

In step one, strategic planning, the strategic plan document is developed to help accomplish the goals of IT in support of the company's business strategies and goals for the three-year planning horizon. According to John, "Our challenge within IT has always been to take a look at how we treat the goals, whether they're well defined or not, and determine what our strategic plan is to help accomplish those goals. Our approach is an evolutionary process. Once you know where you want to go, you take it one step at the time, one program at a time, and each program gets you closer to the goals."

Then, business value/activities, called strategic initiatives and programs, are formulated based on the goals of the business strategy. These are done by the business advisory group and Program Management Competency Center. Tools at the strategic level, which are used to ensure the quality of the alignment, include the strategic plan document, roadmap charts, and alignment charts.

- Strategic plan document is a tentative capital plan that shows the snapshot of all program summaries and types, and estimated headcount and payback. It is based on the business strategy and goals of different business units that IT supports.
- Roadmap charts are used to define where the company wants to be in IT in the next three-year timeframe. They address all of the business goals of different business units that IT attempts to support and their timeframes.
- Alignment charts are the mapping of the business goals, the business values, or strategic initiatives, and the IT programs.

Figure 19.2   The Program to Business Strategy Alignment Process

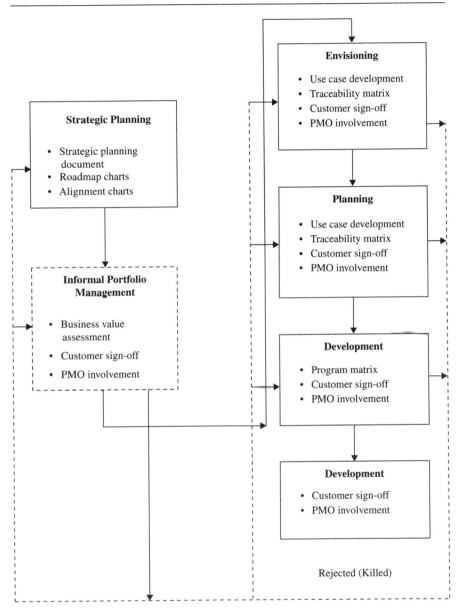

Step two, the informal program portfolio management process, is a significant part of LorryMer's alignment process. Once the portfolio management process is completed, it produces information about a set of candidate programs, their investment opportunities and program priorities. In short, the process reveals

Table 19.1 LorryMer Business Value Assessment

Business Value Weighting	1-Lowest 10-Highest	Business Risk Weighting	1-Highest 10-Lowest
Generates at least $1 million in revenue per year.		Regulatory requirement.	
Cost savings of at least $500,000 per year.		Necessary cost of doing business and key business processes will cease if not done.	
Improves employee communication, culture, or morale, resulting in a 10% increase in the next annual employee survey results.		Required to support new product.	
Product quality that directly affects JD Powers measurement.		Replaces legacy systems that inhibit required business process changes.	
Improves dealer/customer experience, resulting in an increase of 1/2% market share.		Synergy with ZBZ.	
Cost avoidance of at least $500,000 per year.		Affects existing product build capability.	
Payback in less than 18 months.		Affects existing motor vehicle and parts sales capability.	
		Current customer satisfaction level will decrease as measured by JD Powers.	
		Competitive threat will result in the loss of market share and/or net profit.	
TOTAL *		**TOTAL** *	

*Note: the higher the total, the better

the most viable programs and possible risks that could occur in the envisioning, planning, development, and deployment phases of their life cycles. Saul said, "Portfolios help to assess and monitor programs that are the best investment." Currently, this assessment is done in the portfolio process through business value assessment (BVA). BVA consists of a set of questions prepared by value account managers. The program management office (PMO) is asked to first answer the questions regarding business value weighting, which is expressed on a 1-to-10 scale, 1 being the lowest value, 10 being the highest value. Second, it assesses business risk, also expressed on a 1-to-10 scale, 1 being the highest risk, 10 being the lowest risk. An example of BVA questions is shown in Table 19.1.

Programs then are evaluated based on BVA, program type, program size, priority and business area, and are selected accordingly. Once programs are selected, they go through the standard life cycle phases: envisioning, planning, development, and deployment. A program manager is usually assigned at the end of the portfolio process or the beginning of the envisioning phase, which is step three of the alignment process. Saul commented, "Ideally, we like to have a program manager on board when envisioning is just ready to begin. That allows enough time to develop a close relationship with the business system analyst, who is responsible for detailed planning and execution of the program. Sometimes we engage a program manager immediately at the time the portfolio process starts. Therefore, the program manager is responsible, along with the value account manager, for the business value of the program, and solely responsible for achieving the program's other requirements."

In the envisioning and planning stages, the ball is in the program manager's court. Jeff Barrison, a program manager, said, "Use case development and traceability matrixes are used to develop the program plan. A program manager maps the details of the program with the use case, describing a series of tasks that users will accomplish using the software, which also includes the responses of the software to user actions. The program manager also traces program actions back to the initial requirements (traceability matrix) to ensure that the plan is still aligned with the requirements. In addition, the business system analyst and the program manager meet with the business members and end users to get sign-off, as part of a mechanism to make sure that the program is still in line with the business goals and requirements."

After the program plan is approved, the program team starts to implement step five, or the development stage, of the alignment process and later, the deployment stage. According to Jeff, " Program metrics are used as a mechanism to track progress to plan. At the end of each program life cycle phase, a program manager is required to present the program status to the PMO and get customer sign-off to be able to proceed from one phase to the next. Therefore, customer sign-off and PMO involvement are the other control mechanisms to make sure that the program is still in line with the expectation throughout the program life cycle."

The program status report is used as a vehicle to communicate between the program team, PMO, and customers. When a program does not meet the requirements at each of the phase gates, adjustments are required (if the program is not killed). This mechanism is referred to as a corrective approach, in which a new strategy or action emerges from a stream of managerial decisions through time.

CLOSING

Aligning programs to business strategy is new to LorryMer. John remarked, "We are learning the alignment process, and I am happy with its results. For the first time, we are aligning IT programs with the business strategy—and it works."

The LorryMer experience offers a good example of the use of an integrated management system in practice to achieve strategic business objectives. Key elements of the LorryMer experience include the following:

- Alignment starts at the top—strategy drives intended business results and competitive advantage.
- To execute on strategy, a company needs to consistently select, fund, and resource the programs that best align with the strategic objectives and contribute the highest business value.
- Effective program management practices deliver the intended strategic business results and create a competitive advantage for a company.

Discussion item

1. Suggest an approach to align a project/program to the organization's business strategy.

Who Owns the Portfolio?

Dragan Z. Milosevic and Peerasit Patanakul

He knew something was wrong with his project mix after reading a three-month report showing the cumulative number of projects done for the current year. To Ian Plachy, Vice-President of Engineering for Interconnecting Cable, Inc. (ICC), the report showed that his engineers, for the first three months of the year, completed 48 small projects and 11 medium projects. None of the big projects were completed.

ICC, INC.

ICC makes custom-made interconnecting cables for the health, computer, and other industries, with annual sales of $140 million. Typically, ICC would develop platform cables (big projects), per customer requirements. These platform cables could be changed to a minor degree, e.g., color and small changes of material (small projects) or to a great degree, e.g., color, major changes of materials, and length (medium projects). Platform cables were not very lucrative to make because they were costly to develop. In addition, buyers usually do not pay the full price, but share the cost with ICC. This is a main reason why ICC did not want to pursue too many platform cables. They, however, had to have enough of them to be able to adapt to small and medium projects, which would be the sources of their future revenue. In truth, the more small and medium projects were sold, the more would ICC's cost sharing of platform projects be depreciated. Even though margin on small and medium projects was not big, it was certain. Moreover, it was easy to implement these "bread and butter" projects.

OUR FUTURE IS IN PERIL

Ian's business instinct told him that having completed none of the big projects was a mistake. But he decided to wait and research the issue further. When the six-month report arrived, and showed that his engineers did 90 small projects, 20 medium projects, but no large ones, he was ready to make a splash. In the meantime, he called his project management consultant and explained the situation. The consultant told him that this was a frequent issue in companies as a result of some types of projects being overemphasized, eventually leading to the unbalanced project mix or project portfolio. Ian agreed and explained that if this problem persists, the skills of his company's best engineering talent will go rusty. If they constantly worked on small and medium projects, his engineers would not be up for the challenges of the big projects in the future. Without working on the big projects, they wouldn't be able to develop their skills further. The technical competencies of ICC would be in peril.

The consultant said the best approach to the problem would be to talk it over in a four-hour workshop with the engineering and marketing department's leadership. Ian accepted the consultant's idea, and scheduled the workshop in two months, but could not guarantee that marketing leaders would attend, although he'd do his best. Later that day Ian called the vice-president of marketing, Dustin Miller. Dustin agreed with the idea and the date. He confirmed that he and his main people would attend the workshop.

On the day of the workshop, 11 people showed up. This included Ian and Dustin and their direct reports. Because the workshop was not designed for an educational purpose, it was rather short but informative. The consultant discussed the characteristics of the project portfolio and the responsibilities in managing ICC's portfolio. The discussion on the responsibility for managing ICC's project portfolio took an interesting turn.

Asked how that issue is seen, Dustin Miller answered that the marketing department owns the portfolio, but really does not manage it. He explained, "What we do is we have a guy, called a bidder, who reviews customers' requests and strives to pick those with the highest net present value (NPV)." Ian added that he would like to have a say in selecting projects because it had a lot to do with the development of their skill set, ICC's competencies, and market position. Ian, actually, would prefer that ICC form a portfolio interdepartmental committee to manage the portfolio. Now, the power struggle developed as Dustin jumped in and insisted that some changes would be made in marketing to manage project portfolio in the way it was discussed in the workshop, but beyond these no other changes would be made. Dustin saw no need for real changes, nor for any single department except marketing, to be involved.

Discussion items

1. Who should manage project portfolio? Marketing? Engineering? Both? Some other groups?
2. How does project portfolio management influence engineers' skill sets, a company's competencies, and market position?
3. Does the use of only NPV as a project selection method carry certain risks? What are they?
4. Why is it risky if ICC's portfolio consists of 80 percent of the bread and butter projects?

Our Portfolio Stinks

Dragan Z. Milosevic, Peerasit Patanakul, and Sabin Srivannaboon

The Edge Tech business unit within a major electronics company was almost blindsided by its new product portfolio problems. The focus of their strategy was on the development of many new products, which leveraged its core competencies. The business unit developed and put in place a formal new product process, in which the role of a gatekeeper belonged to the newly hired general manager of the unit. After several months on the job, he convened a portfolio review meeting. The meeting's purpose was to provide an overview of the status of product development projects.

PORTFOLIO ISSUES

The meeting went very well and helped the general manager learn a lot about the current direction of the new product development. As he pondered about the learning, he tried to systematize and group the issues he discovered. His groups of issues contained the following:

Projects were taking far too long. To tell the truth, some of the projects were in the pipeline for years. When he asked why, the standard answer he heard was that project teams were working hard, but they were stretched across too many projects. His feeling was that the pipeline was clogged, with too many projects in it.

Apparently, there were some real good projects. In every aspect, they looked like winners. What perplexed him was the large number of subpar projects in the pipeline. Searching for clues about the survival of these marginal-value projects, he asked about the project reviews. The answers he got indicated that the reviews were not doing the job of killing poor projects.

There were 28 projects in the pipeline. No two of them were in the same product performance/market segment area. In all fairness, this approach was as a strategic scattergun. Projects aimed at a large number of different markets—consumer, automotive, computer manufacturers, musical equipment, and so on. Some of these markets the business did not sell to, and some were off-strategy.

It was quite obvious that the proposed products offered a wide variety of benefits in many different performance segments. Practically, there was no real glue to bond all those projects—the total lack of focus was more than evident.

All of the projects held a promise of high reward, obviously a good sign. The problem was that all of them also were highly risky, with a reasonable probability to fail technically or commercially.

Almost all projects were long-term. The absence of "quick hits" to balance the long-term development projects was apparent.

TO SUMMARIZE!

"Okay," the general manager thought to himself. "Now, the groups of issues seem clear. In order to talk to my direct reports, I need to summarize this into a few words, describing where our portfolio is as of now. Would the following description make sense?"

- Overall, our portfolio stinks!
- Our portfolio is unfocused!
- Too many projects, resources spread too thinly!
- Too many poor projects!
- Our portfolio is poorly balanced!
- Our portfolio does not support our business strategy!

Discussion items

1. Do you agree with his verdict? Why, or why not?
2. Suggest a way to improve the current situation.

Chapter 20

STANDARDIZED METHODOLOGIES

This chapter contains four cases with related to the development and use of standard methodologies. The issues discussed in this chapter support several best practices in OPM3 such as Apply Project Management Processes Flexibility and Customize Project Management Methodology. There are two critical incidents and two issue-based cases in this chapter.

1. Standardized Program Risk Management

    Standardized Program Risk Management is an issue-based case. It portrays a standard risk management methodology of one of the world's leading companies. The case discusses the possibility of improving this standard process.

2. Go With the Template Always

    Go With the Template Always is an issue-based case focusing on the misuse of standard project management methodology. The case also portrays an issue of methodology reconciliation where two organizations practice different project management methodologies.

3. We Do Not Need Standard Methodology

    We Do Not Need Standard Methodology is a critical incident discussing the implementation of standard methodology across different directorates. The argument is that "our project is different and standard methodology does not really apply to us." Is it the case?

4. Joy Knows How to Defend

As a critical incident, Joy Knows How to Defend discusses the need to implement the project selection and evaluation process. Although an organization does not really practice project management, establishing standardized project selection may help defend the bottom line—profits.

CHAPTER SUMMARY

Name of Case	Area Supported by Case	Case Type	Author of Case
Standardized Program Risk Management	Standardized Methodology	Issue-based Case	Peerasit Patanakul
Go With the Template Always	Standardized Methodology	Issue-based Case	Murugappan Chettiar
We Do Not Need Standard Methodology	Flexible Use of Standard Methodology	Critical Incident	Peerasit Patanakul, Sabin Srivannaboon, and Dragan Z. Milosevic
Joy Knows How to Defend	Standardized Methodology	Critical Incident	Dragan Z. Milosevic, Peerasit Patanakul, and Sabin Srivannaboon

Standardized Program
Risk Management

Peerasit Patanakul

Rose Martin, a senior program manager of the AB-99 program, sits at her desk and looks at the risk management framework her team currently uses. This framework has been developed over the years and is adopted by her organization as a standard framework. Rose is wondering whether this standard methodology should be revised to address the challenges of future programs. "Our organization is taking more risks these days. If our risk management framework cannot handle this, one day we will be on the front page, which may not necessarily be good."

THE COMPANY AND THE AB-99 PROGRAM

Rose works for a multinational aerospace manufacturer, global security, and advanced technology company. The company is the world's largest defense contractor by revenue and offers a variety of products and services. Systems integration is among these services.

Rose's program, AB-99, is one of the programs under the Systems Integration business unit. The company is the prime systems integration contractor for AB-99 with the total responsibility of overseeing all systems integration efforts. It is a multiyear contract to integrate advanced electronic systems into 99 multimission helicopters to improve their communications capabilities.

STANDARD RISK MANAGEMENT METHODOLOGY

For AB-99, Rose creates the environment that promotes a positive attitude toward risk management. Risk management is considered as an important part of program management. In fact, risk management is ingrained in the company culture. It is practiced in every program by everyone. Since it is typical that a major defense program consists of extended teams, it is important that the program stakeholders are involved in risk management on at least two different levels—the team level and the program level (see Figure 20.1). Each level forms its own risk management board. At the team level, the team takes responsibility for the program risk

Figure 20.1 Standard Risk Management Process

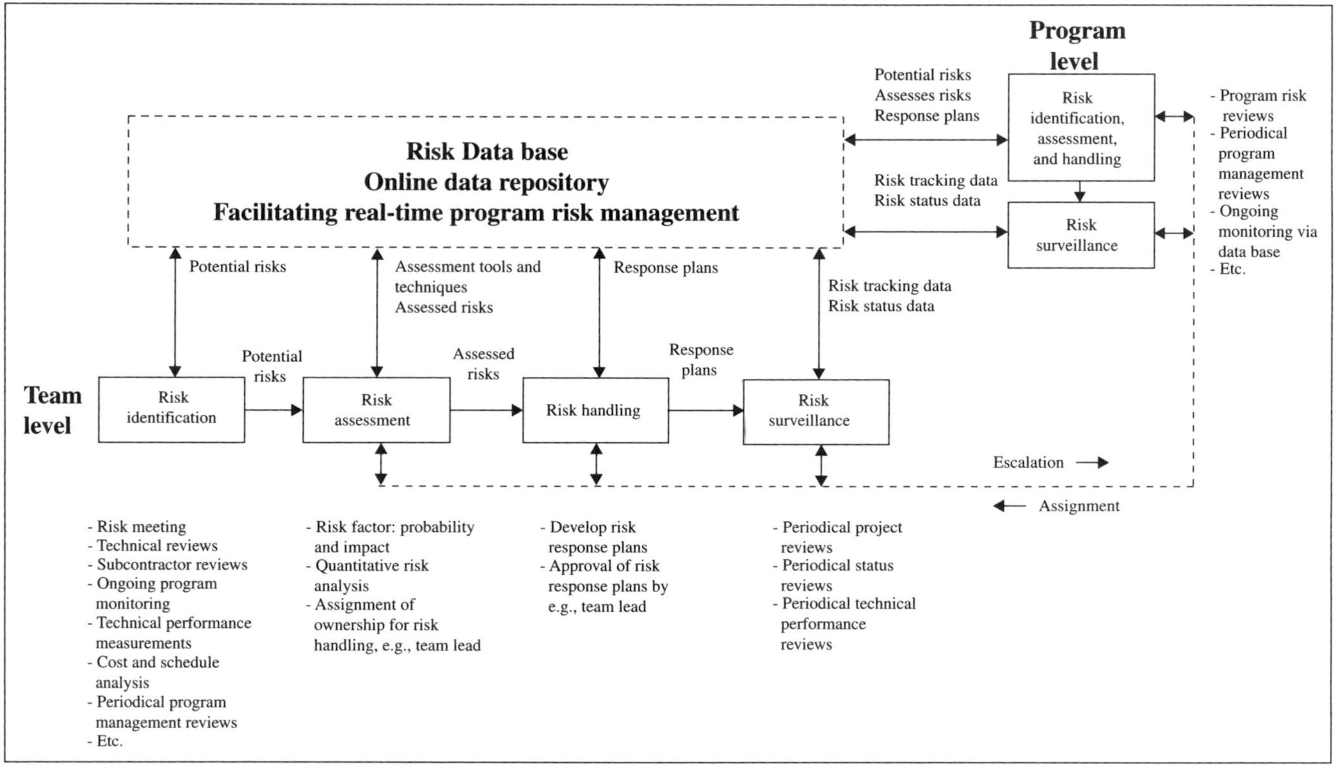

Program level

Team level

Risk Data base
Online data repository
Facilitating real-time program risk management

Risk identification, assessment, and handling

Risk surveillance

- Program risk reviews
- Periodical program management reviews
- Ongoing monitoring via data base
- Etc.

Potential risks
Assesses risks
Response plans

Risk tracking data
Risk status data

Escalation →

← Assignment

Risk surveillance

Risk tracking data
Risk status data

- Periodical project reviews
- Periodical status reviews
- Periodical technical performance reviews

Response plans

Risk handling

Response plans

- Develop risk response plans
- Approval of risk response plans by e.g., team lead

Assessed risks

Risk assessment

Assessment tools and techniques
Assessed risks

- Risk factor: probability and impact
- Quantitative risk analysis
- Assignment of ownership for risk handling, e.g., team lead

Potential risks

Potential risks

Risk identification

- Risk meeting
- Technical reviews
- Subcontractor reviews
- Ongoing program monitoring
- Technical performance measurements
- Cost and schedule analysis
- Periodical program management reviews
- Etc.

in their areas. The team is also responsible for highlighting risks that warrant elevation to the Program Risk Management Board. Risk management at the program level is performed to facilitate the coordination and oversight of the overall program risks.

The company practices a standard risk management process which is aligned with the Department of Defense policy. The typical steps in the risk management process include risk identification, risk assessment, risk handling, risk surveillance, and risk closure.

Risk Identification: Risks are identified at all levels and throughout the entire program life cycle. For AB-99, Rose recalls that, during the early stage of the program, they identified risks by examining the contract requirements, Statement of Work (SOW), specification and other Request for Proposal (RFP) materials, as well as the internally developed work breakdown structure (WBS), Integrated Master Schedule (IMS), and program plans. This initial assessment led to areas where the team focused on improving the performance position to meet technical requirements, cost targets, and schedule goals, while balancing all three elements. The risk identification process also includes the review of the WBS against the internal risk taxonomy matrix. This matrix, in fact, serves as a checklist for risk categories, which include requirements, design, integration and test, management processes, program constraints, production, and logistics (including obsolescence).

Risk Assessment: To identify risk priority, risks were assessed by using both qualitative and quantitative approaches. For the qualitative risk assessment, risks were classified by likelihood and impact levels. The likelihood represents the probability of risk occurrence. For the impact level, the adverse trends in performance-measuring parameters from the impact or risks are measured and predicted. These parameters are the technical, cost, and schedule dimensions. In terms of assessment scales, the AB-99 team uses a scale of 1 to 5 to assess likelihood (remote to near certain) and impact (high to low). The explicit operational definitions for both likelihood and impact were used to facilitate a consistent evaluation standard. After the assessment, risks were added to a matrix. This risk matrix helped facilitate the risk prioritization and the group review and discussion of the risk and corresponding step-by-step mitigation schemes. Rose remembers that the cumulative effect of all of the risks on program cost in dollars was provided by a summation of the individual factored cost exposures. Cost exposure is reviewed on a premitigation and a postmitigation basis, enabling the program to review the predicted reduction of cost exposure.

In addition to the qualitative risk assessment, the AB-99 team also employed the quantitative risk assessment. In particular, the risk likelihood was evaluated in terms of percent and the impact level was identified in terms of dollars (cost) or days (schedule). Factored cost or schedule exposure was defined as the product of the likelihood and impact in dollars or days. Rose recalls that each risk analysis included the determination of a root cause. By categorizing the root cause, potential mitigation actions became more evident and more effective.

Risk Handling: After the program risks were evaluated and prioritized, the action plans were developed for responding to the moderate and high risks. Avoidance, mitigation, and the use of contingency (acceptance) were the common action plans. Low risks were maintained on a watch list, reviewed quarterly for changes, and closed when no longer applicable.

Risk Surveillance: At the team level, the risk management board of each team continually tracked high- or moderate-risk items. On a particular risk item, once the board agreed that the risk level changed from moderate or high to low, that risk item was placed on the watch list and monitored for changes. The board also monitored whether any risk items needed possible additional funding from the management reserve, whether any risks warranted elevation to the program risk management board, and whether there were any newly identified risks. At the program level, the program risk management board met regularly to review and discuss new potential risks and to manage existing risk mitigation efforts.

Risk Closure: Closure criteria were developed to evaluate risk items. According to the criteria, if risks (especially the ones on the watch list) are assessed as no longer a factor, they are closed and removed from the list.

THE USE OF A RISK MANAGEMENT DATA BASE

To make this risk management activity possible, the AB-99 team uses a risk management data base. In the data base, risks are described, catalogued, updated, tracked, and so forth. Rose believes fully that the use of the data base helps them manage risks effectively. The team uses the data base to support (1) the integrated risk management, (2) the risk verification by risk management boards, (3) the risk scoring system, and (4) the establishment of metrics and closure criteria for mitigation tracking. The data base also (1) is the central location for obtaining risk assessment data for the program, (2) provides for the team to respond quickly to emerging risks, (3) facilitates shared monitoring of risks affecting multiple subsystems, and (4) is the tool used to communicate risk areas and status to the chief engineers and program leadership council. Rose loves the automated notification feature of her data base because it gives any team member the ability to enter or update risks and the owner/auctioneer receives an immediate notification so they are aware of the changes and can access the information directly.

Discussion item

1. Discuss the standard risk management methodology of AB-99 and suggest what Rose should do to improve this standard methodology.

Go With the Template Always

Murugappan Chettiar

The sales team at Click Computing Services (CCS) was exuberant upon closure of a multi-million-dollar contract to outsource IT services for a Minnesota-based Fortune 100 company, SafeMining Inc. The deal was going to open doors for CCS in the vertical markets of mining and manufacturing where they had no track record for the last 20 years.

The client management team at SafeMining foresaw the global economic slowdown in early 2007 and wanted to improve operating margins. They picked CCS for being an industry leader and relied on their experience to effectively transition IT services from over 25-year-old technology to state-of-the-art, scalable platforms and also adopt best practices. One request client management unambiguously made clear was "never talk in abstract terms, always provide specifics." They named the project Stella.

The implementation effort was organized into a program with three major IT projects—infrastructure, development, and services. The project team members are both from CCS and SafeMining. Jeff Malta, a CCS program manager was a PMI professional. With his multiyear industry experience, Jeff foresees a problem with the reconciliation of project management standards between the two companies.

CLICK COMPUTING SERVICES

Click Computing Services (CCS) is an industry leader in outsourcing IT services with annual revenue of over $5 billion, and 15,000 people located across 20 offices in North America, Europe, Asia, and Australia. They manage IT functions i.e., computer hardware, computer networks, develop and deploy software, and offer helpdesk support and Internet services. CCS started as a Y2K solution provider in the late 1980s and forayed into outsourcing leveraging offshore methodology.

PROJECT MANAGEMENT AT CCS

Typical Projects

Typical projects would include enhancements to current applications either through product upgrade or custom programming. For example, an ERP procurement module for a pharmaceutical client is being extended for international use with custom programming done by a team of onsite, onshore, and offshore resources. New implementation projects would include migration of applications from client infrastructure to CCS infrastructure, such as hosting and support of an AutoCad system for a manufacturing client and maintenance and support of a billing application for a retailer.

Project Organization

CCS built strong project/process methodologies to support its outsourcing activities. Vertical markets such as government services, healthcare, consumer products, manufacturing, finance, etc. were organized into portfolios under a principal partner. Principal partners had dedicated portfolio managers reporting to them from various offices across the company. Each portfolio manager was responsible for the profitability of accounts (i.e., client) in his/her portfolio. Each portfolio would contain multiple programs either to implement new projects or service existing clients. Every individual client would have an implementation program and/or service program at any given time; and each program would have multiple individual projects. Staffing was managed at an organizational level, and resources were shared across projects/programs based on technical expertise. This included project managers, program managers, technical resources, and subject matter experts.

Project Resources

Project managers typically were drawn from the technical pool of resources and given the additional responsibility of project management based on tenure. Formal on-the-job training was subsequently provided. Program managers, however, were dedicated within the portfolio and were subject matter experts on the industry vertical market. Project managers reported a dotted line to program managers and a solid line to functional managers. The bulk of the project activity was done by technical resources grouped under various functional managers. Project resources, project managers, and program managers were typically onsite at the client's location for the duration of project execution, barring conflict with other priority programs.

Standard Project Management Process

Projects followed the *PMBOK® Guide* phases of Initiate, Plan, Execute, Monitor, and Close. The initiate phase generally included sales and portfolio managers selecting the project team. The planning phase included preplanning internal to CCS and planning with the client. Execution was left to project resources. The monitor and close was the responsibility of program managers along with a steering team that was comprised of members from CCS and client management. All projects had to be approved by a CCS portfolio manager prior to start/termination.

Organizational Support

CCS management supported project management practices and encouraged training; CCS management did not believe in oversight through a central project management office and left the success/failure of each project to its portfolio managers. Tenure of resources played a crucial role in the informal hierarchy; communication flow among project resources was not always open but took place on an as-needed basis. CCS, as an organization, struggled with the adoption of its own project management standards in a consistent way across all projects/programs. This also aggravated existing clients who continued to have different experiences with execution of each project.

STELLA PROGRAM

The program was organized into three major IT projects—infrastructure, development, and services; lasted nine months, used 20 resources, and had almost an open checkbook. Jeff Malta, the CCS program manager with many years of industry experience, was relatively new to CCS. He joined CCS a little over two years ago. Both the development and service project managers had been with the company for less than five years. Dave Wu, an infrastructure project manager, was a veteran. He had the most tenure at CCS with over 15 years.

AT PROJECT ANALYSIS MEETING

Jeff: Dave, based on what we know from sales, in your opinion, what do you think is the complexity we are dealing with here and how long will this take to implement?

Dave: The project is of average complexity, right in the middle of our scale and shouldn't take longer than six months.

Jeff: Really. When can we present a draft schedule to the client team and what process do we follow?

Dave: We can present a schedule when we are done with project analysis. I just plug in the dates to MS-Project templates and we will have a plan.

Jeff: Wait a minute. I presume our project team will also be involved in the analysis prior to finalizing the dates or presenting it to the client, right?

Dave: No, our project team is never involved in schedule preparation. We just change start dates on the template and that's it.

Jeff: I am a little concerned about this approach for two reasons. One, every project is different. It's okay to use a template as the starting point, but to use it as if it is the final plan. Two, not including the project team will not get us the buy-in.

Dave: This is how we have been operating for 15 years and have been successful. I see no reason to change this and do not have the time to do it either.

Jeff: I can help you do it, so we have a quality deliverable to the client based on actual requirements; not blindly using a template.

Dave: Sorry, I have three other projects to manage and this should not be a concern. As long as the client is available, we will do whatever it takes to make it happen.

Jeff: How do we identify when client resources are needed? Don't we have to plan with both project teams instead of assuming resource availability per template?

Dave: Let me take care of it, Jeff.

Jeff: But Dave, you know that SafeMining practices a standard project management process and they want us to be very specific and realistic. We have to take them very seriously. Plus, this is a multi-million-dollar contract. Getting it right will help our company expand our services to the mining and manufacturing industry.

Dave: I will take care of it. Don't worry.

PROJECT ANALYSIS WAS COMPLETED

The client and CCS team met to discuss the schedule; upon review of CCS's infrastructure project schedule the client team was dumb-founded and refused to continue work on the project. SafeMining's project manager requested a steering team meeting to discuss challenges.

Similar to a project management office, SafeMining had an Engineering Management Office with emphasis on engineering subject matter related to mining and manufacturing. All new projects were evaluated so as to not impact their production operations; the company also had a methodology close to *PMBOK*'s five phases. Project Stella had been assigned a representative from the engineering management office to support the deployment and change management as a result of outsourcing to CCS.

Client management reviewed the situation with their project managers and engineering management office representative to find out that CCS never consulted their team members while creating the infrastructure schedule and never provided specifics for the activities to be completed, while assuming client resources would be available almost 90 percent of the time for the project. Any SafeMining resource workload of more than 50 percent had to be explicitly approved by the engineering management office. SafeMining management was concerned about the lack of consultation from CCS and collaboration to create a successful working relationship. SafeMining decided not to pay the next installment of implementation fees and decided to re-review the relationship.

Discussion items

1. What is the role of a template and how was it applied at CCS?
2. How entrenched were current practices and what could have helped with change management?

We Do Not Need Standard Methodology

Peerasit Patanakul, Sabin Srivannaboon, and Dragan Z. Milosevic

Since the organization's restructuring, the project management department of Zeus Inc. has been struggling to implement standard methodologies across all of its directorates. Some directorates have a formal way of managing their projects and require their project managers to be certified. "The first thing I say to someone who wants to be my project manager is: 'go study, take the exam, become certified, and then, you can come work for me,'" said Tim Darby, the director of product deployment directorate. Other directorates, on the other hand, are more flexible in their project management approach. Project management certification is not mandated. This case discusses how they reached a common ground.

PROJECT MANAGEMENT VETERAN

Tim's product deployment directorate is among the four directorates under the program management department. James Higgs is Tim's peer, who is also the director of product deployment directorate. The difference is that James deploys new products related to Zeus's new strategic initiative. Tim's products are those of the existing business initiatives, a bread-and-butter kind.

Tim started his career as an engineer, then project manager, and is now the director. As a project management veteran, Tim has developed a standard project management methodology for his organization. It is a web-based tool called PDX, which includes the five-step product deployment process. Templates, tools, and techniques were created to support the project management activity in each phase. Tim requires his project managers to update the project document, status, and issues every Monday. This web-based tool is also used as a project repository. So far, Tim has the information of over 500 projects that have been implemented.

400

DON'T YOU WANT TO USE PDX?

Tim is very proud of his PDX. He owns it. He has a full-time staff allocated to monitor and upgrade the system. His boss also sees the benefits of PDX and encourages the other directors to use the system. However, it is not mandatory. Tim does not understand why other directorates do not fully adopt PDX.

Tim: James, have you checked out PDX? What do you think about it?

James: Yes. It is great. You have all the project history, all the checklists, and tools. I also like the fact that it provides us a means to check the status of projects 24/7. You even have a function to print out the project status report. This is all great stuff. But I am not sure how much this will apply to my projects. I am not sure if I should require my project managers to do all of that.

Tim: Why?

James: My projects are unique. They are different. Some of those forms may not be applicable to us. My project managers and I look at this and to tell you the truth, we are overwhelmed by it. We decided that we should not commit to this at this time because we should focus on getting the projects done.

Tim: But James, if you guys follow the process in PDX, wouldn't it help you jump-start your projects?

James: Probably not. With the entire document that we have to create, we will focus too much time in documentation rather than leading the projects. My goal for them is to focus the least amount of time on creating the document. It does not mean that we do not do our job. We just go for the necessity.

Plus, I do not need to look at PDX to get the status update. My project managers are collocated. I can go and talk with them, review the project document, check the project status whenever we want. They can come and talk to me when issues arise. You are my office neighbor but your project managers are all in different locations. I think PDX suits your needs perfectly.

Tim: All the excuses, James. You just do not want to try. I cannot see the real reason why PDX is not applicable to you guys. Your product may be unique but it does not mean that the standard deployment process doesn't apply to it. Don't tell me that your project managers are lazy or not capable of using the system.

James: Hey Tim. That's not nice. Even though my project managers are not certified like yours, I can guarantee that they are capable. I know that you are proud of your PDX. Don't take it personally if we don't use it. I told you that in my opinion it is a real good system but it is just not what I need my project managers to focus on at this time. Plus our boss never complains about whether I use it or not. I provide the documents he needs, in timely fashion, and I am on top of all my projects without using PDX.

Tim: Sorry, I did not mean to offend you or your project managers. I just think that you and your project managers will benefit from using PDX. All the forms, templates, checklists, guidelines, tools, and techniques are there in PDX, arranged per phase. You guys don't have to create anything new. They are there for you to use. I think our organization will benefit from this, too. All of the past project records are there. In fact, when we start new projects, I require my project managers to search for similar projects in PDX, read all the documents, critical issues, lessons learned, etc. So far, this helps them jump-start the new projects. The system makes it legitimate for me to set a goal of a 90 percent success rate for my project managers. I know that you are struggling with setting up a 90 percent success rate goal. PDX has a lot of benefits: Efficiency, effectiveness, project success, personal learning, and organization learning are among them.

COMMON GROUND

James knows wholeheartedly that what Tim says makes sense. He agrees to implement PDX slowly. The overall five-phase process is applied to all of James's projects, however, not all the documents, templates, tools, and techniques in each phase will be used. They agree that James's projects are more dynamic and not all cookie-cutter projects, like the majority of Tim's projects. However, having a common place for project repository is very important.

Discussion items

1. To what extent do the following items play a role in implementing a standard methodology: project characteristics, project organization, and readiness of project managers?
2. Suggest an approach to alleviate the resistant to use a standard methodology.
3. List the benefits and shortcomings of having a standard methodology.

Joy Knows How to Defend

Dragan Z. Milosevic, Peerasit Patanakul, and Sabin Srivannaboon

Joy Pechum, CEO of California Anesthesiology Group (CAG), was sitting in her office, preparing the agenda for the meeting of the board of directors to be held next Sunday. She thought about what to offer the board as an explanation for why she would spend $20,000 to deploy standardized project management methodology. The board typically viewed these methodologies as unnecessary red tape—quite the contrary to her own beliefs that having too many methodologies in a company was not productive but saw them as a prerequisite to having orderly business processes.

At that moment the phone rang. She picked it up and heard the voice of Patrick, her VP of finance. He began, "Joy, I have a vendor invoice here. It shows that Dr. Squirrel had bought a computerized patient management system, costing $250,000. Its deployment would cost us another $250,000, i.e., a total of half a million dollars! All I want to know is whether you have approved it." "No, *I have not*," answered Joy hanging up angrily, trying to visualize the face of Dr. Squirrel, a big gun in the company, both as an anesthesiology expert and shareholder-owner, when she tells the board about this purchase.

OUR UNIQUE PRACTICE

CAG is a for-profit company of 200+ medical doctors that provides their services by placing anesthesiologists in hospitals where they are needed; with annual sales of $90 million. CAG is an employee-owned company, where the largest shareholders are the most senior medical doctors in the company. Accustomed to being treated as royalty by the hospitals, these owners–senior doctors often behaved financially irresponsibly and ignored CEOs before Joy.

Then, two years ago, Joy, with the reputation of the best CEO in town for mid-sized health organizations, was hired at CAG. The first order of business for Joy was to introduce financial discipline. She was successful in her mission but only after spending a lot of time and anguish. But, still some of these owners–senior doctors behaved in the same old way. Being owners–senior doctors, they often visited the premiere hospitals in the country, and, staying there a month or two, learned from the best in the business. Also, they often had the opportunity to see the best or newest professional software, usually for managing patient systems. In pre-Joy times, some of them would order the software without consulting anyone. The then CEOs did not object to these purchases because they feared the power of these owners–senior doctors.

EVIDENCE AND EVIDENCE

Then came Joy, who did not tolerate this behavior. She, of course, tried to avoid conflict with the owners–senior doctors, but nevertheless brought two cases of such a purchase to the board's attention. That was an appropriate way, Joy thought, since all owners–senior doctors sat on the board. Basically, she painted such purchases as an undesired attack on company profits. She thought that culprits were not aware of that, viewing the purchases only as CAG's technological improvement, which eventually is passed onto customers. The board was on Joy's side whenever she mentioned erosion of profits with such unplanned purchases. Simply, this was their money, and they did not like one to spend it, whether or not that someone was a colleague.

HERE IS AN IDEA

"Good God," Joy thought to herself, "Half a million dollars! Gone! Wasted!" She remembered the last two software computerized patient management systems they had were never deployed! CAG did not have the skills to deploy the software, and doctors-owners appreciated making money working more than not making it and spending time deploying software. Then, she got an idea. What if she explains to the board that she needs standardized project management methodology to help prevent this purchasing behavior? Actually, she wanted to deploy standardized project management methodology exactly for this reason. Only, she could not tell the board earlier, and now that Dr. Squirrel made that dangerous move on the software purchase, she could.

The first part of standardized project management methodology will be a standardized project selection process, mostly dealing with various patient management software projects deployed in order to improve service. Joy may use ZBB (Zero-Based Budgeting) or some other process that will prevent pet projects like this. The second part will be project life cycle–based project planning to secure that no one rushes into the project without first thinking it through. And, the third part will prescribe the project implementation procedure for Joy to know what is going on. Good idea? She once more scrutinized the idea and imagined how individuals on the board would react to how much of their money the system would save. Then, she concluded, "Joy knows how to defend profits."

Discussion items

1. Which part of the standardized project management system that CAG intends to introduce may help prevent purchasing behavior of owners–senior doctors when buying patient management software on their own?
2. Identify several options to help prevent purchasing behavior of owners–senior doctors when buying patient management software on their own. Do pro and con analysis of each and decide which option is the best.
3. How much does Joy risk conflict with the owners–senior doctors by bringing the attention of the board to the purchasing behavior of individual owners–senior doctors when buying patient management software on their own? Describe this risk.

Chapter 21

COMPETENCIES OF PROJECT MANAGERS AND THE PROJECT MANAGEMENT OFFICE

This chapter contains three cases—one issue-based case and two comprehensive cases, discussing competencies of project managers and the project management office. The issues discussed in this chapter support several best practices in OPM3 such as Establish Project Manager Competency Processes, Facilitate Project Manager Development, Establish Internal Project Management Communities, and Provide Organizational Project Management Support Office.

1. They Are Business Leaders at Spotlight Corporation

   They Are Business Leaders at Spotlight Corporation is an issue-based case. It portrays a set of competency requirements of project managers who typically lead more than one project at a time. This type of project management arrangement is typical in high-tech industries.

2. The Program Management Office

   The Program Management Office is a comprehensive case. It portrays the purpose, contribution, and scope of the program management office in an information technology organization. The case also discusses the placement of the PMO.

3. Progress—One Step at a Time

   Progress—One Step at a Time is a comprehensive case discussing the evolution of Project Management Office (PMO) in an organization.

The case details the journey of PMO from a volunteer-based working group. Roles and responsibilities, structure and staffing, and partnership of PMO are discussed in the case.

CHAPTER SUMMARY

Name of Case	Area Supported by Case	Case Type	Author of Case
They Are Business Leaders at Spotlight Corporation	Competency of Project Managers	Issue-based Case	Peerasit Patanakul and Dragan Z. Milosevic
The Program Management Office	Project Management Office	Comprehensive Case	Sabin Srivannaboon and Dragan Z. Milosevic
Progress—One Step at a Time	Evolution of Project Management Office	Comprehensive Case	James Schneidmuller and Peerasit Patanakul

They Are Business Leaders at Spotlight Corporation

Peerasit Patanakul and Dragan Z. Milosevic

This case focuses on program management competencies as practiced in a technology-driven organization within a high-velocity environment, and on the new product development programs which fuel the organization's growth engine. The organization is Spotlight Corporation, a leader in the liquid crystal display projector industry. The leaders of Spotlight Corp see program management competencies as an essential element of their program managers' performance and continued professional growth, as well as a key factor in whether or not a program will succeed in achieving the desired business goals. Competencies for program managers at Spotlight Corp are driven by and adapted to the business strategy. This principle is widely used in the business world. Spotlight Corp makes sure that the competency-group mix for program managers is in tune with the company needs. This also includes adjusting the various skills needed within a particular competency area as required.

SPOTLIGHT CORPORATION

Having arrived at work a little early, Dave Moskhill reflected on the business of the company he had recently joined, Spotlight Corporation. As the newest member of Spotlight's executive management team, Dave spent the last couple of weeks getting familiar with his new company. Although he collected a lot of information about Spotlight, many aspects of the business were still unfamiliar to him.

For Spotlight, product development is the engine of its growth. To maintain a leadership position in the market, the company emphasizes leveraging its core competencies in advanced research and engineering design to develop new markets, while maintaining customer focus and improving efficiency and effectiveness in product development processes. Rapidly changing technology, customer demands, and aggressive competitors require a clear and well-developed strategy.

Like many U.S. corporations, Spotlight recently outsourced its manufacturing to China. On average, Spotlight implements 40 to 50 product development programs per year. The programs include all types, from derivative to breakthrough, and range from $1 million to more than $5 million in budget, and 9 to 24 months in duration.

Dave scheduled a meeting with Brian Hall, director of the program management office (PMO), to discuss the way things work at Spotlight. Their discussion, focusing on Spotlight's program management function, follows.

PROGRAM MANAGERS AT SPOTLIGHT

Dave had a particular interest in program managers and the competencies they need in order to be real business leaders. He opened the discussion by asking Brian how many program managers there are, who they report to, and what they do.

According to Brian, there are 12 program managers in the PMO, all reporting to him. On average, each of them leads two to three programs at a time—some big, some small. Some of the programs are derivative with added features. Others are new product development programs. Because Spotlight's goal is to consistently make its products physically smaller, it requires new technologies. New product development programs are fast-paced and involve high market uncertainty, and high technological and organizational complexity.

"You need true leaders with significant competencies and a wide range of experience to lead programs like those," Dave said. Brian agreed, and added, "In our program management office, we have three levels of program managers— entry level with very little experience, highly experienced veterans who've seen it all, and those in between."

Brian added that Spotlight uses a competency metric to help determine the level of maturity their program managers possess. "We have a comprehensive list of competencies that we think our program managers need and we rate their competency level against that list," he said.

Brian explained how Spotlight's competency list was developed. "When we first thought about the competency metric we sat down, studied models in the technical press, and benchmarked some companies in our industry. We put the list together, but when we used it to gauge program managers, some mid-level program managers only scored as a level one, so we knew the model had some serious flaws. We went back to the drawing board, testing different models of the competency metric. Finally, it dawned on us that the competency metric we developed corresponded to models from the technical press and benchmarked companies. What we really needed was a model for our own business strategy and our program managers executing that strategy. We picked our five most successful program managers, evaluated their competencies, averaged them out and proclaimed them a temporary competency metric. Then, we refined the metric for almost a year until we polished it to reflect our company, and the competencies of our program managers. We take pride in the fact that it is not just a competency list, but competencies that help Spotlight's program managers improve program results," Brian said.

Brian put the competency metric sheet on the desk and offered its rationale (see Table 21.1). The metric had six groups of competencies on the list: administrative

Table 21.1 Competencies of Program Managers at Spotlight Corporation

Competencies	Competencies
Technical	*Administrative/process*
- Knowledge of product applications	- Monitoring/control
- Knowledge of technology and trends	- Risk management
- Knowledge of program products	- Planning/scheduling
- Knowledge/competencies of tech tools and techniques	- Resource management
- Ability to solve technical problems	- Company's program management process
Interpersonal	*Intrapersonal*
- Leadership	- Organized and disciplined
- Communication	- Responsible
- Team management	- Proactive and ambitious
- Problem solving	- Mature and self-controlled
- Conflict management	- Flexible
Business/strategic	*Multiple Program Management*
- Business sense	- Experience of managing multiple programs
- Customer concern	- Interdependency management
- Integrative capability	- Multitasking
- Strategic thinking	- Simultaneous team management
- Profit/cost consciousness	- Interprogram process

and process competencies, intrapersonal competencies, interpersonal competencies, business and strategic competencies, and technical competencies. In addition, Spotlight program managers need to have multiple program management competencies as well, considering that they lead two to three programs simultaneously. These competencies will help them coordinate multiple concurrent programs.

PROGRAM MANAGERS AND THEIR TECHNICAL COMPETENCIES

Brian explained to Dave that technical competencies constitute the knowledge, skill, and experience of a program manager related to the technical facets of the program product. At Spotlight, the program managers do not have to have extensive knowledge of product technology, know how to use technical tools and techniques related to product development, or have the sophisticated ability to solve technical problems. "According to our experience, our program managers need to know the latest product technologies in the market, but also understand when we are talking about the importance of system architecture being defined and how critical that is to the program," Brian added. "In general, they understand the technological concepts of products and their application, including the knowledge of technology and dominant trends."

"That concept is easy for me to grasp because it is similar to what we used in my previous company," Dave said. "However, the technical aspects of products are very important. How do you know that the technical issues will be taken care of?" he asked.

Brian emphasized that it is not just that the technical aspects are very important, but in some programs there are also a lot of uncertainties. If a program is highly complex and involves high technological uncertainties, an experienced technical lead will be assigned to the program and the team will be staffed with technical people. The program manager focuses on the big picture and the business aspects of the program.

Dave agreed and emphasized his point that since program managers lead two to three programs simultaneously, it is almost impossible for them to have technical knowledge of all of them. "In general, they should have the knowledge of technology and dominant trends," he said.

ADMINISTRATIVE AND PROCESS COMPETENCIES ARE A GIVEN

Dave turned his attention to the administrative and process competencies. Common sense told him that if program managers do not have the knowledge, skills, and experience in planning, scheduling, organizing, monitoring, and control, they shouldn't be program managers. But he wondered to what extent those competencies helped program managers enhance program success.

Brian gave Dave good information about this issue. He emphasized that, besides planning and scheduling, Spotlight's program managers should be proficient in monitoring and control. "This is very important, since the business environment changes rapidly, and those changes often impact their programs directly," Brian said. "In other words, if the changes impact the direction of their program, program managers should be able to recognize those changes, talk to executives about setting new goals, and control the new direction of the program."

Brian continued that risk management is also very important, since most of the programs involve high levels of technological uncertainty. "Program managers know that they have to identify risks in their programs, estimate the probability of those risks occurring, identify the severity of risks, and propose the countermeasure or strategies to prevent or mitigate them," he said. "I have to admit that we are not very good at risk management yet," he added, "but I believe we will be."

Dave was glad to hear that there was a desire to improve risk management and decided to emphasize it. "Good," he said. "Risk management is very important in our environment, and I understand that it is not easy. Maybe we have to provide more training on effective risk management practices." Dave went on to inquire if program managers have the ability to negotiate and allocate resources.

"Yes," Brian replied. "The program managers know how to identify and estimate resources, negotiate resources with the functional groups, allocate them, and monitor and control resources. I also expect our program managers to understand how things work here." He added that program managers understand the program management processes, including policies, procedures, and tools that are used to manage programs. In other words, even though they have other competencies, they have to learn the company's lingo, forms, processes, and get to know people. This will help them manage programs effectively and, in fact, will help them properly employ their program management competencies. "I mean, all in all, I want our program managers to have a solid foundation of program management that they can pretty much tie into any type of program," Brian said. "I call all of these administrative and process competencies, and they consistently help to enhance Spotlight's program managers' results."

Dave agreed and the discussion turned toward people competencies.

SOFT COMPETENCIES DO MATTER

Spotlight learned that having only process competencies is not enough. Their program managers are expected to have interpersonal and intrapersonal competencies as well. The company considers these soft competencies, and Brian was a strong advocate of this approach. Interpersonal competencies of leadership, communication, team building, problem solving, and conflict management were the competencies listed in the table he and Dave were looking at.

"Let me give you the details, Dave," suggested Brian, "Our program managers are proficient in setting direction, delegating authority, and influencing the team with fairness. They also have political competency to be able to set the priority of program activities to be in line with management and the company's goals. In addition, they should be able to influence people and have credibility in the eyes of all the program stakeholders, including senior management. They should also know when to involve me or senior management in program activities," he added.

This was a hot-button issue for Dave, and he jumped in to comment. "I agree. Based on my experience, program managers should have good communication competencies both verbally and in writing," he said. "I think a successful program manager should be a good listener and ask the right questions. I see that you have communication competency on the list. But I want to emphasize that our program managers need to know how to articulate and handle any kind of information, whether it is technical, legal, administrative, or interpersonal in nature. Now, what about team management?" he asked.

Brian commented that team building is very important, especially at Spotlight, where some team members are from Europe and Asia. "There is an expectation that program managers put a team together that is committed and mutually accountable," he said. "Then, they should be able to keep the team motivated in a group setting. They are expected to be good at problem solving and conflict management. As we know, these issues increase the level of complexity when dealing with distributed teams," he added.

Dave asked about intrapersonal competencies, and Brian explained that he expects his program managers to be very organized, thoughtful, and methodical. "By being organized and disciplined, program managers are able to perform their job better in a high-velocity environment," he said. "They should also be proactive and ambitious. Program managers should be action-oriented and self-motivated, so that they can anticipate issues and develop a plan to account for them." He added that he believes program managers are responsible for getting people motivated. "If they cannot get themselves motivated on the program issues, their teams won't get motivated. The other characteristics that they need, as I listed in the table, are being mature and self-controlled. These will help them have emotional stability, patience, poise, and tolerance toward uncertainty. Our program managers need these characteristics, especially at Spotlight where things change quickly," he said.

Dave added that program managers should be spontaneous, adaptive to the working situation, and open to change. "Sometimes program managers have to lead a nontraditional program that requires them to stay up until 2:00 or 3:00 in the morning to have a teleconference with a team overseas," he said. "Therefore, being flexible is important."

Brian agreed, adding that besides the intrapersonal competencies on the list, being entrepreneurial, creative, visionary, and competitive are also important.

BUSINESS AND STRATEGIC COMPETENCIES ARE A MUST

Dave noticed that business and strategic competencies were on the list, and Brian commented that program managers need to understand business and strategic aspects of their programs. "These will help them make the right trade-off decisions on the program. We can tell from Spotlight's experience that having business sense assists them in formulating any program issues in a business context, recognizing fine variations among schedule, budget, and performance needs, and making benefit/cost trade-offs," he said. "In addition, they pay attention to customers in order to understand and respond to their concerns. We believe that these competencies have helped improve our programs over time."

Brian added that the competencies Spotlight believes are important are integrative capabilities because they help a program manager make decisions in the system's context. "They have to take into account the big picture, the revenue associated with the products, and customer involvement, while making decisions," he said.

Dave stared at the list, saying that he was somewhat confused because he thought that the strategic thinking competency was one of top management.

"At Spotlight, people believe that it is very important for our program managers to have this competency as well—maybe not as extensive as what top management has," Brian clarified. "We expect them to understand and adapt to the strategic direction of Spotlight and recognize our competitive components. These will help them manage their programs more effectively. Last but not least, our program managers should always take profit and cost into account when they manage program details. In a nutshell, we are doing business. Therefore, in our experience, our program managers should understand the business aspects. This has made our organization more competitive and helped improve program results," he added.

MULTIPLE PROGRAM MANAGEMENT COMPETENCIES

Not having seen this group of competencies before, Dave asked Brian to explain more about it. Brian answered that the competencies they had previously discussed are very important for managing each individual program. But when it comes to coordinating multiple programs simultaneously, multiple program management competencies are needed.

Brian added that to be good in multiple program management, program managers should have at least two years' experience in managing multiple programs for Spotlight. "The point is that they need time to establish their credibility and network inside the company. Program managers should also be competent in interdependency management to be able to manage interdependencies and interactions among programs related to shared milestones, resources, and technology," he said. "They should see the big picture and not get lost in details. There are an incredible amount of details in any one program, and program managers have to be able to step back and focus on the right things."

Brian continued that program managers need to move out of the mindset that they need to know the task level of details and need to move up to managing the deliverable level across the board for all disciplines. "I think this will help them in problem solving, too," he said. "They have to understand how to solve a problem to the benefit of all the programs they are working on."

Multitasking is also important at Spotlight, according to Brian. He stated that good program managers are able to estimate their own resource capacity (e.g., number of work hours per week or month) in order to set priorities and switch contexts and multitask among different programs. "Multitasking is a significant challenge when managing more than one program because each program often has unique characteristics," he said. "Balancing elements of time, cost, and performance across the metrics of two or three programs is always challenging."

"Sure, they have to get refocused when they move from issues of one program to the next," Dave agreed. "But does being good at multitasking mean losing time when switching from one program's issues to another program's issues?" Dave asked.

Brian nodded. "Yes. Our program managers once told me that they, on average, lose 20 to 30 minutes during context switching. This can be a big loss of their time if they change gears, i.e., programs, many times a day," he said. He added that another issue needing emphasis is that program managers should know how to lead several teams simultaneously. "They should be able to select and use different management styles specifically for each team. Since they have limited time to spend with each team, program managers need to be able to organize the team and have it become effective in a timely manner," he added. "In addition, they have to be expert in communications. The communication has to be concise," he added. "It's hard to find time to have a lot of face-to-face communication with many people."

COMPETENCY METRIC IN ACTION

Finally, Dave and Brian came to the point of how to gauge these competences and determine the competency level of program managers. "The competency metric is used in many ways," Brian said. "First, it is used to measure the competency levels of our program managers. Once that information is available, Spotlight senior management will know what kinds of programs can be assigned to them. This is a part of the program manager assignment process."

Brian continued that, among other things, it is important that the competencies for each Spotlight program align with the skill level of the program manager assigned. "In addition, we consider how important the programs are to Spotlight and issues limiting the assignments, for example, the time availability of our program managers," he said. "We have quite an elaborate model for program manager assignment, but I don't think that we will have time to discuss it now. The idea is that we want to assign important strategic programs to program managers who have sufficient competencies to lead those programs. We also use this metric to determine what kind of training our program managers need for professional development. Additionally, the metric helps us identify qualified candidates for promotion when new positions for program managers are created," he said.

Dave concluded the discussion by saying, "Thanks for a very thorough description of our program management competencies. It is clear to me from our discussion that these competencies are a pivotal factor in Spotlight's competitiveness, and our ability to consistently execute our product strategy."

Discussion items

1. Suggest a list of competencies that a project manager in your industry should possess.
2. Compare the list of competencies you suggest with the list from Spotlight Corporation. Discuss the similarities and the differences.

The Program Management Office

Sabin Srivannaboon
and Dragan Z. Milosevic

Trust Corp is a specialty instrument company in the U.S. cell phone industry. Six years ago, Trust Corp introduced a breakthrough product to the market, leaving the competition in the dust. In the high-technology industry, product advantages over the competition are short-term unless followed by the continuous stream of equally good products. Since Trust Corp has not had another successful new product introduction, the company saw a decline of business in the last five years.

While Trust Corp was developing and introducing a few lackluster products, the competition caught up, the market slumped, and Trust Corp sales dropped significantly. The company's annual sales were $360 million in 1998, but only $130 million in 2004. As a consequence, management began focusing on cost cutting, especially in Trust Corp's information technology (IT) group. The general feeling was that too much money was being wasted on poorly scoped programs that resulted in scope creep, delays, and late cancellations. There was also a feeling that the establishment of a program management office would help put IT's ducks in a row, terminating poor programs earlier and saving money through efficiency improvements.

Saul Cognito, the chief information officer, put together a team led by Barry Senders, a longtime program manager, to study the feasibility of creating a PMO. After the team spent some time working on the problem, it suggested going forward with the establishment of the PMO and hiring an experienced program management officer from outside the corporation. Peter Deerling, a former program management officer in the IT group of a much larger and more successful company named Stellar Corp, was hired.

Saul's plan was to first organize a series of meetings between Peter and major stakeholders to further promote the PMO as a vehicle to increase program success. His second goal was to get input from the stakeholders. A summary of one such meeting, attended by Saul, Barry, and Peter, follows.

THE REAL PURPOSE OF THE PROGRAM MANAGEMENT OFFICE

The first thing Saul planned was to collect data based on Peter's experience with the PMO at his previous company. He wanted to know whether it helped the success of program management and, if so, how. He also wanted Barry to hear Peter's story because Barry was an influential stakeholder who would have a major role in shaping the PMO to be program manager?friendly. Barry was a longtime program manager at Trust Corp and was respected enough to be viewed as the voice of program managers. In addition, Barry helped Saul develop a plan to establish a PMO before Peter came aboard. Therefore, Saul let his ally, Barry, ask the questions.

Barry had already heard a high-level story about Peter's PMO while interviewing him for this job. It was interesting, but Barry wanted more details.

Barry: Can you tell me about the history of the PMO at your previous company, like how it started and was developed?

Peter: I believe that to truly understand a PMO, one has to fathom the organizational context. Therefore, I'll begin with the background of Stellar Corp. The company is one of the leaders in North America in specialty manufacturing and has more than 10,000 employees. It started about 50 years ago and at the time I left, it was operating five major manufacturing plants and one parts manufacturing plant in North America. The company has several business units, which are supported by the IT group, including verticals like the infrastructure division, and the support competency center.

The PMO was established in the company about seven years ago. Overall, it was seen as an important means to create a common and standard language, where program teams could use program management tools, processes, and methodologies. Prior to that, we did not have any standardized ways of using tools or sharing the program management methodologies. Some program managers followed the policies and procedures much more closely, and programs became more successful as a result. However, some program managers didn't follow them closely enough and did things their own way. Sometimes they were successful, but many times programs required rework late in the development cycle because something was missing. We needed the PMO to standardize our practices.

Saul: I feel that I need to interject, because that answer seems commonplace. If I understand it right, one of the PMO's purposes is having program managers adhere to the policies and procedures in order to increase program success. Don't take me wrong, but we want to learn from you what aspects of the PMO bring success and what is just company folklore.

Peter: The purpose is to increase program success. Beyond that, the PMO is viewed as a business function whose job is to develop, implement, and continuously improve program management processes and tools. That solely is infrastructure to improve program performance in order to help the company achieve its business objectives. That was made clear by the senior management in the PMO founding charter—show that improved practices and processes improve business results.

Saul: Okay. But how did you make it happen?

Peter: I'm going to try to approach this from a different angle to help you understand. To achieve its business objectives, the PMO includes working with the community of program managers on how to develop processes (and tools to help enable the process), how to provide training, mentoring, and coaching, and how to implement those things, all with the simple purpose of making our business better. In addition, the PMO is a governing approval body to evaluate and determine if programs should go forward from one program phase to another. In other words, programs need to get approval from PMO committees or officers in every major milestone to be able to move forward. We got very good at terminating poorly defined, planned, and executed programs.

THE TANGIBLE CONTRIBUTION OF THE PROGRAM MANAGEMENT OFFICE

Barry: I am afraid that my thought process needs more details to fully fathom the concept. For example, what were the specific benefits of running the PMO? What success did the PMO bring? What was the PMO's return on investment?

Peter: I want to try to avoid theory and explain it all by means of the company bottom line. I think that the PMO can really make a difference with its contributions. For example, the PMO develops and installs the standardized process. It makes programs shorter, cheaper, and faster, meaning increased productivity, which translates into lower cost, better sales, and higher profits. But management wanted to know the return on PMO investment. So, we had to take a micro view of the operation and spend a huge amount of time trying to provide our executives with an accurate return on investment value for the PMO. I mean, something like how many dollars we get back for every dollar spent on the PMO. That is the language of senior management—the language of money. We proved our case by making money for Stellar Corp. We proved that our PMO increased program success in dollars. But as much as we understood what senior management wanted from us, we figured out it was a one-sided approach. There are so many intangible benefits you simply cannot quantify.

THE INTANGIBLE CONTRIBUTION OF THE PROGRAM MANAGEMENT OFFICE

Peter: Let me give you another example. At times, we needed to make sure that we did use case analysis for programs correctly, and we were struggling with end-user surveys for determining the wants and needs of the infrastructure users.

In my opinion, they were full of leading questions. "Would you like to have this capability that can do this and that for you?" Those kinds of surveys gave us misleading data. Wrong decisions were made based on those answers. "Oh, they like this capability," we heard many times. But when push came to shove, you could not really implement it because, for whatever reasons, customers we had surveyed did not want to use it. The PMO helped drive common end-user surveys and use case development. Stellar Corp's end-user surveys now help make better decisions about the infrastructure feature set and its use, and this increases worker productivity. But for figuring the PMO return on investment, we needed to determine how much money was saved because of the PMO's help with end-user surveys. We were not able to determine that. Sometimes, things do not look black and white, but rather gray. That is the world of the intangible.

Barry: Aren't there too many intangibles that our executives wouldn't like?

Peter: Lots of the PMO benefits I see come from tangible dollars, like I explained, but many of them are intangible, like having a common process, having a common standard language, and having a common understanding of what program management is. We used that in a program business case, for example, so that we were all on the same page, thinking about the same thing. We used that to identify risks before they became problems, so we were much more proactive. One of the real big benefits was improved communication so that we could do more upfront planning and avoid more crisis or reactive management later downstream. All those things create intangible benefits that were difficult to quantify for the PMO return on investment. Also, what looks more quantifiable, for example, gains in time-to-market speed that can be attributed to improvements created by the PMO, is not as tangible as it looks. There is no doubt that they can be roughly estimated, but many question their accuracy or see it as splitting hairs.

So I expect to see that the PMO contributes to the organization as a business solution, looking at the big picture from the business level, and providing tools, policies, and procedures for the individual program level.

THE SCOPE OF THE PMO

Saul: What does the PMO do now and what will it do in the future?

Peter: I've written down the PMO's duties. Essentially, the PMO does what it was originally charted for. I can summarize what the PMO does:

- *Process standardization*: This means that the PMO organizes the company's program management practices by standardizing the program management elements:
  - Processes
  - Procedures

- Metrics
- Tools
- Vocabulary
- *Personal development*: This includes organizing the career development aspects for the program manager:
  - Competency set development
  - Training
  - Mentoring
  - Coaching
- *Consulting*: This means that the PMO organizes delivering program management consulting services to programs to increase the effectiveness of our program management practices.
- *Program reviews*: The PMO is a program governing body that organizes, reviews, and approves:
  - Program phase/gate decision checkpoints
  - Major program milestones
  - Program audits
- *Program information*: The PMO organizes and standardizes data and information to support:
  - Program reporting to senior management and other key stakeholders
  - Information technology system

I want to emphasize two points. One is to pay attention to the word "organize." This means that the Stellar Corp PMO only organized all these tasks. For this, the PMO borrowed human resources from functional groups, because the PMO did not possess its own resources. In fact, the PMO had only two administrative assistants and me. For example, to organize the development, deployment, and improvement of a company's standardized program management process, the PMO would borrow program managers from functional groups, and form a team. The team members were doers—developing, deploying, and improving the process.

The second point was that exactly this approach and the chosen scope of the PMO were also contributors to the success of Stellar Corp.

Barry: You did not mention how you decided what the PMO would do and what the PMO would not do. Can you tell me about that?

Peter: Stellar Corp struggled a lot with that choice. They benchmarked PMOs in several companies and saw different versions of the PMO application. Some PMOs only cared about the standardization. Another managed all programs in the company. Yet another PMO in a big consulting company was virtual in nature because they had sites in different cities, and because their business units (profit/loss centers) were strongly independently minded, they established a PMO on the web.

The more PMOs Stellar Corp benchmarked, the more they believed in what they saw. The major factors influencing the decision about the scope of the PMO are the company's business strategy and organizational culture. For example, the company with the virtual PMO pursued a strategy of independent business units, so what they wanted of the PMO was to develop a common, standardized process and make it easily available to all business units. That's all. The organizational culture was such that the PMO could not even check or ask business units if the process was used. If they tried, the answer would be along the lines of, "If we need further help, we'll call you."

So, in my previous company, we did the same. We asked ourselves two questions:

- What does our business strategy need the PMO to do?
- What kind of PMO will our organizational culture tolerate?

Answers to these questions pretty much shaped our decisions about the scope of the PMO. I think these answers about our PMO scope of activities helped us to develop more successful programs.

ORGANIZATIONAL LOCATION OF PMO

Barry: What organizational structure, location, and size made Stellar's PMO successful?

Peter: The Stellar Corp IT department had about 20 full-time and 20 to 30 part-time program managers. Their jobs were to manage programs in their verticals (functional groups), supporting all business units as assigned per the program charter. Formally, the program managers are not located in the PMO, but in verticals. Depending on their experience and knowledge, some of them are very powerful, exerting full control over the program and making all major decisions. Others, typically less experienced, act more like program coordinators, collecting and bringing information to their vertical manager to make all major decisions.

How is the PMO structured and who does it report to? The PMO is structured in a simple way—the head of the PMO or, as we called it, the program management officer, is responsible for the work of the PMO. That, of course, was my job. The officer had two administrators to support him. That is all. It was very lean, as the organizational culture dictates. Also, part of our business strategy is to be cost leaders in the industry. So, we had to have a cost-effective PMO. Let me show you the organizational chart (see Figure 21.1).

Saul: Okay, let Barry and I read the chart, and if we make any mistake, Peter, jump in.

Peter: Agreed.

Figure 21.1   Partial Stellar Corp Organization Chart

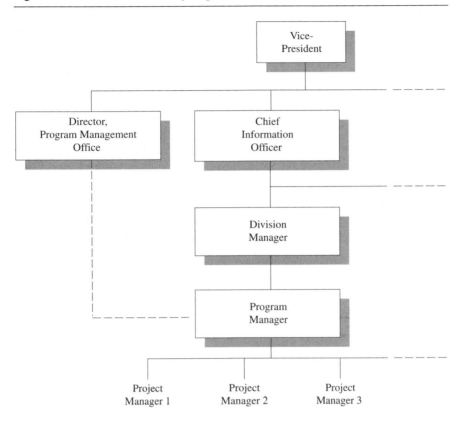

Saul: A program manager reports to his or her division, i.e., a vertical manager. Then, the division manager reports to the CIO, who in turn reports to a VP of the company. A program manager has that direct line relationship with their division manager but, at the same time, there is an informal dotted line relationship to the PMO because the PMO helps develop, deploy, and continuously improve all the processes and tools that they use. So, Peter what you show here on the chart is the organizational structure that helped Stellar do. . . stellar programs, if you will. A lean PMO that does not own any program managers. Verticals own them—pretty simple!

Peter: Exactly.

Barry: Were there any problems in terms of establishing the PMO?

Peter: Resistance to change would definitely come to the top of the list because we never had this role and function before. So there was some tough resistance for some time, especially from the divisional managers who were worried about the division of power. First, some people doubted if the benefits of having the PMO for the organization were significant. So communication and support from the CIO and vice-president who served as the sponsor was key.

Second, I think we were struggling at first with how to establish the standardized process. My job at the PMO was to ask program managers to make sure that they understand our business strategy and to use our procedures to align their programs with the business strategy. At first, I did not know how to ask those questions, and such a standardized procedure did not exist. That caused a lot of dissatisfaction among senior managers and program managers. But a few of us put our heads together and developed the standardized procedure for it. After that, I saw no resistance.

Saul: Okay, I have another meeting in five minutes. We will meet again. In the meantime, I want you to understand that you guys have to have a plan for any anticipated and unanticipated resistance. When it comes time to spend money on building the PMO, we want to look prepared for success, and I want all players in this company to know we are prepared. That is, Peter, why I hired you and that is why, Barry, I chose you to help Peter. Clear?

Peter and Barry: Clear!

Discussion items

1. Based on the business you are in, would a PMO add any value? Why or why not?
2. Suggest a plan for dealing with any anticipated and unanticipated resistance. What can the company do to lessen resistance?

Progress—One Step at a Time

James Schneidmuller
and Peerasit Patanakul

Alex Jordan, a volunteer member of Global Telecommunications' Project Management Community of Interest (PMCOI), just ended a meeting with John Payton, the Senior Vice-President of the Network Business Unit. In the meeting, Alex reported the progress the PMCOI had made with respect to the current year's business plan. He sensed that John was concerned about the lack of progress since the last status review. John challenged Alex to come up with a way to improve the focus on the plan and add a significant value to the company.

GLOBAL TELECOMMUNICATIONS INC.

Global Telecommunications Inc. (GlobalTel) is a business leader in the telecommunications industry. With its global presence firmly entrenched, the company provides a broad range of wireless voice and data services, global backbone networks, broadband, IP, and Wi-Fi to its clients, ranging from individual consumers to global enterprises. In recent years, GlobalTel has had annual revenues of over $100 billion. As a global corporation, GlobalTel is organized by business units (BU). The major BUs are Business, Consumer, Network, Wholesale, and Operations, with support units of Legal, Human Resources, Advertising, Finance, and Executive Operations (see Figure 21.2). In addition, there is a Quality Management office as part of the Network BU that also reports "dotted-line" into the CEO. Each BU has its own independent organizational structure. Even though project management has been established as the standard approach to deliver business results and maximize customer satisfaction, most project managers at GlobalTel operate within their independent entities with no overall cohesion. The consistency in managing projects is lacking at GlobalTel. While some support may exist within the organization, there is no formal, centralized support for the project managers. There is no available source for project management advice, counsel, tools, techniques, processes, metrics, etc.

GlobalTel's largest presence is within the United States, but it has successfully established business within its international territories (Europe, Middle East,

Africa, Latin America, and the Asia-Pacific rim). This has proven to be extremely effective, given its desire to do business with the multinational corporations. This business can originate within the United States, or from within one of the international territories, and can be extended to any or all of the others.

GlobalTel's projects are of varying size, scope, and duration. Its larger projects are typically generated by the Business BU. These projects tend to be significantly complex, with a multitude of its products and services integrated into a total large sale. The Business projects usually focus on a single business customer, but may either be focused on a single site or span multiple domestic U.S. locations and a number of global sites. The Consumer projects, while significant in number, tend to be more limited in scope and breadth of products/services. The Consumer projects have a shorter duration, often with expedite demands, but require more in the way of customer support and service in order to ensure overall satisfaction. GlobalTel's internal projects, typically generated from within the Network or Operations BUs, are also complex with large amounts of capital and expense dollar funding levels. These projects can vary from several months to many years in duration, with another GlobalTel entity needing to be satisfied. Process improvement, network infrastructure enhancements, or other internal demands are typically the basis for these projects.

FLASHBACK

The history of PMCOI started with the recognition of the prevalent global project management issues existing at GlobalTel. To deal with it, a small group of dedicated project managers banded together. This self-initiated, "grassroots" team of volunteers referred to themselves as the GlobalTel Project Management Community of Interest. All of the PMCOI members shared a passion for the profession and wanted to see things get better for the project managers. They faced the existing issues of constrained resources, the lack of a standard project management process or methodology, little or no training requirements for project managers, and inconsistent project performance measures. They wanted to bring attention to the GlobalTel project managers and the important role they served.

When first formed, the PMCOI was loosely structured. There was no real leader and meetings took on an almost social club feeling. It became increasingly clear to most members that a more formal structure, executive support, and recognition of the PMCOI were needed. The executive support was especially important, since without it the group lacked the legitimacy necessary to implement its plan. In addition, new members were reluctant to join fearing that their contribution of time and effort would go unrecognized, perhaps even be challenged, by their local management.

As they worked to identify potential executives, one of the members recommended her senior manager since he had a strong belief in project management. When approached, John Payton, a senior vice-president in the Network business

Figure 21.2   GlobalTel, Inc. Organizational Structure

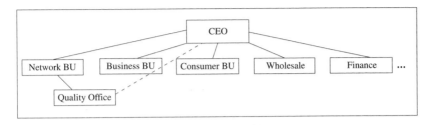

unit, willingly accepted their invitation to become Executive Champion of the PMCOI. This marked a significant turning point, for the PMCOI was able to now accelerate its progress, become a more formalized entity, and attract additional volunteer members. With John Payton's support, the PMCOI planned and successfully implemented GlobalTel's first-ever project management symposium which brought together the project managers, their leaders and executives, and external experts in a forum aimed at spotlighting the importance of project management. John Payton arranged for the CEO to make opening remarks. The event was a sold-out success.

HOW TO MOVE FORWARD

Alex realized that even though the PMCOI had accomplished a lot, it had never operated to its full potential. Although it had an excellent Champion and more recognition, the members were still participating on a voluntary basis. Not being a formal corporate entity, the PMCOI was not funded through the company budgeting process. The members were called upon to use whatever was available to them through their organizations. Alex stops by Sally's office to update her about his meeting with John. Sally is a senior project manager, working under John, and a PMCOI member.

Alex: Knock. Knock.

Sally: Alex, come on in. How was the meeting? Sorry that I could not make it. I had a conference call with one of our clients. Do you remember the order access project I told you about? Now Cortel, the client, wants us to change the web-based application.

Alex: Really. That's going to be a challenge.

Sally: Yes. Anyhow, tell me about the meeting.

Alex: The meeting went fine. I just reported our progress to John. You know what we were expecting to hear, right? We have not done much lately. Everybody has been very busy with work. And with the downsizing, everybody is being asked to do even more. It seems that we are losing the members' interest in the PMCOI little by little.

Sally: You're right. It is unfortunate but what can we do? Our formal responsibility should come first. So, what did John say about it?

Alex: He did not say much. But I could see that he was not very pleased. He asked if we can do something to revitalize the PMCOI. I think that's going to be tough. We can't do much given the downsizing that's going on. People just don't have a lot of time to volunteer.

Sally: Okay. Do you want to talk about this now, Alex? I mean, I am free now and we can also continue over lunch. How is your schedule today?

Alex: I can talk. Let's talk. Now, what can we do so we don't lose the focus on project management that we've been able to create?

ROLES AND RESPONSIBILITIES

Sally: Let's step back a little. I think first we should think about what should be the future roles and responsibilities of the PMCOI.

Alex: Our vision and mission statements are good. John liked them a lot when we shared them and asked him to be our Executive Champion. So if we still want to provide leadership and direction for the advancement of project management here at GlobalTel and we want to advocate and support the implementation of project management practices, processes, and principles across GlobalTel, what should we do? Even though we agree that project management is important to our company, each BU has its own way of doing things. Project managers also do things their own way. We don't have a standard methodology, so we all don't speak the same language when it comes to our project management. Also, our global clients have noted this as well. They want consistency irrespective of the region of the company that they are dealing with.

Sally: We have done some things in that area but I guess it's not enough, right? Project managers in my BU would say that they follow the same methodology. I think it is a good start.

Alex: Right. Things go well in your BU since you have support from John. It is a different story in other BUs. Even though John helps to convince other executives to support us, they still do not take us seriously. They think that their project management approach works well for them or they just don't care. They do not see why they have to put more effort into project management.

Sally: I agree. Besides creating a standard methodology, we also want to continue our project management symposium, right?

Alex: Definitely. It was such a great idea to organize a project management symposium. I think people are more positive about it than the standard methodology. Our first symposium was a great one-day event. The internal and external presentations helped to share knowledge and experiences in project management. John provided great support for the symposium. Personally inviting other executives to attend was a great idea! He attended for the whole day. . . I don't know how he was able to do that with his schedule! We could expand the symposium to be an annual two-, or even three-day event. I have heard from many GlobalTel people that they are interested in presenting next year. The external speakers are interested as well. Even our global regions said they enjoyed it. It has been a big success. I believe we have established a tradition that will last well into the future.

Sally: Definitely. Besides sharing our internal knowledge and experiences, I like the fact that we got to hear so many perspectives in panel discussions led by outside industry leaders. Having outside guest speakers was also great.

Alex: So, we agree that creating a project management standard and organizing the symposium are the major responsibilities for the future.

Sally: Yes. I think that training should be one of our responsibilities as well. There is really no project management curriculum and no consistency with respect to the training that project managers should have. We have a good partnership with the HR department. We can give them a list of the subject areas and then they can organize the training.

Alex: You mean. . . you want the PMCOI to organize the training, too?

Sally: I don't know. I think we have to think more about that. I think it depends on the structure of the future PMCOI. If we are still on a voluntary basis, we will not have time to organize the training. In that case, using the partnership with HR is the best way.

Alex: In the future, besides sharing knowledge and experience through the symposium, I think the PMCOI can play the role of a consulting unit. You know. For example, we can create a review process. Within it, there are sub-processes for data collection, providing consultation and corrective advice, and suggesting additional areas of opportunity. The PMCOI can participate in project audits, too. This review process can provide a linkage between us and the Auditing department. We can work through Kathy for this, since she is from Internal Auditing.

Sally: Wow. This will require a lot of work and people. With the way the PMCOI is structured now, I don't think we can do all of this.

Alex: I know. I'm concerned as well. One more thing, I learned from other companies that they have an organizational entity that helps monitor some major projects. We may be able to do that, too. We can get some information about our project management from those key projects. We can help establish the current baseline and track the progress of deployment. I know Sally, this is huge. The structure of PMCOI must be changed to accommodate these new roles and responsibilities.

STRUCTURE AND STAFFING

Sally: Sure. So, what are our options on the future structure of the PMCOI?

Alex: With all the major responsibilities we just discussed, we may not be able to do everything, but at least I think the PMCOI should increase its visibility here. One way to do that is to have more executive engagement. Maybe we can create something like an Executive Steering Committee. Then, as we interact with them, we keep them in the loop. This way, we may get buy-in from project managers working under them.

Sally: When we get their support, the implementation of the standard methodology should go smoother and we may get more people to be volunteers in PMCOI, too.

Alex: Right. But we can't guarantee that more people will volunteer. In fact, the Executive Steering Committee may be a long shot. It may generate some initial excitement, but in the long run, what will happen if the executives lose their interest? What kind of signal will that send? Executives come and go, too. Would the new executives accept our invitation to join? We can't control that. We can't even control our PMCOI members now.

Sally: Right. Another option is making PMCOI a permanent part of the organization.

Alex: You mean having full-time staff and all of that? But we're in a down-sizing mode. What will executives think about this? They are reducing their headcounts and we are proposing a new organization.

Sally: I know. I also have some concerns about that as well. But we will never know what will happen if we don't try. And if PMCOI is not a permanent entity, you can forget about all of those consulting and monitoring responsibilities, Alex.

Alex: You're right. Let's assume that the possibility exists. How will this new entity be staffed? Where should it report?

Sally: Good questions. Assume that we will call it a PMO for now. Based on its future responsibilities, the PMO should be staffed with experienced project managers. Probably PMPs.

Alex: How many?

Sally: Eight to ten.

Alex: It is downsizing time, Sally. Who will give up their experienced project managers to this new PMO? Eight to ten seems to be too large.

Sally: I know. We have to think more about this. Maybe we should start small. Four to five maybe? It all depends on the responsibilities of the new PMO. It definitely should not limit its responsibilities to what the PMCOI currently does. It should do more. For example, it should definitely keep a partnership with HR, but it should organize its own PM training. It should be sure to keep the symposium running, and implement a standard methodology across all of the BUs. Also, it should consider all of those consulting and monitoring items that you suggested. Don't you think that it needs that many experienced staff?

Alex: I share your view, Sally. I know that during this crunch time, if this new PMO is established, it has to prove to people that it adds value. I know that we need that kind of staff. But I am not sure it will be possible. BUs may not want to give up their best project managers.

Sally: It has to come from the top anyway Alex. We should just try to come up with the options to present to John, right? We are not making an implementation plan here. Now, if it is possible, where do you think it should be placed within the company?

Alex: Since John is our champion, should the new PMO be placed under him? Should it report to him?

Sally: That is possible. The plus side is that John already knows what it's all about. The negative side is this may not help with the implementation of the standard methodology across BUs. People will view the new PMO as just focused on John's organization instead of the full, global focus. You know, focus and support all of the BUs. . . the entire enterprise.

Alex: True. It can be a staff organization, reporting to a senior executive, perhaps even the CEO.

Sally: I have seen that in other companies.

Alex: Or it can work under the Quality Office. The implementation of standard PM methodology after all should be similar to the implementation of the standard processes and methods under TQM and Six Sigma initiatives. Maybe this is a good place for PMO.

Sally: Maybe. In that case then, the new PMO will be funded under the VP of Quality. I wonder what John will think about that.

Alex: Whatever structure it will be, can we have both PMO and an Executive Steering Committee? Oh, can we keep the PMCOI, too?

PARTNERSHIPS

Alex: How about the partnerships we have already formed?

Sally: You mean like the partnership with HR.

Alex: Right. I think we still can keep our partnership with HR. You know what, the new PMO may not need to organize the training itself. It may just need to consult and provide feedback on the content of project management training courses. HR knows what the career path for a project manager is.

Sally: You're right. Now if we go ahead with the auditing responsibilities, we have to have a strong partnership with the Internal Auditing department. When you mentioned about the review process, I think the Internal Auditing department can significantly help us with it.

Alex: You're right. In fact, when we propose the standard methodology, the project review approach should adhere to it. Auditing could help us review that process for some real projects. And you know what, they can provide us with additional areas of focus after they have done each project audit.

Sally: Right. Now, in your dream, Alex, of having a full-time PMO, the PMCOI, and an Executive Steering Committee, what is the relationship between the PMO and the other two entities then?

Alex: I don't know. The PMO will need support and even direction from the Executive Steering Committee no matter what. And for PMCOI, maybe it can structure itself into a set of Steering Committees with some focus on different operational areas. Maybe they can help the PMO review some Education and Training programs. Other parts could be Awards and Recognition, since John mentioned that in our first meeting with him. A Standards and Tools committee could work on the methodology. And we certainly would want a symposium planning committee!

NEXT STEPS . . .

Sally: Alex, let's stop here. I think we have enough ideas. We discussed the future roles and responsibilities, the possible structure and staffing, and the partnerships. I think we can start the dialogs with the other members of the PMCOI. Then, we can present the ideas to John.

Alex: All right. I feel a lot better now after having this conversation with you. The next PMCOI meeting is next Tuesday at noon. It is a lunch brownbag this time. I will send an email reminder to everyone.

Discussion items

1. With the current company's situation, what recommendation should Alex and Sally present to John?
2. If the establishment of the PMO is the way to go, what will be its initial set of responsibilities? What will be the short-term and long-term responsibilities?

3. How will it be staffed: initially, short term, and long term?
4. To where should the PMO report?
5. How should the new PMO be presented so that it will be accepted and embraced quickly?
6. How should the new PMO operate so that it will always have a strong connection with the project management community within GlobalTel, Inc.?

Chapter 22

INFORMATION SYSTEMS, ORGANIZATION, AND METRICS

This chapter contains five cases, discussing OPM in the aspects of information systems, organization, and metrics. The information in the cases is relevant to several best practices in OPM3, such as Project Management Information System and Establish, Adopt, or Institutionalize Organizational Project Management Structure. There are one critical incident, two issue-based cases, and two comprehensive cases in this chapter.

1. Is It Information Systems That We Need?

   Is It Information Systems That We Need? is a comprehensive case. It portrays problems regarding project managers' learning during job reassignment. A question to the readers is that whether or not having an information system would help speed up project managers' learning.

2. Spreadsheet Is Everything

   Spreadsheet Is Everything is an issue-based case, discussing an adoption of information systems in an organization. While information systems supporting project portfolio management were implemented, the user still cannot use the systems at its full potential. This problem is quite common to many organizations.

3. R&D and Operations: How to Make Them Talk

   R&D and Operations: How to Make Them Talk is an issue-based case. It portrays a specific situation where management would like to see more of the

coordination between R&D and Operations. Such coordination is necessary for creating breakthrough products. The case discussed several options that promote internal coordination.
4. Bluedogs USA

Bluedogs USA is a comprehensive case, providing information regarding progressive reorganization of an IT division of an international firm. Since the need for reorganization comes from the bottom, the case also discusses an approach used to promote reorganization to top management.
5. Point of Contact

Point of Contact is a critical incident. It discusses the possibility of a new organizational arrangement of a functional department to support project and program activities.

CHAPTER SUMMARY

Name of Case	Area Supported by Case	Case Type	Author of Case
Is It Information Systems That We Need?	Information Systems, Metrics	Comprehensive Case	Peerasit Patanakul and Sung Han
Spreadsheet Is Everything	Information Systems	Issue-based Case	Peerasit Patanakul, Sabin Srivannaboon, and Dragan Z. Milosevic
R&D and Operations: How to Make Them Talk	Organization Arrangement	Issue-based Case	Priya Venugopal
Bluedogs USA	Organization Arrangement	Comprehensive Case	Nicolas Charpenel
Point of Contact	Organization Arrangement	Critical Incident	Peerasit Patanakul, Sabin Srivannaboon, and Dragan Z. Milosevic

Is It Information Systems That We Need?

Peerasit Patanakul and Sung Han

Scott Hamilton, a manager of the project management group of LGS, walks out of the meeting room with mixed feelings. On the one hand, he is very happy that Andy Smith, his star project manager, got promoted to be a manager of the wireless business unit, a major client of his group. On the other hand, he knows that his project management group will face some challenges because of the job reassignment. Andy is currently managing five projects and they must be reassigned. Even though management gives him enough funds to hire a new project manager, Scott knows that it is not as easy as it sounds. The turnover rate of his project managers is quite high lately. The reasons stem from both internal promotion and exiting because of burning out. This makes it pretty common that one project has been led by several project managers. This practice hurts the projects because of the lack of continuity. Projects are often delayed. Helping his project managers come up to speed quickly when job reassignment occurs is always a critical issue that he still cannot resolve.

SOFTWARE LOCALIZATION IS OUR STRENGTH

MTI Corp is a business leader in research to commercialization of electronic components. Besides doing business locally, providing products and services to international customers generates several concerns to MTI's executives on how those products and services can serve the target customers in a specific country. In addition to developing high-quality products, one of the most important issues that MTI's managers have to focus on is software localization.

For MTI, the localization business is under the responsibility of the LGS group within the ITF business unit. In general, ITF serves clients in the areas of IT consulting and customer infrastructure, web development and software engineering, agency and design service, and language solutions. In other words, ITF is MTI's full-service consulting practice and solution provider. This service is available to all business groups (internal clients) and external initiatives.

437

TYPICAL LOCALIZATION PROJECTS

A localization project starts when the project is assigned to LGS from its clients (other business units). After the client provides the English version of the software, the project budget and schedule are negotiated and approved, the LGS project manager oversees the project by defining its own process of completing the job based on the given resources.

One example can be: LGS receives a wireless network project from the wireless business unit, requiring the software localization in 24 languages including both the translation and the functionality testing of the software. More specifically, the main goal of this project is to localize the product including its user interface (UI), user guide (UG), and related documentations with the functionalities undergone in the testing in 24 different localized versions of the software. To achieve high product quality, the project manager must not only eliminate simple linguistic issues such as mistranslations or typos, but also understand overall features of the software itself to ensure that the software works properly as originally designed.

Typically, the budget for a localization project is approximately $2 million. With an intensive schedule and a tight budget, it may not always be possible for the project team to do the work as planned. This common but significant issue forces the team to develop a list of work priority and use it for achieving the minimum quality acceptable for the project. However, it is also possible that the project budget is renegotiated during the course of the project, especially when the scope of the project was changed. The scope change includes, for example, the product was updated with new features and more languages are required because of more international target markets being added. Figure 22.1 summarizes the relationships among client, end user, and LGS.

PROJECT ORGANIZATION

MTI Corp is a large organization, divided into several business units that have their own suborganizational structure. ITF itself is also divided into several subdivisions (see Figure 22.2). Focusing on the localization business, the specific organizational structure of the LGS group could be described as one of the matrix types. The typical project team includes a project manager, translation team, engineering team, and quality assurance team.

As full-time employees, software engineers in general are responsible for various localized software activities such as recompiling and debugging software. It is common that a software engineer works on more than one localization project at a time. The quality assurance team—most of them are contractors—performs quality assurance activities. Their emphasis is on the improvement of the product quality. The translation team lead manages the team of translators to get translations for specific languages that will be used in the software. In general, the translators are mostly contractors, working remotely around the world.

Figure 22.1   The Relationships Among Client, End User, and LGS

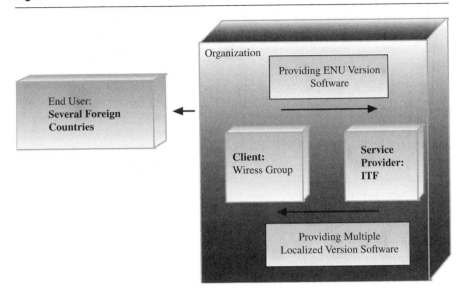

Figure 22.2   Organizational Structure of ITF

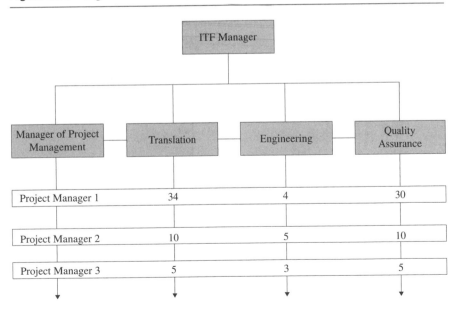

PROJECT MANAGEMENT PROCESS AND TOOLS

LGS follows a standard project management process. Its project life cycle, defined in several process phases, includes discovery, planning, translation, quality assurance, and delivery. Even with its standard phases, a project manager has flexibility to adapt the process to a particular project.

In terms of resource estimation, prior to the project being assigned to LGS, the client typically has estimated the project budget and schedule. Such estimation is made by considering several factors, e.g., the complexity of the software, test, test script, and the number of words and strings of the software. For monitoring and control, tools such as Earned Value Analysis (EVA) are used. In addition, LGS also uses a tool called ITO Timecard to keep track of hours spent by the team members.

LEADERSHIP AND COMMUNICATION

The project managers of LGS have an average of seven years of project management experience. Their backgrounds vary—software engineering, linguistics, etc. In general, the leadership style of LGS project managers is "communicating and getting away." With this type of leadership style, project managers minimize their involvement in the projects unless there is a problem. This style fits well when the team's accountabilities and responsibilities are well clarified and established. The level of trust between the teams and the clients is high since the team members and clients have known each other and worked together for a long time. Leading by example is another preferred leadership style of LGS project managers. In general, project managers provide some information about similar work that has been done. This technique has been used to convince the team members to accomplish their job according to the defined plan. LGS project managers also do their best to create a positive work environment for the team. In addition, they also value communication both by formal and informal means to minimize the barriers between team members and managers.

The importance of communication is recognized by LGS managers as a means of project monitoring and control. A lot of times, when the projects are assigned to LGS, the English version of the software is still in the testing process. During localization, if any versions of the software have issues that should be corrected, it is important to communicate with other versions of the software, which means that the client and the team must be notified.

In general, several communication methods are used. For example, the project team has weekly meetings to communicate project status, discuss problems, plan for the future, etc. The clients usually attend this meeting in order to directly gather information about the current project status from the team. Besides the weekly meetings, the project manager communicates with the team and the clients by email and telephone. Even though some information has been discussed, it is more efficient for the team, especially the quality assurance team, to know the current status of the issues in order to prepare the response plans. In addition, the project manager often has short meetings with specific team members. The meetings can be formal or informal. If new information is generated in those meetings, that information will be shared by email to other team members and discussed in the weekly meetings if it is deemed necessary. Besides formal communication, informal communication, e.g., walking around and asking questions is also very necessary. LGS project managers recognize this as a very useful way to acquire actual project status, problems, and feedbacks directly from the team or communicate the project status to the clients.

PROJECT REASSIGNMENT CHALLENGE

After careful consideration, Scott decided to reassign some of Andy's projects to Steve. This will help Steve get up to speed and possibly become one of the future stars. To do this effectively, he has to reassign some of Steve's projects to other project managers. After looking at the workload of project managers in his group, he also has to hire a new project manager. But before doing anything, he has to talk to Steve about this. He decides to do it over lunch.

Scott: Steve, how is life so far?

Steve: Busy as usual.

Scott: Listen Steve, the reason that I invited you for lunch today is because I want to also update you about issues that have recently happened in our group. Management has decided today to promote Andy to be a manager of the wireless business unit.

Steve: Good for him. We kind of knew about this all along. Andy has done a great job leading all wireless localization projects. He has an electrical engineering degree, so he is a perfect fit.

Scott: Right. I am very happy for him. And it is also great for us because now we will have a client who knows our project management challenges.

Steve: Please do not start. Do you remember the project CAT I am leading? John, our client, just informed me this morning that he wants us to work on the new UI. Gees, we are in the middle of this project. What can I do? I know that we have to do it but the project schedule and budget are not going to be pretty. Some of my engineers are going on vacation. They have already done their job according to our plan, so it is fair for them to take off some days. But now with the new UI, they have to redo their parts.

Scott: Did you negotiate the schedule and budget with John?

Steve: I did, but you know John, he is very aggressive. I told him that I will check the plan and will set up a meeting with him to discuss it. But I know that this project has high priority and we have to hit the deadline.

Scott: Right. Check the project plan. If you need more resources, we can allocate some engineers from other projects to CAT.

Steve: I will. Thanks.

Scott: Let's go back to Andy's promotion.

Steve: I know what is coming.

Scott: You know? Okay. I want you to take Andy's position. Andy is now leading five projects. You will take some and I will reassign some to others. Also, we have to reassign some of your projects to others. I also have enough funds to hire a new project manager. The new project manager will take what is left over. How do you like this idea?

Steve: I think it is good. Thank you for the opportunity. We have to work on the details.

Scott: Right.

SHOULD WE HAVE SOMETHING CALLED "INFORMATION SYSTEMS"?

Steve: Scott, can I talk to you frankly about the situation like this?

Scott: Of course.

Steve: This is not the first time that it has happened. You know that we have a high turnover rate in LGS, especially in a project manager position. Don't you think we should do something to make it easier for us for the transition?

Scott: Do you have anything in mind? We have all the project management training and standard processes and tools to help our project managers.

Steve: I am not talking about those. What I am talking about is what can we do in an organizational level to make the transition period go smoothly? We know that localization is not a rocket science but there are a lot of details that we need to take care of. The project schedule and budget are always tight. Plus, as with my CAT project, it is common that the clients change their requirements in the middle of the projects. Even so, we have to get everything done by the deadline since we do not always have a luxury of renegotiation. All of these force the project managers to come up to speed quickly, which is not always easy. Think about it, if you now reassign CAT to Molly, even though she is a great project manager, she will be bombarded with all the new requirements from John.

How much time does Molly have in order for me to explain the project to her? Not much—she has to act right away.

Scott: I have been thinking about this but have not been able to come across any solution for it yet.

Steve: I am not blaming you. It is not always easy. And it is not just learning about the project. It is about project management, too. I agree that we have a standard process. And it is a CMM process. We also have our flexibility to tweak the process. I bet Andy does something to his projects differently from what I do to mine. I am not sure how much he records. There is no way for me to know about this until I take over his projects.

Scott: Steve, how about we have some guidelines or procedures for the transition?

Steve: Having guidelines or procedures is good but I do not know how much they will help us. I guess what I am talking about is beyond that. It is not only during the transition period. What I am looking for is also what can we do, organization wise, to make our project management easier. What should we have in place so that we can have a record of what we have done? We can learn from each other; we can check our current project status; we know what our teams are doing; the clients know what we are doing; management knows what is going on; etc. See, Scott, you offered me some additional resources if I need for project CAT. How do you know who is available? You may have a way to keep track of things but is it accurate or is it current?

Scott: I understand. So far, we have created our own practices because we agreed that our situation is unique. In each project, we manage a team of contractors around the world. Our process must be flexible enough to accommodate that. But you are right; we should have a central depository or something of that sort to keep our project information, similar to our ITO Timecard that we use to keep track of hours spent by the team members. Maybe I should talk with other groups to see what they have installed. But don't you remember that we had something like this once? It did not work because nobody used it and complained that we should spend more time on project works, not creating these documents and completing them online.

Steve: I remember. But then, we didn't have as many projects to lead at once as we have now. I think having that kind of system now would be helpful. At least, it would help me keep track of my projects. At that time, I also remember that our group just started and we tried to figure out a lot of things. I think if you ask us know, we would all agree that having such a system is necessary.

Scott: Okay. I will see what I can do. We may not have to start everything from scratch.

METRIC SHOULD BE PART OF IT, TOO!

Steve: Another thing, do we have any way to measure that we are doing a good job or a bad job? I know that we have time-cost-performance objectives for our projects. Should I be considered as a loser if I cannot hit the deadline, for example? We tried our best to accommodate our clients, but it is not always easy. In the CAT project, for example, what are the client and management going to think of me if I cannot deliver by the deadline? You know that it is not our fault. It is because of the changes in requirements. And this is common. In the ZABA project, Sandra gave us multiple drops. What I mean is that it took her 10 times to deliver the entire document to us. You would think that once we get the project, the clients will give the entire document to us at once. Then, we will know exactly what we have to do and we can coordinate with the translators. Taking 10 times to deliver the entire document to us will make the translation process a lot more complicated. In such a case, lots of coordination is needed, etc. If something like this happens, and we miss the deadline, are we at fault?

We have the customer evaluation process at the end of the project, when we send the survey out to our clients to gauge their satisfaction with our work. That is a good thing but isn't it too late in the game? I know that the survey results will help us improve our future projects but should we have set measurement criteria upfront so that we know what to expect? You know, the results of the survey may indicate "very satisfied." But it was the result of our hard work to satisfy our clients no matter what circumstance. Are we playing too tough on ourselves? Scott, you know that we are totally burned out. That's why some of us have left the company. I am wondering if it is a good idea to have any standard metric to measure our performance. This metric should help us understand what the management and client expect from us so that we can manage the project to meet that expectation. The metric should also help the clients prioritize their expectations. It will also help us and our clients come to an agreement on those expectations.

Scott: Good idea, Steve. Let's schedule another meeting to discuss this more.

Discussion items

1. What should be implemented to alleviate project management problems at LGS?
2. If Scott decides to implement a project management information system, e.g., online-project management data base, what should be included in the data base?
3. If a standard metric to measure the LGS project performances is developed, what should the measurement criteria be?

Spreadsheet Is Everything

Peerasit Patanakul, Sabin Srivannaboon, and Dragan Z. Milosevic

Avita Corporate and Investment Banking (ACIB) ranks among the top 10 financial institutions on the Euro capital market and remains the global leader in derivatives. Present in the European, U.S., and Asian markets, ACIB offers its client a tailored approach to meet their financial and strategic needs. ACIB is a part of the Avita group, which includes other businesses such as commercial banking.

The Information Technology (IT) division of ACIB just implemented a new information system for its project portfolio management. However, people are reluctant to use it. Work goes on the same way, in a manual and labor-intensive process. Several mistakes are still made. Why aren't they ready for the new system?

PROJECT PORTFOLIO MANAGEMENT IN THE IT DIVISION

The IT division, operated under the Chief Information Officer (CIO), was made up of businesses covering a broad variety of information technology activities for all business lines, accounting and finance services, and the operations services. The IT division uses project portfolio management offices (PPMO) to maintain corporate projects. Generally the PPMO recommends the senior management of project selections, including ways to increase throughputs. The IT division also implements several project management offices (PMOs) to support projects from each business line. The PPMO operates under the umbrella of the office of the CIO. The office just implemented Advance, commercial project portfolio project management software, for the analysis of IT projects. Collecting data for the Advance is done by the PPMO office administrators. The PPMO uses the following labor-intensive steps to collect data.

Step 1: Existing project details from scorecards in Advance are exported into spreadsheets. The project details include the status (open or candidate), project ID number, phase in system development life cycle, start date, estimated end date, description, etc. These spreadsheets are sent to the respective department PMO project managers to verify the data. Changes can be made if need be. After receiving the data back, an administrator loads it back to Advance.

Step 2: Project resource demand data are exported from Advance to a spreadsheet sent to project managers for review. It is again sent back to the PPMO with the updated data, and loaded back into Advance. The resource demand is based on six rolling quarters (the next 1.5 years).

Step 3: Spreadsheets of data regarding resource capacity and availability are collected from functional departments and are loaded into Advance. The IT division uses resource centers to keep track of the resource capacity and availability.

Step 4: Data analysis is performed after all necessary data is collected. The PPMO generates scorecards and graphs based on the configured data format (e.g., quick facts, resource demand versus cost, project scoring map). Presentation slides are made from the analysis results and presented to senior management.

The PPMO executes these steps every month. Common issues are recurrent. A project manager gives a quick call to the PPMO administrator:

MISTAKES ARE STILL BEING MADE

PM: Paul, you send me four spreadsheets for four of my projects. Did you make a mistake on one of them? I do not lead Extreme.

Paul: Wait. Let me check. Extreme was just approved last cycle and in Advance, you are the project manager for Extreme. I just downloaded that information directly from Advance.

PM: Oh, I am sorry. We already changed to project name from Extreme to Vision.

Paul: You should give me an update on this, otherwise I would not know. Go ahead and make the change. I will change it in Advance afterward. You know, we have had a problem like this several times already, especially for new projects. After the first team meeting, the team decided to change the project name to make it sexier. They forgot to let me know. A couple of times, we did not catch those mistakes. The project was shown on the scorecard as two different projects and this led to a confusion and incorrect analysis.

PM: I understand. For the Stardust project, last month you told me that the resource demand spreadsheet I sent you had some conflicts with the resource availability spreadsheet from the functional department. And we spent quite a bit of time figuring that out. What was the problem again?

Paul: It was my fault. I looked at the wrong spreadsheet. I will be more careful this time. You see, we have a total of 224 existing IT projects and some new ones will get approved this time. I cannot keep up with all of those spreadsheets.

PM: Hey, I am with you. We have to send these spreadsheets back and forth. It is time consuming. And if you look at hundreds of spreadsheets, more likely that you will make some mistakes. We have to do a lot of data reconciliation. Can't we set up Advance such that each PM can input data into it directly? We should stop using spreadsheets to exchange data altogether.

Paul: I wish. Not at this point.

PM: Why?

Paul: Advance has not been configured to do that yet.

PM: Why don't we configure it? What's the point of having this powerful tool if we cannot use it?

Paul: We are taking baby steps here. There are a lot of features that we have not used. You see, if we want to use Advance extensively, all of us have to go through trainings. Don't tell me just now that you guys suddenly have time for those trainings.

ABOUT ADVANCE

Advance is a project portfolio management software solution. It provides flexibility and infrastructure for enterprise portfolio management. It has powerful security features and a clean user interface, enabling collaboration among all stakeholders. It is built specifically for proposing, planning, and controlling portfolios, following an objective and transparent process. Advance provides data entry forms for documenting information of each project, scorecards for evaluating a set of projects, maps and diagrams for reviewing and analyzing a portfolio or portfolio of portfolios, etc.

Discussion items

1. For the IT division, does the implementation of Advance help improve efficiency and productivity or create redundancy?
2. What should the PPMO do to improve to way they use Advance?
3. What should the necessary steps be in the successful implementation of enterprise information systems for project management?

R&D and Operations: How to Make Them Talk

Priya Venugopal

As the CEO, John Phillip needs to meet the commitments of his company to sell 5 million units of the solar panels to the OEMs each quarter for the next two quarters. However, the solar technology and industry are growing extremely fast and need rapid improvement in efficiency at lower costs to remain competitive in the market. There are various emerging technologies that are very promising. Some of these include dye-sensitized solar cells, noncrystalline silicon cells, cells based on organic materials, and thin film solar cells using inorganic semiconductors made of elements such as cadmium and tellurium or even copper, indium, gallium, and selenium. Each of these emerging technologies holds the potential to completely destroy Sun Energy's market share by offering various innovative products. It is therefore imperative for Mr. John Phillip to keep the R&D pace extremely brisk in the company. The focus of this case study is organizational structuring, very often used in a process. Its purpose is to define organizational responsibilities and reporting relationships.

SUN ENERGY

Sun Energy Inc. is a manufacturing company that manufactures and markets high-efficiency silicon Photovoltaic (PV) cells. The manufacturing facility operates 24/7 by employing four shifts of technicians. There is a management chain that oversees only manufacturing operations in this plant. In addition there is a significant amount of R&D work that also occurs in tandem. The R&D management employs many engineers to foresee R&D projects and develop next-generation, higher efficiency solar panels. This team of engineers is also responsible for improving the manufacturing technology for enhancing productivity and reducing cost. The manufacturing and the R&D teams often are at logger heads trying to use the same resources in the factory. There is constant friction between the two managers and the technicians are usually torn between the two objectives set by the two managers. There is immediate need to reduce chaos and improve the communication between the two organizations.

448

The manufacturing teams and the R&D teams are two separate management structures which both are responsible for the profitability of the factory. The technicians are direct reports to the operations manager. However, the technicians are trained by the R&D engineers who also are responsible for maintenance of the factory equipment. The operations manager is responsible for improving the productivity of the factory and increasing total output in a shift. However, R&D work often affects the productivity for that shift and this is another constant source of friction between the manufacturing team and the engineering team.

JOB DESCRIPTION

The organizational structure of Sun Energy Inc. can be seen in Figure 22.3.

Operation Manager: The "ops manager" supervises the technicians in each shift and is responsible for the factory production output during this shift. He reports to the factory manager. There are four ops managers for the four shifts in a week.

Factory Manager: This individual is responsible for meeting the factory production output and reports directly to the CEO of the company.

Engineer: The engineers own the production tools and are also the primary drivers of the R&D projects in addition to supporting the performance of the production tools.

R&D Manager: The R&D manager supervises the engineers and the overall R&D program for the factory.

Figure 22.3   Organizational Structure of Sun Energy Inc.

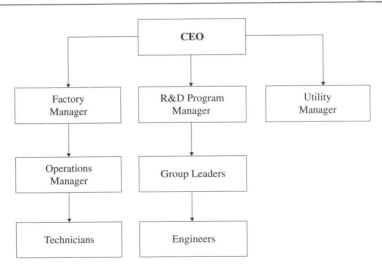

Technician: The technicians report to the ops manager on that shift and are responsible for running the station and achieving production. They are also responsible for conducting some R&D experiments set up by the engineers. In a sense, they also listen to directives from the engineers.

Utilities Manager: The utilities manager is responsible for all the civil, electrical, and chemical facilities in the factory. The contract workers who help demo and/or install the tools are under the supervision of the utilities manager. In addition, any hardware modifications needed to support R&D are controlled by the utilities manager.

ORGANIZATIONAL CHANGES

There were ideas floating around in upper management to merge the manufacturing group and the R&D group to reduce the constant friction between the both teams. However, after many rounds of meetings the project management office in the company effectively put an end to these plans. The PMO's final proposal stated that it was necessary to keep the two groups separate, each with team-specific goals and deliverables. That way, there would be clear understanding as to the owner of each deliverable in the company. The PMO noted that if the organizations could hold effective meetings and communicate their goals efficiently to each other, they could then achieve the necessary objectives for the company.

Sun Energy is currently undergoing an organizational restructuring where the management is trying to improve direct communication between various departments. The change process involved meetings between various managers. These initial meetings involved the following managers: The factory manager, Tim, heads the manufacturing organization; the utilities manager, Josh, heads the factory civil, electrical, and chemical facilities and is responsible for environmental emissions as well; and the technology development program manager, Beth, heads the R&D teams and the various development projects including manufacturing technology development. There are four operations managers working under the factory manager and they are responsible for the four production shifts. The project manager himself has four group leaders working under him that heads the four functional groups in the R&D area.

UPPER-LEVEL MANAGEMENT MEETING

Tim, Josh, and Beth are all invited to the meeting by the CEO in the corporate planning conference room. They know it could be related to the rumors going around about proposed organizational changes in the company, but they do not know the specifics. It is 9 AM and the three managers are waiting for the CEO, Phil, to arrive.

Tim: Beth, I am not sure why we are here. Phil did not mention much in his email. Do you have any idea what this is about?

Beth: Not sure, Tim. But I am wondering if this is related to the issues I raised with Phil during the Program Roadmap Meeting last week.

Tim: What did you both discuss? What were the issues?

Beth: During Q2 of this year, most of our R&D projects had not been completed. Many of my groups complain about the lack of factory resources available to them. There is growing frustration among engineers that they cannot have the technicians complete the R&D projects on time. Also, there are concerns that they cannot get the factory contract workers to execute engineering R&D projects quickly.

Josh: I am surprised this needs an upper-level meeting with Phil.

Tim: Hi Phil. How was the trade show in Vegas? I thought you wouldn't be back until Monday.

Phil: Tim, I curtailed my trip by a few days to address a few outstanding organizational issues. This has been bothering me for a few weeks now.

Tim: So what's bothering you?

Phil: I had a meeting with Beth last month and she briefed me on some of the challenges the R&D team is facing in meeting Q2 goals due to lack of factory resources.

At the trade show, I saw exhibits from our main competitors and their products are achieving an increased level of sophistication and there are rumors that they are able to offer their product at 20 percent lower costs than last quarter due to a breakthrough manufacturing innovation in their product. This really undercuts our price advantage and could affect us seriously if we don't innovate fast enough. Therefore concerns raised by Beth affect our competitive ability directly.

We need to maintain our leadership by enhancing our complementary assets in terms of R&D effectiveness and project management concepts. I would like to ask you all for suggestions about how we can improve coordination between the various groups.

Tim: The operations team currently does not have a formal framework to communicate with the R&D and the utility teams. Our monthly meetings are usually very hectic and don't focus on day-to-day activities.

Beth: I agree. We ought to slightly micromanage everyday activities in the factory to be able to improve operations output and R&D output.

Josh: The utilities team is usually kept in the dark until the last minute as there are no planning meetings. Each morning the factory services contract workers are clueless as to where their services will be needed. After a few hours into the day, requests pour in from operations and R&D and often clash and make it very ineffective. We should have better coordination.

Tim: Can the operations and the R&D teams meet each morning and coordinate their work to allocate utility worker resources and technician priorities? We had tried it once in the past and it worked reasonably well.

Beth: That sounds like a good idea. I think that would give the engineering team a chance to plan their experiments in advance and give a proposal to the operations team.

Phil: In the past what were the reasons why these meetings stopped being effective?

Beth: Often there was no understanding of the R&D roadmap versus The operations roadmap. It was difficult for the teams to understand the priorities of the other team and how it related to the "critical few" mandated by the company. We stopped having daily meetings as it turned out to be an ineffective forum to understand each other's priorities.

Josh: Maybe we can have an upper management planning group that can oversee the long-term priorities and goals of various groups in the company.

Phil: That sounds like a good idea. I recommend creating a project oversight committee involving the R&D manager, utilities manager, and the factory manager. They should meet once every two weeks to discuss project roadmaps in each of these groups. I will also attend this meeting to align the company's long-term goals with the short-term milestones in each of these groups.

Tim: I also suggest having a daily meeting attended by the ops manager for that shift and a few engineers from across various engineering teams to determine daily priorities for the technicians and the factory resources.

Beth: That sounds good. Between the fortnightly meetings and daily meetings, we should be able to coordinate better between the R&D and operations teams. I will set up these meetings and send out outlook requests. Thank you all for your inputs in addressing these issues.

Discussion items

1. What are the pros and cons if Phil decides to merge the operations and R&D units?
2. Summarize Sun Energy's plan to improve coordination between operations and R&D and discuss the pros and cons of such a plan.
3. What should be the roles and responsibilities of the project oversight committee?

Bluedogs USA

Nicolas Charpenel

John, the director of IT project management at Bluedogs USA is on his way to a meeting with Mike, the Vice-President of Marketing and Online Growth Strategy. While walking, John takes a deep breath as he braces himself for what he knows is coming: The business group always has many of those wonderful project ideas, aiming to make the websites better and the business grow, and he will have to tell them, yet again, that he does not have resources available to allocate to anymore projects, unless he is given a budget to hire new project managers.

BLUEDOGS USA BUSINESS

Bluedogs USA is a multichannel retailer, selling its different clothing brands and catalogs through phone, mail order, and online. Bluedogs USA employs about 5,000 people across the United States, with the biggest portion of personnel working on inventory, shipping, and customer-service phone support. The total headcount of the IT group is about 200 employees, half dedicated to the company information systems, half to the development of the brands' different websites.

Bluedogs USA is also part of the international Bluedogs Group, which includes sister companies to Bluedogs USA all around the world. Each sister company has its own set of brands, catalogs, and websites, although they sometimes share brand names and collections. The company generating the most business within Bluedogs is Bluedogs France.

As the online market developed over the last years, Bluedogs USA shifted its strategic focus to online development. A few years earlier, the e-Commerce (or IT) division successfully developed a web platform to host and develop all of their websites. Following the success and the savings from this platform, sister companies to Bluedogs USA and its international counterparts (including Bluedogs France) started to outsource their website and online operations to Bluedogs USA's IT division. As a result, the division tripled in size over the last three years and the platform now hosts over 20 U.S. and European websites generating just about $1 billion a year.

ORGANIZATIONAL STRUCTURE OF THE IT DIVISION

The IT group is a relatively young organization that experienced rapid and recent growth. As such, the division is still trying to achieve maturity in its organizational structure and internal processes. Currently, the IT division is divided into four main teams (see Figure 22.4):

- **Project Management:** This group is led by John, director of IT project management.
- **Quality Assurance:** This team is responsible for verifying that everything is working correctly before releasing a new functionality or version of a website to the public.
- **Engineering:** The engineering team develops new websites or new functionalities to the existing ones (includes developers, network engineers, system engineers, data base administrator, etc.).

Figure 22.4   Initial Organizational Structure of the IT Division 4

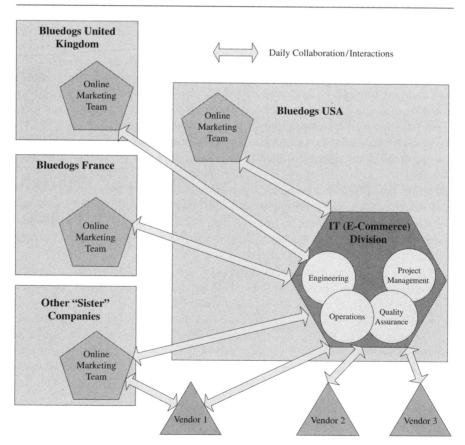

- **Operations:** This group is responsible for monitoring the websites and making sure they are available for business 24/7.

In addition, the IT division—and especially the project management team—tightly collaborates with the following entities:

- **Online Marketing Divisions (The Business):** Although they can share best practices, information, and sometimes resources, every website has its own team of online marketing analysts and web administrators. In the organization, these teams are not part of IT. They belong directly to the brands or sister companies who outsourced their website. Among other duties, the online marketing teams are responsible for analyzing market trends and identifying opportunities to develop the online business. Often, these opportunities take the form of new functionalities on the website developed by the IT division (e.g., the ability to zoom in on a product image, display personalized product recommendations based on the customer purchase history, etc.). The online marketing divisions therefore act as internal clients and project sponsors to the IT group.
- **Third-party Vendors:** When a new functionality is requested by the business, IT does not always reinvent the wheel. The product of a third-party vendor is often implemented and incorporated into the web platform to bring new functionalities to the websites. For example, an internal search engine on a website can be provided by Google or other solution vendors to allow the customer to search for products on the website easily (e.g., typing "red shirt" in the search engine, instead of browsing through shirt categories). The selection of the vendor solutions is either made by the business's online marketing group, the IT division, or sometimes by a combination of the two.

PROJECT ORGANIZATION

Most of the projects coming to the IT division usually fall in one of these categories:

- Implement a new functionality on the platform (either to one or multiple websites).
- Add a new website to the platform.
- Implement a new process, new tool, or new hardware to host the websites (technical projects that do not usually involve anyone other than the IT division and the third-party vendors).

Most projects belong to the first category and are usually short: three months from kick-off to launch. Projects in the second and third categories can have similar durations but can also have much longer lifespan: 6 to 18 months. Depending on how many areas impacted (data base, development, mainframe, third party, etc.), project teams can go from 4 to 15 people.

Projects all start with a request from the business. Following the request, the project management office gathers cost and effort estimates from all groups within IT (Engineering, Quality Assurance, etc.) and negotiates a start and end date for the project with the business.

Once approved, a project manager is assigned to the project and begins working with the business and the third-party vendor(s) (if any are involved and already identified) to define and formalize the scope of the project within the effort hours committed to the project and with the inputs from representatives from the different engineering areas (DBA, Web Development, Network, Systems, etc.). Sometimes the business has a detailed vision of what it wants; other times it only has the general idea figured out. In the latter case, the project manager and the engineering group provide input to help the business detail its requirements and design the final product. In every case, though, it is the project manager who is responsible for formalizing the scope and business requirements into a document and get sign-off from all parties involved.

During this requirements phase, the project team is assembled and the technical approach document (detailing how the scope of the project will be met and tested) is due shortly after. Once IT management signs off on the technical approach, the implementation phase of the project starts. When the development is completed, the project enters the quality assurance phase, followed by the user acceptance test where the business's marketing team verifies that the project meets its scope and expectations. Once approved, a final test is done by the IT operations team to make sure that the project product is stable and compatible with the existing system before releasing it to the public.

OTHER OPERATIONS PERFORMED BY THE IT DIVISION

Besides implementing new projects, the IT group is also responsible for making sure that all of the websites are stable and available for business 24/7. To this end, the operations team has tools to monitor the websites to be sure they are functioning properly. In case the monitoring processes fail to detect a problem, the business usually contacts the project management group during the day. During off-hours and weekends, the business contacts the operations team directly if the problem is critical and prevents sales. Communication on issues to provide status or updates is usually done by a project manager.

IT also develops and provides tools to the business teams to allow them to make specific on-the-fly changes to the website (such as changing images on the homepage, adding new products to the website, creating deals and promotions, etc.). Of course, IT is also responsible for improving these tools and making sure they work properly and continuously.

BACK TO THE STORY . . .

During the meeting, Mike, VP of Marketing and Online Strategy, presents a few upcoming features he would like to be developed on the U.S. websites. One of them is the implementation of a new logic to make smart product recommendations to the customers. This new logic or recommendation engine comes from a third-party solution which demands tight integration with the web platform and data bases developed by IT. As the project will involve a third-party vendor, the business, and multiple engineering teams to complete, John knows that the project manager will have to do a lot of coordinating and micromanaging of the project to successfully see it through. Therefore, John will need to assign someone with a lot of technical expertise and availability. Unfortunately, this is not the best period for such a PM-extensive project as his entire team is already fully allocated to current projects. John starts explaining the challenges he is facing and proposes some solutions to accommodate the business requirements.

> **John:** Basically, if you want the project to be completed in six months we need a project manager to kick-off the project right now, and I don't have anyone available.

> **Mike:** Really? Isn't David available? He just completed the colorization project last week.

> **John:** David still has another project, plus I need Anna to start transferring some of her knowledge on third-party Reportiture to David and let him take over some of the load on that front. Anna has been testing the job market recently and if she leaves, we need to make sure someone knows enough about Reportiture. I know the project is over but we still need someone to follow-up on outstanding issues and evolutions of the tool.

> **Mike:** Why? Isn't there anyone else with sufficient knowledge on this tool and vendor? What about the developer who worked on the project, or maybe one of my analysts who worked with Anna on Reportiture?

> **John:** Well that's the thing. Multiple developers worked at different points of the project and besides, they would mostly know how the tool was technically implemented, not about the business rules and reasons behind the choices made in the requirements. For our guys, it is the opposite. Anna has been there since the beginning of the requirements and she's followed-up on every upgrade since. She is the only one with the full picture right now.

> **Mike:** (Frowns) So basically you are telling me that if she is hit by a bus tomorrow, somebody will have to pick up the pieces from square one. What is your transition plan?

> **John:** I have Anna working on some documentation and she will sit with David to go over the general picture. He will also start following-up on some

upgrades and bug fixes scheduled. That is why I don't want to assign him to another big project right now.

Mike: Alright, fair enough. And you do not have anyone else available?

John: Not really, my team is fully allocated. My guys are already the last ones to leave the office most nights. They are usually clocking more hours than any other team.

Mike: How is that possible? Who do they have to manage and coordinate with during off-hours when the developers are gone? Which reminds me, I talked to the director of engineering and he says he has some resources available. Is the ratio of engineers/PM so bad? We can't have one engineer for one project manager. You know better than I that project managers are expensive resources!

John: The project managers do more than just manage and coordinate efforts of the project team, they do the planning, the communication to stakeholders, etc. But the main reason is that they do a lot of things outside of working on projects.

Mike: What do you mean? Like what?

John: Well, when we experienced production problems on the website, a project manager is usually involved to make sure the right resources are working together to resolve the issue in a timely manner. They also receive and follow-up on daily questions and issues coming from the business brands about how the tools we put at their disposal work. For example, "how come this tool does not work?" or "Can I do this or that through the tool or do I need to ask IT to do it?"

It is similar for third-party functionalities too. When there is an issue or the business needs a new feature, they come to the project manager to ask if it should be addressed by IT or the vendor. To summarize, project managers are basically the first point of contact for the brands whenever there is an issue or a question for IT.

Mike: Now that you mention it, they are also the ones following-up on outstanding bugs and fixes that appear long after the project is over, correct?

John: Absolutely, even though the project is technically over and management focuses on the next, we sometimes fail to realize that there are all those things that project managers still do sometimes months after launch. Those are not projects since they mainly consist of a bug fix and a couple of emails here and there, but they add up and take time away from working on projects.

Mike: Hmmm . . . but why do you think bug fixes always go through a project manager? Can't we have someone in engineering do that?

John: Difficult to say, the project manager is usually involved because, again, he or she has the most understanding of past projects. They were the ones to basically write the requirements, coordinate with the engineers during the implementation phase, were aware of every bug reported by quality assurance and the business before launch. If we want someone else to do it, some serious knowledge transfer will have to take place first.

Mike: I see. But still, it looks like you have the answer to your resource allocation issue. You need to start moving some of those extra-project tasks off the plate of your guys and hand it to someone else. It should help your team focus on project management and take on more projects.

John: I agree and am actually working with other managers in IT to propose a plan to our vice-president of IT to start moving some of these responsibilities away. But you know, it actually comes with the territory of growing so fast. We were only a few project managers and engineers a couple of years ago, and everyone was doing all sorts of things to get the job done and get this web platform started. We are barely figuring out the processes we must put in place.

Mike: Of course, but you know how competitive the market is and how strategic the online part of our business has become. The trend is likely to continue and as an organization we need to reach a new level of maturity if we want to keep growing. When do you think we will be able to see that plan of yours implemented?

John: Not tomorrow if that is what you are asking. We already started gathering data to show the VPs and executives where the problems lie and what would be the ROI of fixing them. We are working on a new organizational structure and the inception of new teams that should help improve our global quality of service to you and the rest of the business. I'm actually hoping it will also help with the retention problems that I had in my team recently. Beside the long hours, spending half the time on support instead of working on projects is not exactly the dream of a project manager.

Mike: (Sigh) It looks like I will either have to reshuffle the priorities to accommodate for the recommendation system project or push it back a little. It will not make the brands or our CEO very happy though.

John: Well, maybe it will make it easier to promote this change in the organizational structure.

ONE YEAR LATER

Is has been about a year after that conversation, and the IT group organizational structure has changed significantly to address some of the issues the department was experiencing and relieve the pressure from the project management team (see Figure 22.5). Three new small teams (two to three persons each) have been introduced:

The Help Desk team: This team is in charge of troubleshooting and communicating with the business on noncritical issues happening on the websites or the internal tools.

The Bug Support Team: This group manages the bug fixes and functionality releases on the websites and management of new or outstanding post-launch bugs projects.

Tool and Third-party Support Team: This team is in charge of the relationship with third-party vendors (negotiations, coordination, etc.). The team coordinates with the business, vendors, and IT engineering to schedule bug fixes or new functionality releases on both in-house and third-party tools.

Figure 22.5 New Organizational Structure

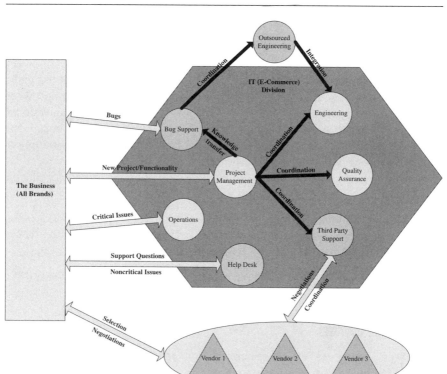

PROMOTING THE REORGANIZATION PROPOSAL
TO TOP MANAGEMENT

The first step toward the reorganization was to prove or demonstrate to executives the issues faced by the IT group and especially the project management team. In order to do so, IT management leveraged the existing time tracking system of the company. Employees were already used to reporting their work hours at the end of every week and divide them between different websites or areas they worked on (maintenance, development of U.S. or U.K. sites, project A, project B, etc.). IT added the following elements to the time tracking system so that the employees could specify how much time they were spending on each function weekly:

- **Production Support/General Operations:** for tracking efforts related to questions coming from the business or production issues on the websites and internal tools.
- **Third-party A, Third-party B:** for tracking coordination efforts with the different third parties.

After a few months of gathering data, the time tracking system showed:

(1) Project managers were spending almost 50 percent of their time doing production support.
(2) While it could not be generalized to the entire project management team, some specific project managers were spending up to 60 percent of their time on the coordination with the third party.
(3) Long after projects were launched, project managers were still clocking some time on the projects to follow-up on outstanding known bugs, questions from the business, and plan improvements or next phases.

The first point served to show executives that the project manager resources could be used more efficiently. Instead of adding more project managers to do more projects, the IT group could free up some project management time by moving production support to dedicated resources somewhat less expensively and who were easier to train. These new resources would share their knowledge base as part of the help desk team and, being dedicated, would provide a better and faster support to the business.

The second point demonstrated that managing the relationship with a third party could be a full-time job. A new position could be created to manage the coordination of a third party full time and perform duties that project managers did not have the time for (perfect the documentation, negotiate contracts, etc.). From there emerged the proposition to create the tool and third-party support team.

The last point highlighted that, in the project management world, the typical life of a project extended much beyond its launch. It led to the proposition of having a bug support team taking the responsibility of all postlaunch follow-up activities, after a knowledge transfer with the project manager. The goal was to free the project manager to take on new projects much faster after a launch.

This analysis was presented to executives along with reports of past and current issues due to project management over-allocation or weak knowledge transfer processes. The reorganization proposal was eventually approved and a budget was allocated to recruit new resources and form the new teams.

A PROGRESSIVE REORGANIZATION

The implementation of the different teams happened independently but each used its own phased or progressive approach.

The help desk team started with only one resource and its manager working on issues on the U.S. websites only. In the first phase, project managers were still responsible for all communication from and to the business, leveraging the help desk team to do the troubleshooting and involving the right engineering teams. Eventually, a second help desk engineer joined the team, project managers were removed from the loop of communications (leaving the help desk team to interact directly with the business), and the team started covering the issues of all websites on the platform.

The tool and third-party support team initially had one resource managing one or two of the most critical third-party applications. The team later developed into a three-person team as it took on the responsibility of internal (in-house) tools while managing an increasing number of third-party applications.

Two bug release managers were recruited at the same time to form the bug support team. Right from the start, they managed the bug fix releases for all the websites on the platform. Eventually, the engineering development team decided to follow the same approach of separating resources working on bugs and resources working on new functionality. The development of bug fixes was outsourced to a consulting company while the Bluedogs developer focused on building new functionalities. The bug release manager's responsibility became the coordination of the in-house engineers and the consulting development team to release the bug fixes in production.

This progressive approach allowed Bluedogs' IT division to prove the merits of its reorganization and make some modifications when necessary before going full speed on the implementation of each team. It also gave the division some time to recruit resources for the newly created positions.

ROLES AND RESPONSIBILITY IN THE NEW ORGANIZATION

The help desk team's role is to troubleshoot and communicate on the status of noncritical issues reported by the business. Any brands experiencing a problem on the tools managed by IT or on their website contact the help desk team directly. The help desk team does a first level of troubleshooting and attempts to solve the issue. If the team cannot resolve or identify where the issue comes from, the help desk provides the right person or team to troubleshoot and solve the issue (Development, Project Manager, Network, etc.). Throughout the process, the help desk team is responsible for communicating status on the incidents to the business and other stakeholders regularly, and making sure the teams involved troubleshoot and resolve the issues in a timely manner.

The tool and third-party support team's responsibilities include the negotiation of new contracts (or renewals) with third parties, making sure that third parties respect their end of the contract, and answer business questions from the brands on both in-house and third-party tools. Members of the team are responsible for maintaining a detailed documentation on third-party contracts and implementation and then sharing it in the team's knowledge base. The team works with project managers to implement new functionalities or new versions of the third-party software, with the bug support team to fix outstanding bugs and with the help desk team to troubleshoot new issues (usually configuration or hardware related) reported in production.

The bug support team manages the outsourced consultant team to fix known outstanding bugs and releases them in production. The process starts with a knowledge transfer between the project manager and the bug release manager once a project is launched or a short time before. The bug release manager then interacts with the business to prioritize which bugs should be fixed first. Once priorities are defined, the bug release manager coordinates with the outsourced consultant team to fix as many bugs as possible in time for the next scheduled release (every three to five weeks). The new version of the website that includes the bug fixes is tested by Bluedogs' quality assurance team and hands it over to the engineering team for a publish in production.

The responsibilities of the quality assurance, engineering, and operations teams did not change. As before, the operations team remains the first point of contact and communication for critical issues (websites not available, impossible to check out and make sales, etc.). The development team in engineering, however, is able to focus on the development of new functionalities and less on bug fixing.

The project management team is still responsible for all the projects—implementing new functionalities, improvements, or launching new websites—but can now focus more on applying standard project management best practices and coordinate a project from the requirements phase to its launch. Project managers can now rely on the new teams to assist on their projects during the implementation cycle or after. On the counterpart, project managers are now responsible for transferring their knowledge to other teams to ensure proper postlaunch support of a website, tool, functionality, or third-party application.

Discussion items

1. What was the role of a project manager before the organization changes described in this case? Based on the discussion, what extra duties were project managers performing in addition to straight project management?

2. Without the reorganization, what were the issues and risks the company was facing?
3. What are the benefits and drawbacks of implementing this new organization structure? How does it help/hurt the project management activities specifically?
4. What would have been the alternatives to solve the issues faced by the organization without changing the organizational structure? What would be the drawbacks and benefits?

Point of Contact

Peerasit Patanakul, Sabin Srivannaboon, and Dragan Z. Milosevic

Jane Malone is contemplating on a way to help workflow better in her organization. Being promoted to Director of Laboratory not long ago, Jane is quite a bit of frustrated with her organization's workflow. Her goal is to provide a better service to both internal and external customers. "What should I do?" thought Jane. "Should I reorganize?"

JANE'S LABORATORY

Jane runs a laboratory that is part of the Technical Center—the world's leading aviation research and development and test and evaluation facility. The center operates under the umbrella of the National Aviation Agency (NAA), a government agency that is responsible for the safety of civil aviation. The role of Jane's laboratory is to perform the test and evaluation, and verification and validation in air traffic control, communications, and navigation. Within the laboratory, different test and evaluation methods are run by different teams.

SUPPORTING NAA PROGRAMS

The NAA's mission is to provide the safest, most efficient aerospace system in the world. To support this mission, the NAA implements several programs. Among them are the development of a new aviation technology, the improvement of a system of air traffic control and navigation for both civil and military aircraft, and research and development of the National Airspace System and civil aeronautics. Jane's laboratory will have to support most of these programs in the areas of system testing and evaluation.

EFFECTIVE OR INEFFECTIVE

Depending on the test and evaluation methods they need, the program teams usually contact the test team leads, who are Jane's direct reports. Obviously, these mega programs consist of multiple subsystems that require different test and evaluation methods. "Contacting team leads directly has worked so far but I am wondering if it is effective and will serve us well in the future," Jane thought. "In addition, we have a lot more programs than we used to have several years ago. With multiple programs, contacting team leads directly will definitely create organizational complexity. I can imagine more chaos in the near future. What should I do?" Jane thought.

Should I have something like one-stop shopping? Should I create a position called test integrator, such that the program management can come and talk to this person? The test integrator can then help the programs arrange the test and evaluation methods they need. The program can also get the status of test and evaluation activities from the test integrator. Our test leads can then focus on test and evaluation instead of worrying about direct communication with the customers. The communication with program management can be more effective through a test integrator. Would this help us be more effective? Are there any disadvantages if I go with this?

Discussion items

1. Would creating the test integrator position help Jane's laboratory be more effective? Why or why not?
2. If Jane decides to go with her "one-stop shopping" idea, what competencies should her test integrator possess?
3. Is there anything else Jane should do to support program management of NAA?

Chapter 23

ORGANIZATIONAL CULTURE AND PROGRAM CULTURE

This chapter contains two cases—one issue-based case and one comprehensive case, discussing organizational and project or program culture. The issues discussed in this chapter support several best practices in OPM3 such as Establish Organizational Project Management Policies and Engage Program Stakeholder Processes.

1. What Helps Us Come This Far?

   What Helps Us Come This Far? is a comprehensive case. It tells a story of a successful program. Elements contributing to program success are discussed in the case. Is program culture a basis of a successful program?

2. Is It Standard Methodology That We Need?

   Is It Standard Methodology That We Need? discusses project management problems in one organization. While some project management methodologies have been implemented, the organization seems to lack a culture that supports project management.

CHAPTER SUMMARY

Name of Case	Area Supported by Case	Case Type	Author of Case
What Helps Us Come This Far?	Program Culture	Comprehensive Case	Peerasit Patanakul
Is It Standard Methodology That We Need?	Organizational Culture	Issue-based Case	Peerasit Patanakul

What Helps Us Come This Far?

Peerasit Patanakul

Sally Stonewell arrives at a meeting early. Today is the day that Flight 23 will be launched. It has been a while since Flight 23 was shipped to Cape Canaveral. There have been a series of launch system issues that the teams have been working very hard to address anticipation of a near-term launch. As a program manager, Sally is always appreciative of the dedication of her team. In the back of her mind, she always wonders what she did right to have her team's dedication so high. Or, is it pure luck? On the back of her ID badge, there is a list of six corporate values. The first on the list is product quality, followed by customer satisfaction. Could these values have something to do with her team's success?

OUR BUSINESS ENHANCES OUR NATION'S SAFETY!

DeTech is a global defense and technology company. It provides innovative systems, products, and solutions in information and services, electronics, aerospace, and shipbuilding to government and commercial customers around the world. The company is organized by several business sectors, for example, Information Technology (DIT), Space Technology (DST), Technical Services (DTS), Electronic Systems (DES), and Shipbuilding (DSB).

In the Space Technology sector, DeTech is considered as a leading producer of space systems including satellites, payloads, and ground segments. The company has an unparalleled portfolio of advanced technologies. Defense Support Program (DSP) is among the major programs of DeTech's DST sector. Since it won the first government contract in the 1960s, 23 satellites have been built and 22 of them have been launched under the program. Flight 23 is its recent satellite. Similar to the satellites previously built, its major mission is to detect and report all missile attacks in near-real time. The secondary mission includes detecting space launches, nuclear events, and tactical missile launches.

Flight 23 will be a new addition of the Integrated Tactical Warning and Attack Assessment System, consisting of a geosynchronous constellation of infrared sensing satellites, associated ground stations, and relays. Integrated by DST, each satellite includes a spacecraft, built by DST; a primary infrared payload, built by DeTech's Electronic Systems (DES); and secondary payloads, built by the subcontracted laboratories. After Flight 23 is launched, the program team expects to support the operational sustainment of Flight 23 and previously launched satellites for more than 15 years.

MISSION SUCCESS AND STRONG PARTNERSHIP ARE THE KEYS

During the previous stages of the program, depending on the key deliverables of Flight 23, several Integrated Product Teams (IPTs) were formed. The teams generally consist of the contractors and government teammates. This helps enhance mutually supportive government-industry partnership, leading to the high level of customer satisfaction. It is typical that the teams meet everyday, at their respective locations. Today, the prelaunch meeting is held at Cape Canaveral Air Force Station's launch complexes and processing facility.

> **Sally:** Today is the day that we will finally launch Flight 23. I got a report from the launch pad last night that everything is set up for the 2 PM launch time.

> **Tom:** That sounds great. We have a couple of hours this morning to do the final walk-through.

Lt. Col. Tom McFee is the DSP program manager from the Air Force Space and Missile Center. Throughout the program, the Air Force personnel have been assigned to work alongside the DSP contractors at several Air Force Bases/Stations and DeTech locations in Florida, Colorado, and California. The partnership is bound with the common imperative for mission success and it has produced a truly integrated team with technical and management excellence.

> **Sally:** It is early to celebrate our success but I have a good feeling about it.

> **Tom:** It is fair to say that Sally. We have worked so hard on every aspect of the program. We put our best effort out there at all times. So, to me, we could not have done any better.

> **Sally:** I'm glad you brought that up. From your perspective, what encouraged all of us to do that?

> **Tom:** Good question. I think as government personnel, we value mission success. We know that if this program is successful, our country and our people will benefit from it. We are talking about the security of our nation and also the world. We are talking about saving the lives of our friends in the frontlines, saving lives of our families and friends from the attack of our enemies, and so forth. I think this is the bottom line. We all are proud to be a part of this significant program.

Sally: You are right. This is the bottom line. And this applies to us, the contractors and co-contractors, too. Working in the defense industries, besides the job challenges, we all know that the mission success is so important, especially the national missions. We are also doing this for future generations, too.

Tom: We also appreciate partnership. We all do. And I think this is a great thing.

Sally: Thank you for recognizing this. At DeTech, one of our corporate goals is to become the trusted supplier of systems and products while, of course, achieving significant, profitable, and sustainable growth. To become the trusted supplier, we value strong partnership with our customers and co-contractors. Thanks to you, also, Tom, for being a strong supporter of this concept. It is great that we have the luxury of having DeTech and government personnel work on collocated IPTs at the Air Force bases and at our DeTech facilities.

Tom: You are welcome. It was difficult at first but we all benefit from it in the end. I think having such a partnership is great. It encourages effective communication at all levels. You and I, we talk a lot since we are the program managers, no doubt about that. But having collocated teams helps the information flow at all levels. We can reinforce our needs and the significance of the mission to you guys, on a daily basis. You respond to our needs and let us know what is going on. We identify and solve problems together. I think all of us appreciate such a working environment that enhances full disclosure and mutual trust.

Sally: You are right. We have nothing to hide. If we have any problems, you know. If you have any concerns, we know. Can I go so far as to call it an open and complete communication process across all program participants?

Tom: You name it.

Sally: One more thing. I can speak on behalf of all of us, the contractors, that having a strong partnership with you is a major source of inspiration. We know firsthand that our actions to assure mission success are critical. Working side-by-side with you guys, who could get deployed to the dangerous areas that our satellites watch, gives compelling motivation to us to get it right. This becomes very personal.

PROGRAM MANAGEMENT PROCESSES

Key to DSP's joint management team process is open, effective communication at all levels through truly integrated IPTs. The IPTs operate as inclusive teams where all disciplines partner across organizational and company lines. The teams are vested with applying knowledge, discovering, and implementing best practices, and capturing lessons learned. DSP uses web-based central repository for process descriptions. Each process has a process owner who is in charge of documenting and sharing best practices and lessons across the program.

Since the mission success is extremely crucial, DSP develops a proactive risk management mentality. Both technical and program management risks are identified and assessed in the technical, cost, and schedule dimensions. Root cause analysis is used rigorously as part of developing risk response plans.

In terms of monitoring and control, DSP uses metric-based, forward-looking predictive techniques throughout all program phases. These methods are part of the established program management process. Key technical, schedule, and cost indicators are monitored and analyzed. Future trends are studied. Proactive course corrections are made to avoid much more costly down-stream discoveries.

Key metrics from program management processes are reported both to corporate management (monthly program reviews) and customer leadership (program management reviews). Key parameters that are tracked and reported as dashboard include technical performance metrics (TPMs), cost metrics, and schedule metrics.

- TPMs are used in the development stage to assess reserves in technical parameters such as weight, electrical power, and software size.
- Cost metrics, e.g., Cost Performance Index, and variance metrics are used regularly to monitor program spending status against the plan. Cost savings resulting from, say, Six Sigma activities are also tracked and reported.
- Schedule metrics, e.g., the number of tasks that do not start or complete on time, are used to avoid schedule delay. Schedule slack of key events is also monitored since it is an early indicator of an overall schedule issue. These early warning metrics allow the management team to institute cost-effective, early corrective actions. Key program milestones and slack to these milestones are addressed in reviews through the company president level.

TEAM RECOGNITION IS IMPORTANT

Tom: So, Sally, what will you award the teams with after we successfully launch Flight 23?

Sally: Me? Are you kidding? This should be a question to our president! Our company's standard motivational practices include success and profit sharing, annual salary adjustment, etcetera. I hope we will get a big raise.

Tom: Congratulations in advance.

Sally: Specifically to DSP, we typically get performance bonuses. This is beyond my control. But what I did along the way as additional retention and morale-building activities are some award recognition activities, bi-annual off-sites, celebrations on major events, picnics, pool tournaments, commemorative pins, patches, outstanding performer awards, a drawing for a trip to a launch, and so forth. You know all of that and you are part of them.

Tom: I know. I just wanted to tease you.

Sally: Thanks. By the way, thanks for pitching in the cash benefits for our key staff who monitored DSP through critical events. I think the Employee Critical Skills Retention Program that we jointly developed is great. We should recognize the need to motivate and reward people.

Tom: You are welcome. It is a win-win, Sally. We want to keep those skilled people who work well with us. The program is not ended after Flight 23 is launched. We still have to work together on the operational sustainment phase of the program for many years. By the way, I heard that Steve will move on to join other programs, is it true?

Sally: Yes. It is a good career move for him. He will be one of those leaders across DeTech who are DSP alumni. Over the past 40 years, DSP is a source of senior leadership across DeTech.

Tom: So, who will take his place?

Sally: Luke. We always promote from within. Employee development is our corporate mandate. Other functional supervisors and I jointly monitor performance and develop goals with each employee that address both program objective and employee development. For the past several years, we are working hard on renewing the program organization. While we want to maintain our skilled employees, we also want to bring in someone new to refresh our knowledge and skills. We have to do this carefully because we do not want to jeopardize the DSP's legacy and culture. Luckily, we have a strong core of DSP-assigned staff who maintain our legacy and culture and also orchestrate education and transition of new team members.

Tom: Promotion is always a good way to recognize the members. But I can also see the difficulty of maintaining program culture. It is good that you find a way to integrate new team members to the program, by education and training.

Sally: Yes. Talking about recognition and motivation, we also have a mentorship program, education options, and career planning for the teams. DST also has a daycare facility onsite at our head office.

Tom: Didn't we answer your question—what encourages the team to be so dedicated to DSP?

FLIGHT 23 ON ORBIT

Flight 23 thundered into the clear afternoon sky around 2:10 PM. It successfully separated from the Alpha III heavy launch vehicle 6 hours and 20 minutes later. The launch marks the beginning of Flight 23's operational service, extending the services of a satellite constellation that has been the United States' eyes in the sky for nearly four decades. The launch was broadcasted throughout DeTech, drawing more than 400 DeTech and Air Force Personnel and their families to the festivity.

Lt. Col. Tom McFee commented at the press conference: "Thanks to the thousands of people who have worked so hard on this program, Flight 23 was successfully launched today. Thanks to the entire team for getting us to mission success."

Discussion items

1. What are the key program cultures ingrained in the Defense Support Program?
2. Does the company culture have an influence on the program culture Provide some examples from the case?
3. Would projects or programs in other industries benefit from having a project or program culture? Discuss specific project or program cultures in specific industries, e.g., IT, pharmaceutical, and telecommunication.
4. Suggest a process to develop the project or program culture.

Is It Standard Methodology That We Need?

Peerasit Patanakul

It is the time of year when the performance evaluation will soon be carried out. Matt Garner, the director of the project management department, sits in his room contemplating all the initiatives he had implemented in the past year. Some of them were successful, but the majority of them were not. Before joining Clement, Matt had many years of experience in project management and also worked as a project management consultant. He joined Clement several years ago in the quality assurance department and was promoted to the director of project management department at the beginning of this year.

COMPANY AND ITS PRODUCTS

Clement Corp is a provider of integrated financial and operations software for small and mid-sized businesses in the construction and real estate industries. The company also provides a range of support services. It operates predominantly within the United States; and also licenses products in other countries, including Australia and Canada. Clement's customers include commercial and industrial contractors, residential builders, specialty contractors, electrical and mechanical contractors, real estate managers, real estate developers, corporate owners, and the government.

Through its integrated family of construction and real estate management solutions, Clement Office construction software gives its customers highly developed cross-functional capabilities to pull everything together for streamlined, single-source control. The software suite includes:

- Accounting: accounts payable, accounts receivable, billing, cash management, contracts, equipment cost, financial statement designer, job cost, pay roll, remote time entry, etc.
- Estimating: tools, data base, etc.
- Procurement: buy-out, inventory, purchasing, etc.
- Reporting and productivity: document management, information assistant, inquiry design, report designer, etc.

PROJECT MANAGEMENT AT CLEMENT

Before Matt assumed the capacity of the director, project management at Clement was in a state of disrepair. Because of the internal promotion, technical staff were promoted to project manager positions with limited experience. In fact, no one had formal project management background and little training was provided. However, project managers did the job as best as they could. Matt knew this problem and he was convinced by the CEO of the company to take this challenging task. He recalled his conversation with the CEO at the beginning of this year:

> **CEO:** I know this is a big issue that we have to address as soon as possible. I want you to develop a methodology, document it, train the company on it, implement it, maintain it, and get these project managers up to speed.

This sounded simple but Matt knew it would not be easy to do. He knew that a previous director could not develop and implement the standard methodology. He had to figure out a way to do it successfully. He thought, at least, he had the support from the CEO.

WHO ARE THESE PROJECT MANAGERS?

Matt recalled the meetings with project managers. He arranged those meetings to get to know who they were. He found that besides not having formal project management training, most of them did not have any experience outside Clement that they could relate to. Also there was a huge lack of internal credibility. Employees in other departments did not understand why they needed project managers. What project managers were really doing was the administrative function; posting project documentation on the website, etc. They knew that they needed to maintain project schedules but they were not trained how to put the schedule together to begin with. They picked up little tidbits of issues. Since they did not have a real project schedule, these project managers could not create any expectation and make any trade-off decisions. They did not know whether decisions they made impacted the rest of the project.

FIRST STEP: CREATING TEMPLATES

With his project management experience, Matt could provide the internal training to the project managers. However, projects were still going on and new ones were continuously coming in. He could not take project managers out of their job and conduct a week or two classroom training. Matt decided to develop templates to help the project managers in the meantime. He started off with maintenance release types of projects since all project managers were familiar with them. He also did some coaching and mentoring.

The template for maintenance release projects seems to have been a success. As Matt anticipated, these project managers were familiar with these projects. They could relate to them right away. The project managers understood key deliverables and dependencies. Matt started to feel a bit relieved that things seemed to be getting along fine. Then one day, he helped one project manager develop a schedule for a new product development project.

Matt: You know what to do, right? We have a standard life cycle phase. So, you can start from there.

PM: Okay.

Matt: Start with the work breakdown schedule. This is the new product development project, so we have to do prototyping. We have to develop iterations and so on.

PM: So, the work with the schedule template does not apply to this project.

Matt: Not all of them. Look at this deliverable; is it going to be suitable for this project?

PM: No.

Matt: Do you need some help? It is obvious that you need some help. Let's go write things down on the board.

PM: You know how to create a work breakdown schedule for this project, right?

Matt: Yes.

PM: Well, could you create one of those for me?

Matt insisted that he would help the project manager create the WBS. They started brainstorming and finally they put together a decent WBS for this project. However, along the way, he realized: Even though these project managers show the progress on the schedule of maintenance release projects, when it comes to new product development projects, things go downhill pretty quickly. They could not identify deliverables. They could not set up dependencies. No wonder they could not track the project progress. He felt so sad for them and also for the company. He wondered how the company could maintain its businesses all these years.

WHAT IS THE PRIORITY HERE?

Besides developing standard methodology, Matt had the typical responsibility of a director, e.g., assigning projects to project managers, solving resource conflicts, and keeping track of projects and their status. He met with one project manager to discuss her new assignment:

Matt: I checked the status of your projects. Every one of them is progressing well and they are almost completed. We are not involved in any implementation. I think you have enough capacity to take on this new project.

PM: Oh no. I do not have capacity. Just today an issue came up. I sent you an email this morning explaining the issue to you.

Matt: I know but I do not think you need to be involved in that.

PM: No, they can't resolve them. I need to be there. I also have to set up a conference room for that meeting.

Matt: You don't need to do that. Do you have anything to contribute?

PM: No, but.

Matt: So, stop doing such things like those. Take on this project. It's good for you.

PM: But I am busy.

Matt: You are not. You are pretty much at 60 percent capacity right now. As I said most of your projects are going to be completed soon. You have enough capacity to take on this project. Be focused. Set your priorities. Also, pull yourself off the administration stuff.

This is a typical issue, Matt recalled. These project managers seemed to be busy all the time but they were busy on activities that they should not have been doing. They seem to get themselves into the situations that they do more of the administrative stuff. Helping them come up to speed with the methodology wouldn't help, Matt thinks. "I have to also help them have their priorities straight."

Talking about priorities, management seems to have an issue with this, too. Matt recalled his conversation with one of the senior vice-presidents.

Matt: We do not have enough resources to implement all of these projects. What is your priority? How do you want me to assign resources?

SVP: Everything is a priority.

Matt: What?

SVP: I just said everything is a priority.

Matt: Okay. But you have to understand that if you want to implement all of these projects simultaneously, nothing is going to complete by the set dates. All of these projects will be delayed.

SVP: But Matt, you have to understand also that if we prioritize, we are limiting ourselves. We cannot say that we are going to do these but not those. We have to respond to what our customers want.

Matt: So, you would rather just get everything out despite the delay than have to choose.

SVP: You work it out, Matt.

ONE-WAY COMMUNICATION: GOES DOWN BUT NOT UP!

Another important issue, Matt recalled, is communication. It did not seem that project managers had sufficient communication with upper management. This caused a lack of support and commitment from management. On the other hand, management did not know what was going on in the projects. He recalled what the CEO told him once:

CEO: I just don't know why we cannot get the product out of the door. Alpha project has been going on for two and a half years and I hardly have an idea how complete it is. What they told me is that we are not going to be able to get it out this year and I do not understand why. I have half of my company work on this project. Matt, from now on, I need you to tell me where we are and where we should be.

Matt thought, great. So what he did shortly after that was set up a project management steering committee and started reporting on all projects' status on a monthly basis. The committee included the CEO, senior vice-presidents, and vice-presidents. Besides listing the status report, he also defined the role of the committee as a decisionmaking group to make decisions on major project management issues, e.g., resource conflict, risks, and future directions.

The committee started off nicely. But all of a sudden, Matt recalled, all engineers were sitting there. Some executives stopped attending. They got frustrated because they would not be able to talk about the decisions that needed to be made. The forum then turned to be a project status reporting. Even though all projects were behind, people did not talk about what they could do, what the risks were, what the response plans should be, or what the next steps were. Then, one of the senior vice-presidents decided that they did not need this anymore. The committee lasted about six months.

MATT'S NEXT MOVE . . .

After thinking through all of these, Matt talks to himself: "I cannot be the only one to be blamed about all of these. I did my best. So, what is wrong here?" Matt is not sure if people recognize the value of project management. "The CEO thought that having a standard methodology would help. But that cannot solve everything. Are we lacking any mechanism to make project management successful? Do we have the right culture? What should I do next?"

Discussion items

1. What is the basis of the project management problem at Clement?
2. What should Matt have done differently? What should he do next?
3. Outline some organizational culture that supports project management.
4. What kind of project management culture should Clement create?

Chapter 24

ORGANIZATIONAL PROJECT MANAGEMENT IN ACTION

This chapter contains two comprehensive cases, discussing the implementation of organizational project management. The issues discussed in this chapter support several best practices in OPM3, such as Establish Organizational Project Management Policies, Standardize Portfolio Select Component Process, and Provide Organizational Project Management Support Office.

1. Let's Go All the Way

    Let's Go All the Way is a comprehensive case. It details an implementation of project management office (PMO). The case discusses the structure of PMO, its authority and responsibilities, process improvement, and the use of information technology.

2. Are We Ready for Project Portfolio Management?

    Are We Ready for Project Portfolio Management? is a comprehensive case. It illustrates some organizational project management problems in one not-for-profit organization. The underlying question in the case is: Would the implementation of project portfolio management help?

CHAPTER SUMMARY

Name of Case	Area Supported by Case	Case Type	Author of Case
Let's Go All the Way	Standardized Methodology, PMO, Information System	Comprehensive Case	James Staffan and Peerasit Patanakul
Are We Ready for Project Portfolio Management?	Portfolio Selection, Alignment	Comprehensive Case	Peerasit Patanakul

Let's Go All the Way

James Staffan and Peerasit Patanakul

It has been almost a year since the Global Infrastructure Group, a business unit of Global Finance Corporation, implemented their program management office (GIG PMO). The main goal of the PMO is to help improve efficiency and empower management to make better and more informed decisions through the use of the processes, tools, and standards. Prior to the implementation of the PMO, a working group was formed to assess the current state of GIG project management and the needs of its project management community. GIG management realized that project and program management added major value to the organization, and through the recommendations of the project managers (PMs), GIG felt it could achieve better efficiencies across the organization with a PMO in place. Several recommendations were presented to the head of GIG on the appropriate placement of the PMO and its authority and responsibility.

In the case study, we take a look into one of the PMO weekly meetings. On the agenda of the meeting today, two major items will be discussed. The first item is the implementation of the standardized project management process. While the team has worked hard on this for many months, the process is still a work in progress. One of the major issues that the PMO has to deal with is the customization of the process. "Standardize with flexibility" and "One size does not fit all" are the frequent comments from the project managers when they review the proposed process. "When I lead small projects, I will not spend time following all the steps in the process. Twenty procedures, three tools, and seven templates for the planning phase? This is ridiculous," said one of the Global Finance Corporation (GFC) project managers.

The second item is the way to improve the collaboration between the GIG PMO and the GIG project management community. The PMO implements the Link for Synchronization program to drive the implementation of the PMO initiatives, to involve GFC project managers in the PMO activities, and to communicate the training and mentoring programs. With the ambitious objectives of Link for Synchronization, PMO personnel anticipate the resistance from the GIG project management community.

GLOBAL FINANCE CORPORATION

GFC is among the international leaders in wealth management, capital markets, and financial advisories. With offices in 40 countries and territories across six continents, GFC serves the needs of both individual and institutional clients with a diverse range of financial services, including: personal financial planning, securities underwriting, trading and brokering, investment banking and advisory services, trading of foreign exchange, commodities and derivatives, banking and lending, insurance, and research.

Among other business units at GFC, Global Infrastructure Group (GIG) is responsible for developing technologies and solutions that serve GFC businesses. Each year, GIG initiates hundreds of projects and programs. Some of them are small (budget less than $50,000) and some of them are large (budget over $10 million). GIG has project management offices support its project management activities.

Project Examples

1. Corporate-wide Global Procurement System
   a. Description: The project is focused on implementing a corporate-wide Global Procurement System. The procurement system can be purchased as a product from another vendor or grown in-house, as long as it meets all requirements of the firm. The procurement system will be used for provisioning any assets.
   b. Duration: 2.5 years
   c. Budget: $11.2 million
2. Add Bloomberg Connectivity to 40 Branch Offices
   a. Description: 40 brokerage offices need connectivity to have Bloomberg stations at their locations. The connectivity needs to be set-up and terminals connected.
   b. Duration: six months
   c. Budget: $1.7 million
3. Move Equity Derivatives Group from the 6th Floor to the 5th Floor
   a. Description: Move the Equity Derivatives Group including all technology and physical equipment from the 6th floor to the 5th floor.
   b. Duration: one month
   c. Budget: $150,000

GIG PROGRAM MANAGEMENT OFFICE

GIG PMOs are organized with respect to the organizational structure of GIG (see Figure 24.1). GIG's business unit consists of two divisions: the Global Technology Group and Architecture and Engineering. Each division is composed of several groups. PMOs are set up to assist the groups within GIG (e.g., Global Ops PMO and Architecture PMO). The PMOs do not report into the group they support. They report to the Division PMOs (GTG PMO and A&E PMO), operating under the

Figure 24.1 Structure of GIG's PMOs

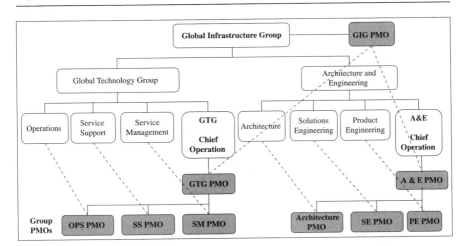

Chief Operation Officers. This helps PMOs promote synergies among the PMO groups and across the organization, which also helps more efficiently maintain standards across the groups. The division PMOs also report to the GIG PMO.

The GIG PMO is led by a managing director, who is also a peer with the heads of the GTG and A&E divisions. While the staffs of the group PMOs are mostly senior program managers who are PMP certified, several members do not have project management background and are tasked to develop project management tools and reporting standards. The head of the GIG PMO holds a weekly meeting where the leads of all the division and group PMOs attend and have a voice in the decisions. During the meeting, the team discusses the issues related to reporting and metrics for team management, future tools analysis, financials and demand management, and many other items.

PMO AUTHORITY AND RESPONSIBILITIES

The GIG PMO is authorized to make policies pertaining to a large variety of initiatives. The core policies are focused on projects that are completed by the GIG organization. The PMO is responsible for improving the process and tools which will support the GIG project management community and provide key management reporting to all levels of GIG management. The focus is to provide consistent and efficient project management methods, processes, and documentation:

- Driving, coordinating, and communicating with users, internal teams (development, logistics, and testers), and vendors
- Providing best in class project management techniques for engaging infrastructure and software development projects
- Providing objective project metrics with standardized reporting

- Constructing, managing, tracking, and reporting progress of project work
- Navigating GIG gate-keeping/governance committees for project initiation and approvals
- Managing and mitigating risk, scoping changes, and project issues
- Ensuring appropriate documentation is constructed for audit compliance

Some of the responsibilities of the GIG PMOs are:

- Managing the division PMOs
- Assigning PMs to projects over $10 million
- Final approval of any demand management, financial management, or project management tools/standards used in GIG

Some of the responsibilities of the division PMOs are:

- Managing the group PMOs
- Assigning PMs to projects between $1 million and $10 million
- Leading all development and research associated with the tools and standards being approved or researched by the GIG PMO

Some of the responsibilities of the group PMOs are:

- Assist in assigning PMs to projects under $1 million
- Assist in project prioritization with the group managers
- Spread the use of the tools/standards among the PMs within the groups
- Recruit PMs to join the programs within the PMO
- Manage and develop needed reports and metrics for the group managers

Typically, PMOs are not tasked to lead any GIG projects. Those projects are led by individual project managers who reside in their respective groups, reporting to their functional manager. If a project (program) has a budget over $1 million or requires extensive cross-organization collaboration, it will be led by a PMO program manager. The project managers from the respective groups who are involved in such an initiative will report to both the PMO program manager and their functional manager.

PROCESS IMPROVEMENT IS A CENTRAL THEME

GIG has a need for a transparent, consistent, and sustainable end-to-end project management process and tool set. The need can be elaborated into the following core areas:

- Process: Standardization and interoperability with processes of other business units
- Tools: Simplification and common platform

- People: Professional development, consistency of roles and responsibility, and qualification criteria
- Metrics: Transparency, management control, and decision support, and people performance and process optimization

To address the need, GIG PMOs developed a list of issues under the core areas that should be addressed. These issues are presented in Table 24.1. PMOs also identified the potential tools that should be implemented. The tools should help increase system throughput, reduce administration, promote standardization, and enhance communication.

Table 24.1 Core Process Improvement Areas

		Tools
Process	1. Project selection - Project prioritization - Reduced channel for requests - Workflow for all governance steps - Reduced administration	- Workflow Manager - Project and Portfolio Management Center
	2. Management of project timeline - Consistent framework/standardization - Collaboration via workflow updates - Project template - Better communication	- Enterprise Project Management System (MS Project professional 2007/ Project server)
	3. Communication and project documentation - Best practice - Continuous process improvement - Standardization	- Enterprise Project Management System - SharePoint 2007
	4. Management of project related expense - Total project expense - Budget alignment - Prioritization	- Project and Portfolio Management Center - Financial Tracking Tool
	5. Better identification and use of resources - Resource forecasting - Better utilization of resources	- Resource Forecasting Module (Time Manager (TM)) - Project and Portfolio Management Center
People	6. Professional development and PM competencies - Structured training - Career development pathway - Professional certification - Forums for sharing best-practice techniques - Mentoring program - Standardized communications platform	- Link for Synchronization
Metrics	7. Project reporting and metrics - Real-time project reporting - System/process/people/performance metrics	- Enterprise Project Reporting System

STANDARDIZED PROJECT MANAGEMENT PROCESS

GIG PMO put together a standardized project management process. Figure 24.2 illustrates the high-level process which includes four major phases. In each phase the more detailed steps are also proposed.

While many of the processes and standards are documented, there is a concern whether every GIG project should follow such processes and standards. Since the typical GIG projects have a short duration with a need to be completed with an as-soon-as-possible mentality, many of the project managers argue that with the methodologies in place, a project can be slowed with the amount of paperwork and tools that need to be involved.

BACK TO THE AGENDA . . .

Dan: It seems that all of us have the same concerns about the process that we've put together.

Dave: Yes. When I presented the process to the project managers in my division, they definitely agree that the process will help them do their job better, but they showed some concern about its application. They don't think that every project should follow all the steps in the process, especially in the planning phase.

Figure 24.2   A Standardized Project Management Process

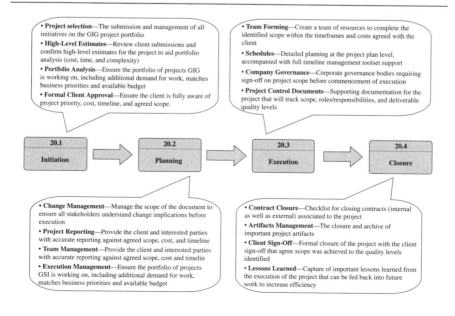

Mary: I've heard the same things. I believe that we agreed when we put this process together that our process should be flexible. A lot of projects we have are small and need to be executed fast to meet the client's needs. For example, the smaller projects may not need to follow all of the 20 steps in the planning phase.

Dan: Okay. Now the difficult part begins. First off, we need to find a way to categorize our projects. The large ones will follow the full-blown process. The smaller ones can follow a shorter version. Now, we just need to decide which procedures and tools are appropriate for the smaller ones.

Dave: Let's think about this carefully. Let me throw in a number so that we can discuss the pros and cons. How about a $100,000 threshold? We could consider a project as "small" if its budget is below $100,000.

Mary: I think that the threshold should depend on the composition of projects we have in our GIG portfolio. You should be able to tell us this, Dan.

Dan: The majority of our projects are below the $100,000 budget.

Mary: So, I don't think that a $100,000 threshold is appropriate.

Dan: Should it be higher or lower?

The discussion continues. Before moving on to the next topic in the agenda. Dan, head of the GIG PMOs, asks all group PMO leads to go back and look at their data with their teams. Group PMOs will discuss with the heads of their division PMO, Dave or Mary respectively. Dan also asks both Chief Operation Officers to be involved in the division discussions. The final decision will be made together in the next meeting.

LINK FOR SYNCHRONIZATION PROGRAM

Recently GFC has placed a tremendous amount of time into recognizing project management as a critical discipline across the firm, yet there has never been a forum that has facilitated knowledge exchange between various business units. As GFC moves toward a more holistic global model, it is becoming increasingly necessary for best practices to be shared across the firm. Recently, GFC has begun to sponsor PMP certifications, developing internal and external professional development training curriculum, and reimbursing other PMP-specific study expenses. To help address this issue and to get all GIG project managers more in sync with the PMOs, including the standards and tools available, the GIG PMO runs a program called the Link for Synchronization.

Link for Synchronization will become a focal point for professional standards (e.g., PMI Body of Knowledge), the driver of tools and techniques (e.g., Microsoft Project Server 2007), the driver of standardized approaches (e.g., Demand

Management and Portfolio Management), and the precedent for extending this across the entire organization. The program is building in four areas: career framework and training, recruiting and mentoring, event coordination, and communications and web content.

Dan: Have any of you had a chance to check the new website of our Link for Synchronization program? I want to make sure that we are familiar with this website and communicate it to our project managers. It should be used to ensure timely communications among us. The website also features key project highlights, events, and lunch box learning schedules.

Dave: I saw it. I think that it looks great. The website is a backbone for our Link for Sync program. It seems to address all four areas we plan to implement.

Dan: Almost. We plan to use it as a platform and to present the outcomes. For example, in Career framework and training, our focus is on career framework updates and alignment. We want our project managers to participate in updating roles and responsibilities for project and program management. They can use the website to share their views.

Mary: This is great. So, we have all the information there. I suppose that the website will inform us about the training and lunch box learning programs too, right?

Dan: Yes. Let me make it clear. The focuses of lunch box learning and training are different. Lunch box learning will focus on key areas of project management and allow for project managers to participate in learning events and share experiences. With this, we will look for feedback on areas in which individuals or teams would like us to focus. The training will have formal curriculum. We will develop a standardized set of developmental programs to support core competencies, toolsets, and the GFC policies and procedures. Training will also focus on soft skills required to enhance the performance of project managers and their teams.

Mary: Should we bring in outside speakers once in awhile?

Dan: Definitely, it is part of event coordination. We will form a committee to create a yearly events calendar, which will be posted on the website. The committee will focus on regional, local, and global networking event planning with key topics and speakers from within GFC and the project management industry.

Dave: The topics that come up from lunch box learning can feed both the training and the events.

Dan: Right. Now since we already talked about three other areas, let's discuss a bit about recruiting and mentoring. Our focus in this area is to identify opportunities to attract new members and look for innovative and creative ways to recruit project managers within the firm. We also plan to coordinate the mentoring program. Our goal is to increase the level of ex-perienced project managers who are available to manage the increasingly complex projects necessary to support the business. We want this to be a competitive program.

Mary: How?

Dan: This program is not for everyone. We are thinking that the project managers who join the program should be nominated from their manager. Then, they will be assigned as an apprentice to a more experienced project manager who will serve as their mentor. This will provide the opportunity for the mentees to work closely with the mentors and to apply the techniques they are learning in a structured, supported environment. The mentors can support, reinforce, and elaborate on the lessons. This is very formal.

Mary: This sounds intriguing. Do you think it will work?

Dan: Why not?

Dave: Are we going to lose productivity during the mentoring? If a mentee, who first must be a star project manager to get nominated, works as an apprentice, who will work on their projects. With the limited resources we have now, do you think that people will have time to go through this program?

Mary: That's my concern, too. All project managers in my division are very busy. They are juggling so many projects at once. But I think the program would work if, at the end of the day, the project managers do not have to worry about their regular job but instead can focus on the mentoring program.

Dave: But we cannot afford that. The functional managers will say no to this right at the beginning.

Mary: Right. And who will be the mentors, the senior project managers in each group or the PMO project managers? If mentors also come from the functional group, then you lose the productivity of two experienced project managers. How can we sell this to the functional managers?

Discussion items

1. Discuss the placement of the PMOs for GIG. List the pros and cons of this placement?
2. Discuss the importance of having the head of the GIG PMO as a peer of the heads of the GTG and A&E divisions.
3. Discuss if GIG PMOs are set up for success or failure. Explain specifically at each of their initiatives.

4. Discuss why the $100,000 threshold is inappropriate if the typical range of project budget in the GTG division is from $30,000 to $220,000 and more than half of the projects have a budget below $100,000.
5. How important is the GIG Link for Synchronization program to the success of the GIG PMO and GIG project management practices? Is the program too ambitious? How should such a program be implemented?
6. Who should be in the committee for event coordination?
7. Suggest a way to implement the GIG mentoring program. Is it possible? If not, suggest an alternative mentoring program.

Are We Ready for Portfolio Management?

Peerasit Patanakul

Star Media is a typical nonprofit organization. While several projects were implemented in the company over the year, Star Media's focus has been its operations. Projects are typically run by functional heads; no real project managers; and teams have limited formal project management training. On a high level, Star Media does not practice project portfolio management per se. This seems to have been sufficient over the past history of the company. But in the last couple of years, the company has been in the midst of a big shift in the media industry. To stay afloat, Star Media has to implement many initiatives both strategically and tactically. It does not seem that its limited project management competence will serve the company well any longer. People have started complaining louder and louder. Top executives must get their acts together.

STAR MEDIA

Star Media is an internationally acclaimed producer and distributor of noncommercial news, talk, and entertainment programming. A privately supported, not-for-profit membership organization, Star Media serves a growing audience of 16 million Americans each week in partnership with more than 360 independently operated, noncommercial public television and radio stations. Each member station serves local listeners with a unique combination of national and local programming. With original online content and audio streaming, StarMedia.org offers hourly newscasts, special features, and 10 years of archived audio and information.

The mission of Star Media is to work in partnership with member stations to create a more informed public—one challenged and revitalized by a more thorough understanding and appreciation of events, ideas, and cultures. To accomplish the mission, Star Media produces, acquires, and distributes programming that meets the most stringent standards of audience service in journalism and cultural expression; it represents member stations in matters of mutual interest; and it provides satellite interconnection for the entire public media system.

493

Figure 24.3 Organizational Structure of Star Media

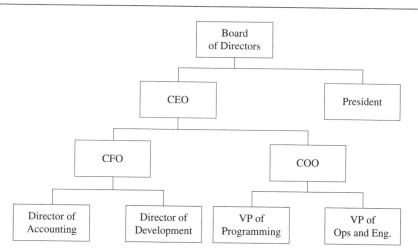

Star Media's annual revenue is approximately $250 million. Each year any profit (about. $1 million) is put toward Star Media's endowment, which is currently close to $90 million. Approximately 50 percent of Star Media's budget comes from subscription fees that member stations pay in exchange for programming, 30 percent comes from corporate underwriting, and the remaining 20 percent comes from foundation and government grants. Simply put, the primary "product" of Star Media is programming and its primary programming is news.

A 23-member board of directors governs Star Media. Sixteen of the directors are station managers who are elected to the board by member stations. The seven remaining directors include the president of Star Media, the chairman of the Foundation, and five prominent members of the public selected by the board and confirmed by the member stations. The organizational structure of Star Media is shown in Figure 24.3.

THREE MAIN PILLARS OF STAR MEDIA

Development, programming, and Operations and Engineering are the three main pillars of Star Media.

- The primary function of the development department is to establish relationships with high-quality businesses, individuals, and organizations in order to enhance Star Media's image, and to develop long-term financial support through underwriting, grants, and other opportunities so as to ensure Star Media's future.
- The primary function of the programming department is to provide comprehensive, and diverse, compelling news and cultural affairs programming

for a significant, diverse audience for distribution to member stations and through new media.

- The primary function of the operations and engineering department is to maintain, operate, improve, research, and implement systems, new and existing technologies, and infrastructure in order to promote an efficient, reliable, and high-quality workplace.

PROJECT MANAGEMENT AT STAR MEDIA

While Star Media focuses on its operations, a variety of projects have been implemented. Typical projects are under the responsibility of three directorates: Development, Programming, and Operations and Engineering. Examples of projects are:

- The implementation of new software for desktop production: The software will be implemented on the desktops of over 300 employees (operations and engineering department)
- Podcasting: Leverage audio production abilities by producing low-cost, high-impact, and quick-response programming (programming department)
- Digitization of Archive: Re-recording reel-to-reel tapes to digital format, over 40,000 hours of archived programming (operations and engineering department)
- Multi-Protocol Layered Switching: Providing standardized and improved connectivity (operations and engineering department)
- Reporter training: Training reporters on multimedia applications, audio and video visuals and web pages development (operations and engineering department)
- New bureau: Setting up the equipment and training staff in a new facility (operations and engineering department)
- Establish a three-tiered elite membership club (e.g., Cornerstone Society) for major donors (development department)
- Create and implement a new company logo (development department)
- Produce a television show of the annual Major Donors' party with high-profile acts (development, programming, and operations and engineering department)
- Establish a scholarship fund so that Star Media can help enterprising low-income students attend the college of their choice (development department)
- Create a partnership program with colleges and universities to expose students to Star Media's educational outreach program (development department)
- Produce a new program for television, radio, or podcast (programming department)

- Establish an interactive online news quiz (programming and operations and engineering department)
- Develop a StarMedia online store where listeners can buy logo items (hats, t-shirts, etc.) and make contributions (development department)

Similar to typical organizations that focus on operations, management at Star Media realizes that its project management practices need to be improved. With functional focus, the majority of projects are led by the respective functional managers. A recent survey of the organization indicated that more standardization, more training, and centralized data bases along with increased accountability would help improve project management at Star Media.

The senior management team also understands there is a lack of project management focus in the company. They realize that the approach they use to select projects is too reactive. Sometimes, projects are approved without the full understanding of desired goals or how these projects are aligned with the company's strategy. In order for the company to be competitive they must commit to improve their project management practices starting from the project evaluation and selection process.

WHAT CAN WE DO BETTER?

In the meeting between the Chief Operating Officer (COO) and the Vice-President of Operations and Engineering, the following discussion occurred:

VP: We have so many projects going on right now and new ones are coming. We do not have enough resources.

COO: I know. But you have to understand that with the recent movement in the media industry, we have to change the way we operate and deliver. We cannot rely on the old model of media anymore. We cannot produce radio and television shows without an Internet component and think we are reaching our full potential audience. We have to use all forms of media. And we have to upgrade our equipment that we have been using for decades. That's why we have so many initiatives recently.

VP: I do understand. But we need to make sure we are putting the horse before the cart. If we can't hire new people, shouldn't we find a better way to pick projects or to know which ones are more important, more effective? That should guide the way we allocate resources. The projects that are less important should be put on hold. Now it seems that we are trying to do too much, which I think is less productive because we may not get them done by their deadlines; not to mention that our people have little project management experience.

COO: I agree. What is your suggestion?

VP: Well, maybe we should concentrate on improving the way we select projects? Maybe we can have something like a list of criteria that will help us select projects. We want to make sure that the projects we implement help our business.

COO: Business is always an interesting term. It has to be clearly defined.

VP: I agree. We are not in it to just make money. It's also about enriching people's lives, entertaining society, keeping people informed, those sorts of things. Remember last year when Bob from programming suggested we produce that new morning news program? That was crazy. We already have a very successful morning news program. I don't work for the programming department, but even I can tell you we need to reach out to a younger audience. We already have seven news programs. I think we need to get back to more of the cultural programming we used to do, some of the music programming we used to do. We've been reaching the same aging audience for decades!

COO: You're absolutely right. The same goes for the development department. Right now we have six initiatives that aim to generate revenue 10 to 15 years from now. We need revenue this year; we need revenue next year, and the year after that. Sure planning for the future is great, but we need to pay salaries this week. Do you know how much that partnership with Duff beer cost us? We spent $6 million over three years . . . and after all that work we only generated $1 million in profits. We can't spend all our investment dollars on low-return investments. We need to find projects where we can double our investment, and we need to find them soon!

VP: What about exposure? Didn't we get a lot of exposure from that Duff partnership?

COO: Yeah, but that was pure luck. Look, what we need to do is develop a list of criteria. We want to make sure that they are applicable to all kinds of projects we have. But programming and operations and engineering projects are very different. The variety becomes greater if we include projects from development. Can we have a set of criteria that is applicable to all?

VP: We have to put more thought into it. Maybe we do not have to have a set of universal criteria. Maybe we should have different sets of criteria to judge different kinds of projects. But we want to make sure that every project we select enhances our business.

COO: Okay. We have to talk more about this. Let's form a working group. I will invite the CFO and his directors to join us.

Discussion items

1. Discuss whether a scoring method would be an appropriate methodology for project selection at Star Media.
2. Discuss whether several scoring models should be used with respect to the major departments of Star Media.
3. What should be the criteria for the scoring models? Should the criteria be weighted?
4. To check the balanced mix of projects in the portfolio, what criteria should be used to represent key dimensions of the bubble diagrams?
5. How does management ensure that projects are selected in alignment with the company's strategy?

Index